ELASTOMER-BASED COMPOSITE MATERIALS

Mechanical, Dynamic, and Microwave
Properties and Engineering Applications

ELASTOMER-BASED COMPOSITE MATERIALS

Mechanical, Dynamic, and Microwave Properties and Engineering Applications

Nikolay Dishovsky, PhD, DSc
Mihail Mihaylov, PhD

CRC Press
Taylor & Francis Group
Boca Raton London New York

CRC Press is an imprint of the
Taylor & Francis Group, an **informa** business

First published 2018 by Apple Academic Press, Inc.

Published 2019 CRC Press
Taylor & Francis Group
6000 Broken Sound Parkway NW, Suite 300
Boca Raton, FL 33487-2742

ISBN-13: 978-1-77463-058-7 (pbk)
ISBN-13: 978-1-77188-620-8 (hbk)

Library and Archives Canada Cataloguing in Publication

Dishovsky, Nikolay, 1953-, author
Elastomer-based composite materials : mechanical, dynamic and microwave properties, and engineering applications / Nikolay Dishovsky, PhD, DSc, Mihail Mihaylov, PhD.
Includes bibliographical references and index.
Issued in print and electronic formats.
ISBN 978-1-77188-620-8 (hardcover).--ISBN 978-1-315-15958-4 (PDF)
1. Elastomers. 2. Composite materials. I. Mihaylov, Mihail, 1981-, author II. Title.
TA455.E4D57 2017   620.1'94   C2017-907434-2   C2017-907435-0

Library of Congress Cataloging-in-Publication Data

Names: Dishovsky, Nikolay, author. | Mihaylov, Mihail, author.
Title: Elastomer-based composite materials : mechanical, dynamic and microwave properties, and engineering applications / Nikolay Dishovsky, PhD, DSc, Mihail Mihaylov, PhD.
Other titles: Elastomer based composite materials
Description: Toronto; New Jersey : Apple Academic Press, 2017. | Includes bibliographical references and index.
Identifiers: LCCN 2017055543 (print) | LCCN 2017056734 (ebook) | ISBN 9781315159584 (ebook) | ISBN 9781771886208 (hardcover : alk. paper)
Subjects: LCSH: Elastomers. | Composite materials.
Classification: LCC TA455.E4 (ebook) | LCC TA455.E4 D57 2017 (print) | DDC 620.1/94--dc23
LC record available at https://lccn.loc.gov/2017055543

Visit the Taylor & Francis Web site at
http://www.taylorandfrancis.com

and the CRC Press Web site at
http://www.crcpress.com

ABOUT THE AUTHORS

Nikolay Dishovsky

Nikolay Dishovsky (born 1953) is a Full Professor who has been Head of the Department of Polymer Engineering at the University of Chemical Technology and Metallurgy, Sofia, Bulgaria since 2003. He obtained his PhD in 1983, working on the synthesis and application of elastomer-based microgels. In 1991, he became an associate professor in macromolecular chemistry. In 1997, Dishovsky earned his Doctor of Science degree defending a thesis on rubber composites containing specific functional fillers. Before his appointment as a Full Professor in rubber technology in 2000, he held DAAD and NATO Science Committee fellowships in Germany and Spain. He is particularly interested in rubber magnetic modification, rubber-based composite materials, absorbing electromagnetic waves, and functional fillers. He has more than 200 publications in IF- and SJR-indexed journals and holds 41 patents. In 2001, his name was inscribed in the Golden Book of Bulgarian Inventors. In 2015, he was chosen as Inventor of the Year in Bulgaria, category "Chemistry and Biotechnology." He has authored two books (Iv. Ivanov, N. Dishovsky, *Patent Analyses*, Sofia, IP Bulgaria, 2004; N. Dishovsky, G. Tsenkov, *Rubber Handbook*, Sofia, EC Print, 2006) and six textbooks (N. Dishovsky, II. *Radulov, Transport Phenomena in Rubbers*, Sofia, EC Print, 2007; N. Dishovsky, I. Radulov, R. Dimitrov, *Reinforcing of Elastomers*, Sofia, EC Print, 2005; N. Dishovsky, *Ingredients for Rubber Compounds*, Sofia, UCTM, 2004; N. Dishovsky, *Ageing and Stabilisation of Elastomers*, Sofia, Informa, 1998; St. Andreev, R. Peshleevsky, N. Dishovsky, *Materials and Coatings Absorbing Ultra High Frequency Electromagnetic Waves*, Sofia, VMT, 1991; IV. Mladenov, N. Dishovsky, *Photooxidation and Stabilization of Polyolephines*, Sofia, Technika, 1987) for students of the BSc, and MSc and PhD programs in polymeric engineering and polymer materials. He has supervised 20 PhD students and more than 70 BSc and MSc students. He has been project leader of more than 60 national and international projects. The results from a considerable part of those projects have found application in real-life practice.

Mihail Mihaylov

Mihail Mihaylov (born 1981) has been an Assistant Professor in the Department of Polymer Engineering at the University of Chemical Technology and Metallurgy, Sofia, Bulgaria since 2011. He obtained his PhD in 2010, working on the possibilities for recycling waste tires and rubber goods. He is particularly interested in rubber modification, modification of fillers, recycling of waste tires, rubber-based composite materials, absorbing electromagnetic waves, and functional fillers. He has more than 25 publications in IF- and SJR-indexed journals and one patent.

CONTENTS

LIST OF ABBREVIATIONS

AES	Auger electron spectroscopy
AGU	anhydroglucopyranose
BE	binding energies
BET	Brunauer–Emmett–Teller
BJH	Barrett–Joyner–Halenda
BR	butadiene rubber
BSS	Bulgarian State Standard
CB	carbon black
CBS	N-cyclohexyl-2-benzothiazolesulfenamide
CCB	conductive carbon black
CCD detector	charge coupled device detector
CNT(s)	carbon nanotubes
CR	chloroprene rubber
CSM	chlorosulfonated polyethylene
CTAB	cetyltrimethylammonium bromide
DBP	dibutylphthalate
DBPA	dibutylphthalate absorption
DFT	density functional theory
DMTA	dynamic mechanical thermal analysis
DSC	differential scanning calorimetry
DTA	differential thermal analysis
DTAB	dodecyltrimethylammonium bromide
EDX, EDS	energy dispersive x-ray spectroscopy
EMC	electromagnetic compatibility
EMI	electromagnetic interference
EMI SE	electromagnetic interference shielding effectiveness
EMI SM	electromagnetic interference shielding material
EMW	electromagnetic wave(s)
ENR	epoxidized natural rubber
EPDM	ethylene–propylene diene terpolymer
EU	European Union
FGS	functionalized graphene sheets
FMR, NFMR	ferromagnetic resonance, natural ferromagnetic resonance
FTIR	Fourier transformed infrared spectroscopy

GNP(s)	graphene nanoplatelets
GNR	Guayule natural rubber
HAADF	high-angle annular dark field
HNBR	hydrogenated acrylonitrile–butadiene rubber
IEP(s)	isoelectric points
IIR	isobutylene–isoprene rubber
IN	iodine number
IPPD	N-isopropyl-N′-phenyl-p-phenylenediamine
IR	isoprene rubber
IRS	infrared spectroscopy
LDA	Vulkacit LDA (zinc diethyldithiocarbamate)
LDPE	low-density polyethylene
MBT	2-mercaptobenzothiazole
MBTS	dibenzothiazyl disulfide
MCC	microcrystalline cellulose
MFA(s)	multifunctional additives
MWCNT	multiwall carbon nanotubes
NAFTA	North American Free Trade Agreement
NBR	acrylonitrile–butadiene rubber
NLDFT	nonlocal density functional theory
NMR	nuclear magnetic resonance
NR	natural rubber
NTCC	negative temperature coefficient of conductivity
OAN	oil absorption number
PCB	pyrolysis carbon black
PCSF	pyrolysis carbon–silica filler
PDF	powder diffraction file
PDMS	polydimethylsiloxane
PET	polyethylene terephthalate
phr	parts per hundred of rubber
Pl	plastikol
P_s–isoprene–P_s	polystyrene–isoprene–polystyrene
PTCC	positive temperature coefficient of conductivity
PTFE	polytetrafluoroethylene
QR	siloxane rubber
R	resistance, Ω
REACH	Registration, Evaluation, Authorization and Restriction of Chemicals
RF	radiofrequency
R_H	Hall coefficient

RI	reinforcing index
SAED	selected area electron diffraction
SBR	styrene–butadiene rubber
SE	shielding effectiveness
SEM	scanning electron microscopy
SFMR	spontaneous ferromagnetic resonance
Sh	Shore A hardness
STEM	scanning transmission electron microscopy
STP	standard temperature and pressure
SWCNT(s)	single-walled nanotubes
TBBS	N-tert-butyl-2-benzothiazolesulfonamide
TEM	transmission electron microscopy
TEOS	tetraethoxysilane
TESPT	bis[3-(triethoxysilyl)propyl] tetrasulfide/silane/
TGA	thermogravimetric analysis
TMTD	tetramethylthiuram disulfide
UHF	ultra-high frequency
VSWR	voltage standing wave ratio
XPS	x-ray photoelectron spectroscopy
XRD	x-ray diffraction
ZDEC	zinc *diethyldithiocarbamate*
$Zn(R)_2$, Rs	zinc resinate
$Zn(St)_2$, St	zinc stearate

LIST OF SYMBOLS

B	magnetic induction
B_m	maximum magnetic induction
B_r	residual magnetic induction
B_s	saturation magnetic induction
d	thickness of the sample
D_{AV}	average pore diameter
D_P	pore diameter
E	electric field
e	elementary charge of the electron
E*	complex dynamic modulus
E'	dynamic storage modulus
E"	dynamic loss modulus
E_a	activation energy
E_{ah}	activation energy for hopping transition
E_{az}	activation energy for zone transition
f	frequency
f_c	frequency without sample
f_r	resonance frequency of an empty cavity resonator
f_s	frequency with sample
f_ε	resonance frequency of cavity resonator with sample
H	magnetic field
h	thickness of the sample between electrodes
H_a	intensity of intrinsic magnetic anisotropy field
H_o	intensity of external magnetic field
I_H	current when determining Hall effect
K	wave number
K_{Sh}	aging coefficient regarding the Shore A hardness
K_ε	aging coefficient regarding the elongation at break
K_ρ	aging coefficient regarding the volume resistivity
K_σ	aging coefficient regarding the tensile strength
L	attenuation
L_r	reversible losses
L_T	induced attenuation
M:HF	magnetite:xeferrite mass ratio

M_{100} stress at 100% elongation (modulus 100)
M_{300} stress at 300% elongation (modulus 300)
M_C molecular weight of the segment between two cross-links
M_H maximum torque
M_L minimum torque
$ML\,(1+4)$ Mooney viscosity
n   quantity of charge carriers
n_C average number of cross-links
P_A absorbed power
P_a adopted power at the end of coaxial line without sample
$PC\rho_V$ compression coefficient of the volume resistivity
P_I incident power
P_p adopted power at the end of coaxial line with sample
P_R reflected power
P_T transmitted power
q   carrier charge
Q_C Q-factor of the cavity without sample
Q_S Q-factor of the cavity with sample
R   resistance
R_L reflection losses
r_P pores radius
R_P Pearson correlation coefficient
R_S resistance (surface)
R_V resistance (volume)
S   cross-sectional area of the measuring electrode
S_{11}, S_{12} complex scattering parameters
S_{BET} BET specific surface area
SE_A absorptive shielding effectiveness
SE_R reflective shielding effectiveness
SE_T total shielding effectiveness
S_{EXT} external specific surface area
S_{MI} specific surface area of micropores
S_r cavity resonator cross-section
S_ε sample cross-section
T_{50} cure time at 50%
T_{90} cure time (optimum)
$\tan\delta$ mechanical loss angle tangent
$\tan\delta_\varepsilon$ dielectric loss angle tangent
$\tan\delta_\mu$ magnetic loss angle tangent
tand@MH   mechanical loss angle tangent at maximum torque

tand@ML	mechanical loss angle tangent at minimum torque
T_s	scorch time
T_{s1}	scorch time $(M_L + 1)$
T_{s2}	scorch time $(M_L + 2)$
V	vulcanization (cure) rate
V_C	volume of the cavity
V_H	voltage when determining Hall effect
V_{MI}	volume of micropores
V_{MES}	volume of mesopores
V_r	volume fraction of the swollen network
V_s	molar volume of the solvent
V_S	volume of the sample
V_t	total pore volume
W_0	total volume of micropores
Z	wave resistance
T_g	glass transition temperature

Greek Symbols

α	attenuation coefficient
α_r	temperature coefficient of resistance
α_t	temperature coefficient of volume resistivity
Γ	reflection coefficient
$|\Gamma|$	magnitude of the reflection coefficient
δ	bending degree
ΔM	$\Delta M = M_H - M_L$
ε	permittivity
$\bar{\varepsilon}$	complex permittivity of the medium
ε_e	equivalent permittivity
ε_r	relative permittivity
ε_r'	real part of permittivity (dielectric constant)
ε_r''	imaginary part of permittivity
ε_{rel}	elongation at break
ε_{res}	residual elongation
ε_S	static dielectric constant
ε_∞	dielectric constant at frequencies so high the molecular orientation does not have time to contribute to polarization
λ	thermal conductivity coefficient
μ	permeability
μ_m	charge carrier mobility
μ_r	relative permeability

μ_r'	real part of permeability
μ_r''	imaginary part of permeability
v	density of the vulcanizates network
ρ_0	resistivity without pressure
ρ_r	rubber density
ρ_s	surface resistivity
ρ_v	volume resistivity
σ	tensile strength
σ_{asd}	average squared deviation
σ_e	effective electrical conductivity
σ_V	electrical conductivity
τ	relaxation time
$\varphi_{(H\alpha p)}$	δ-function characterizing the probability distribution of the electromagnetic field
χ	parameter characterizing rubber–solvent interaction
ω	angular frequency

PREFACE

A composite material (commonly known as "a composite") is a material comprising two or more constituents with significantly different physical or chemical properties that, when combined, produce a material with characteristics different from the individual components. The individual components remain separate and distinct within the finished structure.

Traditionally the composite material is a multi-phase system consisted of a matrix and reinforcements. The matrix is a continuous phase, and includes metals, inorganic non-metallic matrices and polymer matrices. The reinforcing material is a dispersed phase usually in the powder state. In the composites, the effectiveness of the reinforcing material (also called reinforcing filler), the matrix and the interface between them impact directly the performance and possible applications of composite materials.

We discuss only elastomer matrix composite materials in this book.

The book is focused on elastomer based composite materials comprising different types of reinforcing fillers. New hybrid fillers have been synthesized by different techniques, for example impregnation of different substrates (carbon black, conductive carbon black, activated carbons, etc.) with silica or magnetite. The fillers thus obtained have been characterized thoroughly by standard techniques and by up to date methods—energy dispersive X-ray spectroscopy in scanning transmission electron microscopy (STEM-EDX), atomic absorption spectroscopy (AAS), and inductively coupled plasma-optical emission spectroscopy (ICP-OES). The effect of those fillers upon the curing properties, mechanical and dynamic mechanical parameters, electrical conductivity, dielectric and microwave characteristics of elastomer based composites is discussed in details. The book also covers the influence of various types of ceramics (SiC, B_4C, and TiB_2) and barium and strontium hexaferrites upon the aforementioned properties of rubber composites as well as the opportunities for solution of the environmental problems caused by waste tires. The book shows that the pyrolysis cum-water vapor is a suitable and environmentally friendly method for conversion of the waste green tires into useful carbon-silica hybrid filers. The properties of elastomer based composites comprising different types of nanostructures (fullerenes, carbon nanotubes, graphene nanoplatelets), modified activated carbons, calcined kaolin are also discussed. Special attention is paid to the composites with a lowered level of zinc oxide.

The book provides the reader with an idea of the possibilities for broadening the engineering applications of elastomer composites through using various types of hybrid fillers, ferrites and ceramics as well as for their synthesis and characterization. Due to the detailed characterization of these fillers as well as the extensive research and discussion of the curing, mechanical, dynamic mechanical, dielectric and microwave properties of the elastomeric composites, the book provides knowledge on the properties and main applications of these composites.

The book is important and different from the other books similar to it because:

- It describes a new, easy-to-accomplish, and cheap method for synthesis of hybrid fillers for rubber composites by impregnation.
- The fillers obtained have been characterized in details by standard and-up-to-date methods, for example STEM-EDX, AAS, ICP-OES and so forth.
- The effect of those fillers upon the curing properties, mechanical and dynamic mechanical parameters, electrical conductivity, dielectric and microwave characteristics of elastomer based composites is discussed in details.
- Much attention is paid to the electromagnetic interference shielding effectiveness, attenuation and reflection of electromagnetic wave by the elastomer based composites comprising various types of hybrid fillers, ferrites and ceramics.
- The book also has an environmental aspect. It has been shown that the pyrolysis cum-water vapor is a suitable and environmentally friendly method for conversion of the waste green tires into useful carbon-silica hybrid filers. Opportunities for lowering the level of zinc oxide in rubber compounds are also discussed.

The book surveys the most recent research activities of the authors that will, unquestionably, make it a vital reference source for scientists in both the academic and industrial sectors, as well as for individuals who are interested in rubber materials. It will be very useful for students, especially PhD students, scientists, lecturers and engineers working or doing research in the field of polymer materials science, elastomer based composites and nano-composites and their engineering applications in the production of micro-wave absorbers and electromagnetic waves shielding materials, materials for electronics devices and telecommunications.

Professor Nikolay Dishovsky, PhD, DSc
Assistant Professor Mihail Mihaylov, PhD

INTRODUCTION

It is known that at least 3000 years before the first Europeans saw natural rubber, the Mesoamerican communities had developed ways of collecting and forming it into a variety of objects such as toys, domestic products, and items related to ritual sacrifice and to tribute payments.[1] However, rubber became an indispensable factor of industry and of daily life after Charles Goodyear and Thomas Hancock made one of the most effective discoveries and developments of mankind, namely the vulcanization technology, in 1839–1844. The invention consisted in transformation of a material—sticky when warm and brittle when cold—into a high-deformable chemically crosslinked elastic solid. Since that time the rubber industry has been established and shown a markedly progressive development over the last period of more than 100 years. Developments in the field of synthetic elastomers have progressed so rapidly that the whole concept of rubber technology has changed several times adapting to the altering demands from society.[2]

Everyone is aware of the fact that the development of modern techniques and technologies is impossible without using intensive elastomeric materials and composites based thereon. It is difficult to name even one area wherein those materials do not find wide application.

Mass production of elastomers and related products requires very good knowledge of their properties and qualities of materials with potential engineering applications. The properties of the end products are not clearly determined by the chemical composition and structure of elastomers as the latter are depend to a great extent on the fillers properties. The relationship between the composition of the "elastomeric matrix–filler" system and the properties of the product is very complex. Studies on that relationship act as theoretical basis of elastomeric materials processing. The manifestation of some new and unusual properties of elastomer composites makes them not only attractive, but defines them as a material of the future. All this drives the studies on the possibilities to use fillers of different chemical nature and structure in order to manufacture elastomeric materials of unique properties able to afford high performance. On the other hand, it has been of marked interest to study and trace their "composition–structure–properties" relationship with regard to filling the gaps in the knowledge about elastomeric materials science. In that sense, it is essential to discuss issues related to "elastomer–filler" interfacial phenomena, fillers dispersion, their compatibility with the elastomer, the impact that the concentration of individual components has upon composites structure and properties.

REFERENCES

1. Loadman, J. *Tears of the Tree. The Story of Rubber—a Modern Marvel*; Oxford University Press: Oxford, 2005.
2. Gert, H. Ed. *Advanced Rubber Composites*; Springer: Berlin Heidelberg, 2011.

CHAPTER 1

ELASTOMER-BASED COMPOSITE MATERIALS: OPPORTUNITIES FOR BROADER ENGINEERING APPLICATION

CONTENTS

ABSTRACT

Elastomer composites are complex multicomponent system whose properties depend on the type of elastomer (or mixtures of elastomers) as well as on the type of fillers and ingredients used. This chapter briefly presents the main approaches for widening the application areas of rubber composites by chemical modification of already-known elastomers. The focus is mostly on hydrogenation of elastomers, given that one of the oldest possibilities for modification of elastomers is taken into consideration. Attention has been paid to the possibilities for using renewable sources for obtaining elastomers with new properties by epoxidation of natural rubber latex for grafting antioxidants onto the elastomer macromolecule by means of grafting copolymerization.

1.1 INTRODUCTION

The term "composite material" refers to materials consisting of two or more starting materials with different physical and/or chemical properties. The properties of a composite material differ completely from those of its constituent components. Elastomer-based composites can be defined as materials composed of two or more components with distinct boundary surfaces or interfaces. Typically, elastomeric composites consist of a rubber matrix, fillers, and a number of ingredients that have a key role in their vulcanization, exploitation, storage, and so forth. In other words, elastomer-based composites are a complex multicomponent system whose properties depend on the type of elastomer or elastomer blends and on the type of fillers and ingredients used.

Due to unique properties, elastomer-based composites have been the subject of many studies and developmental investigations. As a result of their durability, high strength, light weight, corrosion resistance, design, and process flexibility, these are applicable in many fields, such as medicine, car industry, space research, defense, sports,[1] and so forth.

Over the past two decades, with the development and introduction of nanomaterials in polymer systems, elastomer-based composites are entering a new era of nanostructured polymeric materials.[2] To classify an elastomer-based composite as nanostructured, at least one of its phases should be in nanosized region.[3] Obviously, in the case of elastomer-based composites this phase is the filler.

1.2 OPPORTUNITIES FOR BROADER ENGINEERING APPLICATION OF ELASTOMER-BASED COMPOSITE MATERIALS

Typically, a large number of multicomponent parts and assemblies used in industry contain rubber components whose main function is to assist the precise operation of articles.[4-6] The main application of elastomers is in automotive industry, especially in the production of tires. Another important application is the production of technical rubber goods—various belts, hoses, gaskets, oil seals,[7] and so forth. However, the applications of elastomers are not limited to these areas. The development of so-called conducting rubbers, obtained by formation of interconnected charge carrier paths in the dielectric rubber matrix by loading it with various conductive fillers, such as metal powders, conductive carbon black, graphite, and so forth, significantly widens its application areas. The usage of conducting elastomers as electrostatic charge dissipating materials, pressure sensors, transducers, electromagnetic shielding materials, antennas,[8-12] and so forth, could be seen as an example. The above discussion clearly shows that selection of an appropriate filler during formulation is as important as selecting the type of rubber as it determines the performance specification of a product.[13,14]

Therefore, demand for enhanced properties of elastomers and elastomer materials increases every year. There are several options to meet this high demand and to broaden the fields of applications of elastomeric materials:

- Synthesis of new types of elastomers having new and improved properties
- Chemical modification of already-known elastomers
- Developing and investigating new ingredients for the rubber industry
- Synthesizing new types of reinforcing and functional fillers
- Modification of the conventional fillers, and so forth

1.3 MODIFICATION OF ELASTOMERS

As mentioned earlier, one way to broaden the application fields of elastomer composites is through synthesis of new polymers possessing the desired properties. In most cases, however, the synthesis of novel elastomers is a complicated, laborious, and especially cost-inefficient process. Therefore, most often the solution to this problem is a purposeful change of the properties of already-known elastomers, by means of chemical modification.

Diene elastomers such as natural rubber (NR), butadiene rubber (BR), isoprene rubber (IR), chloroprene rubber (CR), isobutylene–isoprene rubber (IIR), acrylonitrile–butadiene rubber (NBR), and styrene–butadiene rubber (SBR) are some of the most commonly used rubber types. All these types of rubbers have a common trait, that is, the presence of olefin structures in the rubber chains that favors a number of chemical transformations. The latter in their turn are the basis for the chemical modification of elastomers.[15,16]

The main types of chemical transformations of elastomers can be divided into three groups:

1. Reactions that take place without change in the degree of polymerization (intramolecular and polymer-analogous transformations)
2. Reactions leading to an increase in the degree of polymerization (cross-linking and curing of polymers, the preparation of block and graft copolymers)
3. Reactions leading to lowering the degree of polymerization (destruction of polymers)

Polymer-analogous transformations are chemical reactions that take place between the functional groups of the elastomers and low molecular weight substances, leading to modification of the functional groups that brings about changes in the properties of the elastomer used as starting material. These reactions are among the most frequent routes for chemical modification of elastomers. Hydrogenation, halogenation hydrohalogenation, epoxidation, and so forth are typical examples of polymer-analogous reactions of elastomers.

The team of G. Rempel and Q. Pan[16–18] has a significant experience in the field of hydrogenation of elastomers. According to them, catalytic hydrogenation is the most commonly used method for modification of elastomers. When NBR has two types of unsaturated groups—the olefin group ($C = C$) and nitrile group ($C \equiv N$), the challenge in the preparation of hydrogenated NBR (HNBR) is in performing the modification so that the nitrile groups remain unaffected, as that would result in the reduction of NBR oil resistance. The authors achieved a high degree of hydrogenation of NBR latex by means of in situ synthesized $RhCl (PPh_3)_3$ catalyst, which was prepared using water soluble salts of rhodium ($RhCl_3$) and triphenylphosphine (PPh_3) as precursor.[16] The advantages of HNBR over NBR are considerable. The mechanical properties (tensile strength and elongation at break), heat resistance, weather resistance, and the resistance of composites containing HNBR to aggressive fuels and chemicals are significantly improved. In all cases, the processability and oil resistance of the starting elastomer are retained.[17]

The epoxidation of NR is one of the most promising ways to obtain new polymeric materials from a renewable resource. The epoxidation of NR is carried out in stages of latex by means of hydrogen peroxide and formic acid. Epoxidized natural rubber (ENR) is a polymer with high potential, used widely in industries. It not only has good elastic properties like other elastomers but also has specific properties of its own; namely, high resistance to oils and solvents, high glass transition temperature, and so forth. In addition, the chemical and physical properties of this polymer changes depending on the content of epoxy groups, that is, these properties of the ENR can be easily controlled by changing the degree of epoxidation. As a result of epoxidation, the polarity of NR increases, hence it becomes compatible with a variety of polar elastomers. The increase in degree of epoxidation leads to changes in the following properties of NR:

- Lower swelling degree in hydrocarbon oils
- Reduction of permeability
- Better absorption of microwave energy
- Improved dispersion of silica in rubber matrix, hence a stronger reinforcing effect,[19–23] and so forth.

Another interesting approach to improve the properties of elastomers and their composites as well as to widen the application areas is grafting of antioxidants. It is known that olefin groups are thermally unstable; therefore, at high temperatures aging of the elastomers takes place. Low molecular antioxidants, such as N-phenyl-β-naphthylamine, are widely used for solving the problem of thermal oxidation aging of elastomers. A disadvantage of these antioxidants is that they are often released from the elastomer via migration, evaporation, or extraction. In this regards, A. I. Al-Ghonamy and coworkers successfully grafted o-aminophenol onto NR through grafting copolymerization. Their results showed that prepared NR-graft-o-aminophenol can protect NBR vulcanizate against thermal treatment better than the commercial antioxidant.[24]

The above techniques are some of the possibilities that can be used for modification of elastomers to improve their properties, respectively, to widen their fields of application. A number of researchers have been working on the possibilities to broaden the application fields of elastomeric materials through synthesis of new ingredients for the rubber industry. Large amount of work has been done exploiting the opportunities to use ingredients and raw materials (mainly fillers and plasticizers) from renewable sources in the rubber industry. Other researchers are working on opportunities for changing purposefully the properties of elastomer composites using the newly synthesized or modified reinforced and functional fillers. These options are discussed in Chapters 2 and 3.

KEYWORDS

- **elastomer-based composites**
- **modification of elastomers**
- **chemical transformations**
- **grafting antioxidants**
- **hydrogenation**
- **epoxidation**
- **reinforcing fillers**

REFERENCES

1. Papageorgiou, D. G.; Kinloch, I. A.; Young, R. J. Graphene/Elastomer Nanocomposites. *Carbon* **2015**, *95*, 460–484.
2. Bhowmick, A. K.; Bhattacharya, M.; Mitra, S.; Kumar, K. D.; Maji, P. K.; Choudhury, A.; George, J. J.; Basak, G. C. Morphology–Property Relationship in Rubber-Based Nanocomposites: Some Recent Developments. In *Advanced Rubber Composites;* Heinrich, G., Ed.; Springer Berlin Heidelberg: Berlin, Heidelberg, 2011; p. 1–83.
3. Camargo, P. H. C.; Satyanarayana, K. G.; Wypych, F. Nanocomposites: Synthesis, Structure, Properties and New Application Opportunities. *Mater. Res.* **2009**, *12*, 1–39.
4. Sittiphan, T.; Prasassarakich, P.; Poompradub, S. Styrene Grafted Natural Rubber Reinforced by in situ Silica Generated via Sol–Gel Technique. *Mater. Sci. Eng.: B* **2014**, *181*, 39–45.
5. Schneider, M.; Pith, T.; Lambla, M. Toughening of Polystyrene by Natural Rubber-Based Composite Particles: Part I Impact Reinforcement by PMMA and PS Grafted Core-shell Particles. *J. Mater. Sci.* **1997**, *32*, 6331–6342.
6. Schneider, M.; Pith, T.; Lambla, M. Toughening of Polystyrene by Natural Rubber-Based Composite Particles: Part II Influence of the Internal Structure of PMMA and PS-grafted Core-Shell Particles. *J. Mater. Sci.* **1997**, *32*, 6343–6356.
7. Stephen, R.; Thomas, S. Nanocomposites: State of the Art, New Challenges and Opportunities. In *Rubber Nanocomposites*. John Wiley & Sons, Ltd: Singapore, 2010; p. 1–19.

8. Sahoo, B. P.; Naskar, K.; Tripathy, D. K. Conductive Carbon Black-Filled Ethylene Acrylic Elastomer Vulcanizates: Physico-mechanical, Thermal, and Electrical Properties. *J. Mater. Sci.* **2012,** *47,* 2421–2433.
9. Sau, K. P.; Chaki, T. K.; Khastgir, D. Electrical and Mechanical Properties of Conducting Carbon Black Filled Composites Based on Rubber and Rubber Blends. *J. Appl. Polym. Sci.* **1999,** *71,* 887–895.
10. Sridhar, V.; Choudhary, R. N. P.; Tripathy, D. K. Effect of Carbon Blacks on Relaxation Phenomenon of Chlorobutyl Vulcanizates. *J. Appl. Polym. Sci.* **2006,** *102,* 1809–1820.
11. Mahapatra, S. P.; Sridhar, V.; Chaudhary, R. N. P.; Tripathy, D. K. Relaxation Behavior of Conductive Carbon Black Reinforced EPDM Microcellular Vulcanizates. *Polym. Eng. Sci.* **2007,** *47,* 984–995.
12. Rahaman, M.; Chaki, T. K.; Khastgir, D. Development of High Performance EMI Shielding Material from EVA, NBR, and Their Blends: Effect of Carbon Black Structure. *J. Mater. Sci.* **2011,** *46,* 3989–3999.
13. Donnet, J.-B.; Custodero, E. Reinforcement of Elastomers by Particulate Fillers. In *The Science and Technology of Rubber,* 4th ed.; Mark, J. E., Erman, B., Roland, C. M., Eds.; Academic Press: Boston, 2013; pp 383–416.
14. Wolff, S. Chemical Aspects of Rubber Reinforcement by Fillers. *Rubber Chem. Technol.* **1996,** *69,* 325–346.
15. Puskas, J. E. Diene-based Elastomers. In *Handbook of Elastomers,* 2nd ed.; Bhowmick, A. K., Stephens, H. L., Eds.; Marcel Dekker, Inc: New York, 2000; pp 817–833.
16. Liu, Y.; Wei, Z.; Pan, Q.; Rempel, G. L. Hydrogenation of Acrylonitrile-Butadiene Rubber Latex using in situ Synthesized RhCl(PPh₃)₃ Catalyst. *Appl. Catal., A* **2013,** *457,* 62–68.
17. Wang, H.; Yang, L.; Rempel, G. L. Homogeneous Hydrogenation Art of Nitrile Butadiene Rubber: A Review. *Polym. Rev.* **2013,** *53,* 192–239.
18. Wei, Z.; Wu, J.; Pan, Q.; Rempel, G. L. Direct Catalytic Hydrogenation of an Acrylonitrile-butadiene Rubber Latex Using Wilkinson's Catalyst. *Macromol. Rapid Commun.* **2005,** *26,* 1768–1772.
19. Gelling, I. R.; Porter, M. Chapter 10. In *Natural Rubber Science and Technology;* Robert, A. D., Ed.; Oxford University Press: Oxford, 1988.
20. Chapman, A. V. Natural Rubber and NR-Based Polymers: Renewable Materials with Unique Properties. In *24th international H F Mark—Symposium on Advances in the Field of Elastomers & Thermoplastic Elastomers.* Vienna, Austria 2007.
21. Bac, N. V.; Huu, C. C. Synthesis and Application of Epoxidized Natural Rubber. *J. Macromol. Sci.,* Part A **1996,** *33,* 1949–1955.
22. Bac, N. V.; Terlemezyan, L.; Mihailov, M. Epoxidation of Natural Rubber in Latex in the Presence of a Reducing Agent. *J. Appl. Polym. Sci.* **1993,** *50,* 845–849.
23. Heping, Y.; Sidong, L.; Zheng, P. Preparation and Study of Epoxidized Natural Rubber. *J. Therm. Anal. Calorim.* **1999,** *58,* 293–299.
24. Al-Ghonamy, A. I.; El-Wakil, A. A.; Ramadan, M. Enhancement the Thermal Stability and the Mechanical Properties of Acrylonitrile-Butadiene Copolymer by Grafting Antioxidant. *Int. J. Polym. Sci.* **2010,** v. 2010, article ID 981690.

CHAPTER 2

INGREDIENTS FOR ELASTOMER-BASED COMPOSITE MATERIALS: REQUIREMENTS AND ECOLOGICAL CONCERNS

CONTENTS

ABSTRACT

A number of ingredients used in the manufacture of rubber composites are of petroleum origin. Eventually, after use, they are converted into carbon dioxide—the greenhouse gas responsible for global warming. Therefore, more researchers are focusing on the opportunities to obtain elastomers and ingredients from renewable sources. This chapter presents briefly the main groups of ingredients used in the manufacture of elastomer composites, as well as opportunities for synthesis and use of new polymers, fillers, and plasticizers from renewable sources.

2.1   INTRODUCTION

Elastomers are amorphous polymers whose application in practice is near to impossible without the use of different ingredients. Usually, rubber compounds are manufactured using the following ingredients: vulcanization activators, processing aids, extenders, reinforcements (fillers), anti-degradants, cross-linking agents, some specific ingredients like anti-pyrenes, colorants[1,2], and so forth. Choosing the type of elastomer, as well as the type and dosage of the necessary ingredients, is a complicated and demanding task. The choice depends entirely on the requirements for the properties of the final composites, their purpose, and also on the exploitation and storage conditions. The preparation of composites with desired properties often requires implementing a combination of two or more elastomers, or two or more fillers, accelerators, antiaging agents, and so forth. Unfortunately, many of the organic ingredients used today are of petroleum origin. According to the American Chemical Society, about 32 L of crude oil is consumed to produce approximately a billion car tires every year worldwide. In fact, upon being used they transform into carbon dioxide, the greenhouse gas responsible for global warming. Therefore, the latest tire technologies have been developed specifically to reduce their consumption. It is essential to replace crude oil sourced technologies (for production of polymers, additives, and reinforcements) with non-petroleum derived materials. Consumers are also growing aware of the need to protect our fragile planet and buy products that are environmentally friendly.[3,4] In Europe, production and usage of chemical substances as well as their potential hazards to human health are regulated by one of the most stringent systems for control over the ingredients in rubber industry— Regulation 1907/2006 of the European Parliament and of the Council concerning the Registration, Evaluation, Authorization and Restriction of Chemicals (REACH).[5]

Generally, researchers focus on renewable resources and examine precisely the possibilities of using the latter for obtaining raw materials for rubber industry. A fact could do better than any conviction: it has been estimated that the world vegetable biomass amounts to about 10^{13} tons and that solar energy renews about 3% of it per annum (i.e., 3.10^{11}, 100,000,000,000 tons).[6]

This chapter aims at presenting briefly the main groups of elastomers and ingredients as well as some of the possibilities for using polymers and ingredients from renewable resources in the preparation of elastomer composites.

2.2 TYPICALLY USED INGREDIENTS

2.2.1 RUBBER MATRICES

The elastomers used in practice could be divided into two main groups: general purpose elastomers and specialty elastomers.

Elastomers of general purpose are composed of hydrocarbon chains. The group consists of: natural rubber (NR), isoprene rubber (IR), butadiene rubber (BR), and styrene-butadiene rubber (SBR). A common property of all these elastomers is being unsaturated, hence they are prone to thermal oxidation and ozone aging. Meanwhile, they are non-resistant to fuels, oils, acids, and so forth. Nevertheless, each one possesses valuable properties and therefore finds wide application in manufacturing of a great number of technical goods, car tires[2], and so forth.

There are a number of polymers belonging to the group of specialty polymers, namely: chloroprene rubber (CR), acrylonitrile-butadiene rubber (NBR)—hydrogenated acrylonitrile-butadiene rubber (HNBR), isobutylene-isoprene rubber (IIR), ethylene-propylene diene terpolymer (EPDM), silicone rubber (QR), chlorosulfonated polyethylene (CSM), fluorocarbon rubbers, urethane rubber, and so forth. Unlike the elastomers of general purpose, the specialty ones possess a valuable complex of more specific properties like good thermal stability, resistance to oils, low gas permeability, resistance to various fuels, acids, and so forth. A part of those elastomers are saturated or weakly unsaturated what ensures their good resistance to thermal oxidation or ozone aging.[2]

2.2.2 CROSS-LINKING AGENTS

Sulfur vulcanization in the presence of various types of accelerators and activators is performed in elastomers containing double bonds. The name of this vulcanization type itself shows that the vulcanizing agent in the case is elemental sulfur. Sometimes it is possible to use organic substances containing sulfur in their molecule, which serve as donors of sulfur. Such a substance, for example is tetramethylthiuram disulfide (TMTD). Sulfur vulcanization is a process whereby mono-, di-, and polysulfide cross-links are formed between the elastomeric macromolecules. The advantage of composites wherein polysulfide cross-links dominate is their elasticity. These cross-links, however, are thermolabile, which determines the low resistance of the composites to thermal oxidation and ozone aging. The problem could be overcome using the so-called efficient or semi-efficient vulcanization system. The method requires using a sulfur amount in the range of 0.5–1 phr (parts per hundred rubber), and an accelerator (usually TMTD) at about 2.5–3.5 phr. Thus, significantly more thermostable mono- and disulfide cross-links prevail in the vulcanization network of the elastomer composites built through an effective vulcanization system,[7–10] unlike in a conventional vulcanization system, wherein the amount of sulfur is 1–2.5 phr, and the accelerator about 1.2–2 phr.

Sulfuric vulcanization cannot be applied to saturated elastomers; therefore, various vulcanization methods are used, such as peroxide,[11] resin,[12] in the presence of metal oxide,[13] and so forth. Peroxide vulcanization is applicable to both saturated and unsaturated elastomers, and is mainly applied to cure EPDM and QR. Dibenzoyl peroxide, di-tertbutyl peroxide, dicumyl peroxide, and so forth are commonly used. Resin vulcanization applies to IIR, which makes use of Phenol-formaldehyde and alkylphenol-formaldehyde resins. Vulcanization by metal oxides is applied to CR using a combination of zinc and magnesium oxide. In all cases the advantage of the above sulfur-free vulcanization systems, is the preparation of composites that are resistant to heat and thermal oxidation and ozone aging. This is due to the carbon–carbon or ether bonds in the composite that these types of vulcanization give rise to.

2.2.3 VULCANIZATION ACCELERATORS AND ACTIVATORS

Vulcanization of elastomers without using accelerators and activators is an extremely slow process. The accelerators are organic substances that to a significant extent increase the vulcanization rate and influence the final properties of the composites. They can be classified by several attributes—most often according to their chemical structure and composition. According to the chemical classification, the vulcanization accelerators are divided into thiazoles, sulfenamides, tiuramsulfides, dithiocarbamates, guanidines, xanthates, and so forth. Each of these groups of accelerators possesses characteristic properties; therefore, their selection in the development of an elastomer composition is a complex task.[7,8]

Vulcanization activators are some of the ingredients used in the rubber industry, whose mechanism is subject to the most contradictory theories. However, the benefits from the presence of activators on the properties of the elastomer composites have been proven repeatedly. As already mentioned, in order for an effective rubber vulcanization to proceed in the presence of sulfur or substances that act as donors of sulfur, other substances that act as accelerators in the vulcanization process should be also present. The accelerators, in turn, exert their full effect in the presence of third substances called activators. A large number of metal oxides, such as ZnO, MgO, CdO, CaO, PbO, and so forth, act as activators of accelerated sulfur vulcanization of rubber. The most important representative of this group is zinc oxide. Almost every rubber compound vulcanized with sulfur comprises this oxide.[8,14]

The mechanism of action of accelerators and activators in the vulcanization process is a complex and rarely studied process. It has been found that an increase in the concentration of free zinc ions leads to an increase in the vulcanization rate in the initial stage. However, the rate of cross-linking decreases, but the density of the resulting vulcanization network is higher, which can be explained by the assumption that the chelating zinc complex formed between the accelerator and the activator changes the position of the S–S bond, mainly ceasing it. The destruction of these bonds takes time, hence the rate of cross-linking decreases, but the formed cross-links are of a higher density and a lower sulfidity.[8]

2.2.4   REINFORCEMENTS (FILLERS)

Fillers are highly dispersible substances used for targeted modification of the properties of elastomer composites. The introduction of a filler in rubber compounds leads to an improvement of the technological, mechanical, and performance properties of the finished goods. To a large extent, their use leads to a significantly lower price. The classification of fillers can be made according to their chemical nature (organic and inorganic), reinforcing effects (reinforcing, semi-reinforcing, and inert), size of their particles (micro- and nano-sized), and so forth. The effects that the filler has on the properties of elastomer composites are due to many different factors. Most often these are the size and shape of their particles, specific surface area, ability to be dispersed in the polymer matrix, specific structure, the presence of functional groups on their surface, capability to adsorb elastomer macromolecules on their surface, and so on. Carbon black (CB) and silica are the two most used reinforcing fillers in rubber industry. They provide good mechanical and dynamic properties to the elastomer composites, but are used mainly in the production of technical rubber goods and car tires.[15] A large number of fillers have been used with a main purpose to improve the processability of elastomers (extrusion, calendering, and so forth.) and lower the cost of the latter; such fillers are kaolin, chalk, talc, and so forth. Multi-wall carbon nanotubes (MWCNT), graphene nanoplatelets, different types of carbides and borides (SiC, B_4C, TiB_2), and so forth., are used as fillers in the development of elastomeric composites with specific properties.

Carbon nanotubes have attracted enormous attention for their fundamental behavior and are used in a wide variety of applications in nano-electronic devices,[16–19] probe tips for scanning probe microscopes,[20–24] or in the automotive and aero-space industries for the dissipation of electrostatic charges.[25,26] Due to their structural characteristics and electrical and mechanical properties, one of the most important opportunities in the future is the emergence of a new generation of composite materials. Such materials with a relatively low carbon nanotube loading (<10 wt.%) on polymeric matrices, are required for various applications.[27–29]

Graphene is an allotrope of carbon, whose structure is one-atom thick planar sheets of sp2– bonded carbon atoms packed in a honeycomb lattice. It is the basic structural element of some carbon allotropes including graphite, charcoal, carbon nanotubes, and fullerenes.[30,31] Compared to carbon nanotubes, as well as its high aspect ratio and low density, graphene has attracted considerable attention because of its unique and outstanding mechanical, electrical, and electronic properties (e.g., exceptional in-plane electrical conductivity—up to ~ 20 000 S/cm and the highest thermal conductivity—up to ~ 5 300 W/m.K). All these unique properties in a single nanomaterial have raised the interest of physicists, chemists, and material scientists in the potential of graphene.[29]

SiC, B_4C, and TiB_2 are fillers finding application in manufacturing elastomers designed for thermistors and piezoresistive sensor materials.[32–34] Fillers possessing high dielectric and/or magnetic loss values (such as conductive carbon black, γ-Fe_2O_3, Ni-Zn-Fe_2O_4, Fe_3O_4, $SrFe_{12}O_{19}$, and so forth.)

are suitable for production of conductive composites, microwave shielding materials, and microwave absorbers.[29, 35-40]

2.2.5 PROCESSING AIDS AND PLASTICIZERS

Technological additives are substances favoring the processing of a rubber compound. They facilitate the manufacture of compounds of sophisticated forms via extrusion, injection molding, calendering, and so forth. The additives lower the molecular weight of the elastomer used (peptizers), increase its plasticity (plasticizers), decrease the viscosity of the compound (softeners and plasticizers), increase its productivity, help overcome the reversion occurring during the vulcanization of NR (anti-reversion agents), improve the dispersion of powdered ingredients in the polymer matrix (dispersing agents), improve the interaction between the elastomer and the filler (coupling agents—bifunctional organosilanes), and so forth. Typically, processing additives are used in small amounts and do not change the mechanical properties of the composites. A variety of substances are used as processing additives. Most often these are hydrocarbons (mineral oils, paraffin waxes), fatty acid derivatives, synthetic resins,[41-44] and so forth.

Plasticizers are low molecular weight, oligomeric, or polymeric substances that improve the technological properties of the rubber compounds and the technical properties of their composites as a result of the increasing mobility of polymer macromolecules. Elastomer composites containing plasticizers are characterized by a lower glass transition temperature, that is, have better cold resistance. The addition of plasticizers to rubber compounds, allows using large amount of filler. A large number of substances act as plasticizers, mainly products derived from petroleum processing such as mineral oils (paraffin, naphthenic and aromatic), fuel oils, asphalt, bitumen, paraffin, and so forth. Certain products of plant and animal origin such as rosin, factice, and fatty acids also have a plasticizing effect. There are also synthetic plasticizers that are esters of certain carboxylic acids (phthalic, adipic, sebacic, and so forth.).[45]

The most serious drawback of most of the plasticizers used in the rubber industry is the fact that they are not chemically bound to the elastomer matrix and therefore easily migrate to the surface of products containing them. Given their chemical composition, this is a serious environmental problem. Phthalate plasticizers are considered carcinogenic and therefore their application in Europe is prohibited by Regulation 1907/2006 of the European Parliament and of the Council concerning the REACH.[5]

2.2.6 ANTI-DEGRADANTS

The presence of double bonds in the elastomer predetermines their ability to be vulcanized with sulfur. However, that makes them susceptible to various factors such as heat, light, UV-radiation, oxygen, ozone, aggressive chemicals, and so forth. Some undesirable changes in the structure of elastomers occur under the influence of these factors, namely, destruction or further cross-linking. This set of physical and chemical changes in the polymer may occur during their processing as well as operation and storage of the final composites, and leads to a deterioration of their mechanical properties or to the so-called aging of the elastomers.[46,47]

Aging of the elastomers can be prevented or at least considerably reduced by implementing polymers of lesser unsaturation, as well as by accurately selecting the vulcanization system. Unfortunately, that is not always possible, hence specialists resort to the use of stabilizing additives (anti-degradants). Depending on their mode of action they are divided into antioxidants, anti-ozonants, and anti-fatigue agents. However, the most common classification is according to their chemical nature. According to this classification they are amine, phenolic, heterocyclic nitrogen-containing, and so forth.

Amine anti-degradants such as p-phenylenediamine, diphenylamine, N-isopropyl-N'-phenyl-p-phenylenediamine N-(1,3-dimethylbutyl)-N'-phenyl-p-phenylenediamine, and so forth are efficient antioxidants, antiozonants, and anti-fatigue agents. Phenolic anti-degradants such as 2,2-methylene-bis-(4-methyl-6-tert butyl phenol) and 2,6-di-tert-butyl-4-methyl phenol are efficient antioxidants with a weaker antiozone effect. Heterocyclic nitrogen containing anti-degradants such as 4- and 5-methyl-2-mercapto-benzimidazole and 2,2,4-trimethyl-1,2-dihydroquinoline are powerful antioxidants and antiozonants. Microcrystal waxes, whose function is forming a protective film via their migration over the surface of polymer composites are also used as an effective antiozonants. [46,47]

Like plasticizers, a number of anti-degradants (so-called staining anti-degradants) are prone to migrating to the polymer surface. Although being toxic the application in manufacturing medical goods or such having contact with food and drinking water is limited.

2.3 ELASTOMERS AND INGREDIENTS FROM RENEWABLE RESOURCES

Over the past two decades, scientists and researchers have been paying great attention to polymers and ingredients from renewable resources, mainly for two reasons: concern for the environment and limited oil resources. There-fore, this chapter addresses several options for obtaining elastomers from renewable resources and usage of ingredients (reinforcing fillers, plasti-cizers, and so forth.) of the same origin.

2.3.1 ELASTOMERS FROM RENEWABLE RESOURCES

Most often synthetic elastomers are made from petroleum products. NR and ENR are the only bio-based materials among rubbers. Unlike synthetic elas-tomers it is obtained via an environment -friendly process, namely through coagulation of the latex extracted from the tree *Hevea brasiliensis.* The epox-idation of NR is one of the most promising ways to obtain a new polymeric material from a renewable resource.[48-52] Nevertheless, the consumption of NR increases each year. That makes a greater number of investigators to put efforts in finding new sources for production of elastomers from natural substrates.

Parthenium argentatum Gray (Guayule) is a low-input perennial shrub native to Mexico and southern Texas that has received considerable atten-tion as an alternative source of NR. It has the potential to replace the most common types of rubbers, including synthetic rubber derived from petro-leum. The guayule plant produces NR in parenchyma cells of the bark and the shrub is processed to extract the latex.[53,54] According to literature, guayule natural rubber (GNR) is concurrent to that extracted form *H. brasiliensis.*[55] Much more research should be done in order to confirm the suppositions and give an answer to what extent GNR could replace the elastomers used in current practice.

2.3.2 FILLERS FROM RENEWABLE SOURCES

2.3.2.1 SILICA FROM RICE HUSK

Global production of rice, the majority of which is grown in Asia, is approxi-mately 550 million tons/year. The milling of rice generates a waste mate-rial—rice husk, which is generated at a rate of about 20% of the weight of the product rice, or some 110 million tons/year globally. The husk in turn contains about 25–35% cellulose, 18–21% hemicellulose, 26–31% of lignin, 2–5% soluble compounds, 7.5% moisture, and 15–20% mineral matter, the majority of which is amorphous silica.[56]

Rice husk is currently being used for energy production through direct combustion or gasification in many areas of the world. Unfortunately, in almost all of these installations, the ash produced is not suitable for use as a silica fume substitute. Generally, there are two shortcomings in the ash by-product obtained by the rice husk to energy technology: first, it can contain unacceptably high concentrations of residual carbon and second, a portion of the amorphous silica has been transformed into crystalline silica, cristobalite.

Due to the proven effectiveness of silica in the production of tires (guaranteeing lower rolling resistance, lower fuel consumption, respectively, and less environment pollution and safer travel under winter conditions), silicon dioxide gets more and more widely used in the manufacture of tires.[57–59] Encouraged by the results from using silica, in 2009 Pirelli & C. SpA was the first company in the world to open a plant for the production of standard and easily dispersible silica from rice husks in Brazil. The technology developed by the company is based on the efficient and optimized utilization of energy generated and consumed at various production stages—burning the rice husks to ash, dissolution of the ash, precipitation of silica, drying, silanes treatment, and so forth. Currently, many companies offer silica from rice husk as a commercial product.

2.3.2.2 LIGNIN AND ITS USE AS A BIO-FILLER

Lignin, together with cellulose and chitin, is one of the most common polymers on earth. It is a complex organic compound, a biopolymer, which is derived from wood, wherefrom, comes its name (Lat. -Lignum- "wood"). Structurally, lignin varies depending on the plant source and method of isolation. In recent years, considerable research has been carried out on using lignin as a filler in rubber composites as a substitute for carbon black, particularly in composites based on NR.[60]

It has been found that lignin affects the vulcanization kinetics, improves adhesion to textile, and has an antioxidant effect on thermal aging. The starting material for the preparation of lignin comprises polysaccharides in addition to lignin. The latter impacts the interaction with the elastomer, and purification process is carried out in sodium carbonate. In the process, all impurities are removed eventually to obtain pure lignin. However, the physical and mechanical performance of vulcanizates filled with standard lignin is worse, as compared to that of vulcanizates containing carbon black. In addition, lignin dispersion in the rubber matrix is problematic and the problems are solved by modification of the purified lignin, resulting in improved dispersion, as well as in a better "elastomer-filler" interaction, respectively, and in improved physical and mechanical parameters of the vulcanizates. This is why in recent years, the efforts have focused on developing methods for targeted surface modification of lignin, which guarantees its better dispersion and interaction with the elastomer.[61]

2.3.2.3 STARCH AND ITS USE AS A BIO-FILLER

Polysaccharides are good candidates for renewable nano-fillers because they have partly crystalline structures conferring interesting properties. Starch is a natural, renewable, and biodegradable polymer produced by many plants as a source of stored energy. It is the second most abundant biomass material in nature, and is found in plant roots, stalks, crop seeds, and staple crops such as rice, corn, wheat, tapioca, and potato. Starch industry extracts and refines starches by wet grinding, sieving, and drying. It is either extracted from the plant and is called "native starch," or undergoes one or more chemical modifications to reach specific properties and is called "modified starch." Worldwide, the main sources of starch are maize (82%), wheat (8%), potatoes (5%), and cassava (5%) from which tapioca starch is derived.[62] However, starch has drawbacks when used as a reinforcing filler for elastomers, namely its large particle size, highly polar surface, high cohesive energy, and high soft temperature.[63] The challenge here is the achievement of good compatibility between both polymers (starch and rubber) because of their different chemical affinities—starch is hydrophilic and NR is hydrophobic. Therefore, often starch has to undergo chemical modification, mostly surface modification, so that the disadvantages are eliminated. An appropriate method for modification of starch particles is their esterification by a method that is relatively easy. Starch is mixed with water to form a paste, and then sodium hydroxide is added. Thereafter, carbon disulfide and hydrogen peroxide solutions are added. The resulting final product is a starch xanthate. It has been found that the modification results into a transformation of the crystalline starch structure into an amorphous one. The obtained starch particles have a size of about 200 nm and are readily dispersible in the rubber. Addition of 20 phr to NR improves linearly the physical and mechanical properties with the increasing filler concentration.[64]

2.3.2.4 MICROCRYSTALLINE CELLULOSE (MCC)

Cellulose is the most abundant natural polymer. Commercial cellulose fibers are commonly produced from cotton and wood. Native cellulose is amorphous and crystalline in nature. Crystalline cellulose includes microcrystalline and nano-crystalline cellulose and can thus be derived from the native cellulose fibers. As crystalline cellulose has high strength and high stiffness, it has a great potential to be used as a reinforcing material in the rubber matrix.

From a chemical point of view cellulose is a large and linear-chained polymer with many hydroxyl groups. Its basic unit is anhydroglucopyranose (AGU) having three hydroxyl groups per AGU except the terminal ends.[65] Microcrystalline cellulose consists of cellulose particles with a high degree of crystallinity. These are mainly cellulose aggregates obtained by removing the amorphous domains from the raw cellulose material purified by hydrolytic degradation (typically with a strong mineral acid HCl). The acid hydrolysis process affords coarse aggregates with particle size in the 10 to 50 μm range.

MCC fillers have hydrophilic surface and are not very compatible with hydrophobic rubbers. Coupling agents or compatibilizers are often required for improving the bonding between MCC and the rubber matrix. Surface modification with three-aminopropyl-triethoxysilane has been described.[66] The potential for partial replacement of silica with surface-modified nano-crystalline cellulose and the effect of the latter on the properties of NR composites has been studied. It has been established that modified cellulose activates the vulcanization process, suppresses the Payne effect, improves the stress at 300% elongation, tear strength, and hardness, meanwhile the modification decreases heat build-up and residual deformation.

2.3.3 PLASTICIZERS FROM RENEWABLE SOURCES

The tendency of using ingredients from renewable sources as plasticizers is particularly pronounced. The plasticizers implemented are sunflower, soybean, rapeseed, coconut, hazelnut oil, cashew nuts oil, and so forth. Some of these oils are subjected to a preliminary treatment, mostly epoxidation. Many studies have been devoted to the comparison of the properties of rubber compounds and vulcanizates based thereon containing phthalates and vegetable oils. The conclusions of these studies indicate that some of these plasticizers have already become commercial products, for example,

Pionier TP130 (modified sunflower oil produced by the company Hansen & Rosenthal KG), which a suitable alternative to phthalate ones in nitrile and other elastomers.[67,68] This product ensures the same level of improved processability and resistance to aging, as phthalate plasticizers, being at the same time environment friendly and having no adverse effects on the latter.

Bio-based fillers have several major advantages over conventional fillers such as low density, easy biodegradability, and pollution free. They have good processability and the production of rubber compounds with their addition is energy saving.

A drawback of the fillers from renewable energy sources, however, is their hydrophilicity and need to chemically modify their surface in order to make them hydrophobic and compatible with the elastomeric matrix. However, environmental problems related to the conservation of clean nature, and problems with the gradual depletion of global oil reserves will force more manufacturers of rubber products and their users to turn to fillers from renewable sources.

Therefore, we believe that if not in a short, then in the medium term (15–20 years) use of ingredients from renewable resources will increasingly become widespread in rubber technology in accordance with global standards for keeping the environment clean.

KEYWORDS

- **elastomers**
- **reinforcing fillers**
- **vulcanization activators**
- **curing agents**
- **processing aids**
- **ingredients from renewable sources**

REFERENCES

1. Hertz Jr, D. L. Introduction. In *Engineering with Rubber*, 3rd ed.; Gent, A. N. Ed.; Hanser: Munich, Germany, 2012; pp 1–9.
2. Hamed, G. R. Materials and Compounds In *Engineering with Rubber*, 3rd ed.; Gent, A. N. Ed.; Hanser: Munich, Germany, 2012; pp 11–36.

3. Matlack, A. S. *Introduction to Green Chemistry*, 2nd ed.; CRC Press, Taylor & Francis Group: USA, 2010.

4. Job, K. A. Tendencies in the Green Tire Production. *Rubber World* **2014**, *249*, 32–38.

5. Regulation (ec) 1907/2006 of the European Parliament and of the council. *Off. J. Eur. Un.* **2006**, *L396*, 1–849.

6. Gandini, A.; Belgacem, M. N. The State of the Art. In *Monomers, Polymers and Composites from Renewable Resources*; Belgacem, M. N., Gandini, A., Eds.; Elsevier: Amsterdam, 2008; pp 1–16.

7. Engels, H.-W.; Weidenhaupt, H.-J.; Pieroth, M.; Hofmann, W.; Menting, K.-H.; Mergenhagen, T.; Schmoll, R.; Uhrlandt, S. *Chemicals and Additives. Ullmann's Encyclopedia of Industrial Chemistry*; Wiley-VCH: Berlin, 2000.

8. Coran, A. Y. Vulcanization. In *The Science and Technology of Rubber*, 4th ed.; Mark, J. E., Erman, B., Roland, C. M., Eds.; Academic Press: Boston, 2013; pp 337–381.

9. Formela, K.; Wąsowicz, D.; Formela, M.; Hejna, A.; Haponiuk, J. Curing Characteristics, Mechanical and Thermal Properties of Reclaimed Ground Tire Rubber Cured with Various Vulcanizing Systems. *Iran. Polym. J.* **2015**, *24*, 289–297.

10. Rattanasom, N.; Poonsuk, A.; Makmoon, T. Effect of Curing System on the Mechanical Properties and Heat Aging Resistance of Natural Rubber/Tire Tread Reclaimed Rubber Blends. *Polym. Test.* **2005**, *24*, 728–732.

11. Dluzneski, P. R. Peroxide Vulcanization of Elastomers. *Rubber Chem. Technol.* **2001**, *74*, 451–492.

12. Lattimer, R. P.; Kinsey, R. A.; Layer, R. W.; Rhee, C. K. The Mechanism of Phenolic Resin Vulcanization of Unsaturated Elastomers. *Rubber Chem. Technol.* **1989**, *62*, 107–123.

13. Miyata, Y.; Atsumi, M. Zinc Oxide Crosslinking Reaction of Polychloroprene Rubber. *Rubber Chem. Technol.* **1989**, *62*, 1–12.

14. Garreta, E.; Agullo, N.; Borros, S. The Rle of the Activator During the Vulcanization of Natural Rubber Using Sulfenamide Accelerator Type. *KGK-Kautschuk Gummi Kunststoffe* **2002**, *55*, 82–85.

15. Donnet, J.-B.; Custodero, E. Reinforcement of Elastomers by Particulate Fillers. In *The Science and Technology of Rubber*, 4th ed.; Mark, J. E., Erman, B., Roland, C. M., Eds.; Academic Press: Boston, 2013; pp 383–416.

16. Bachtold, A.; Hadley, P.; Nakanishi, T.; Dekker, C. Logic Circuits with Carbon Nanotube Transistors. *Science* **2001**, *294*, 1317–1320.

17. Derycke, V.; Martel, R.; Appenzeller, J.; Avouris, P. Carbon Nanotube Inter- and Intramolecular Logic Gates. *Nano Lett.* **2001**, *1*, 453–456.

18. Rotkin, S. V.; Zharov, I. Nanotube Light-Controlled Electronic Switch. *Int. J. Nanosci.* **2002**, *01*, 347–355.

19. Iijima, S. Helical Microtubules of Graphitic Carbon. *Nature* **1991**, *354*, 56–58.

20. Akita, S.; Nishijima, H.; Nakayama, Y.; Tokumasu, F.; Takeyasu, K. Carbon Nanotube Tips for a Scanning Probe Microscope: Their Fabrication and Properties. *J. Phys. D: Appl. Phys.* **1999**, *32*, 1044–1048.

21. Cheung, C. L.; Hafner, J. H.; Odom, T. W.; Kim, K.; Lieber, C. M. Growth and Fabrication with Single-Walled Carbon Nanotube Probe Microscopy Tips. *Appl. Phys. Lett.* **2000**, *76*, 3136–3138.

22. Wilson, N. R.; Cobden, D. H.; Macpherson, J. V. Single-Wall Carbon Nanotube Conducting Probe Tips. *J. Phys. Chem. B* **2002**, *106*, 13102–13105.

23. Yenilmez, E.; Wang, Q.; Chen, R. J.; Wang, D. W.; Dai, H. J. Wafer Scale Production of Carbon Nanotube Scanning Probe Tips for Atomic Force Microscopy. *Appl. Phys. Lett.* **2002,** *80,* 2225–2227.

24. Ye, Q.; Cassell, A. M.; Liu, H. B.; Chao, K. J.; Han, J.; Meyyappan, M. Large-Scale Fabrication of Carbon Nanotube Probe Tips for Atomic Force Microscopy Critical Dimension Imaging Applications. *Nano Lett.* **2004,** *4,* 1301–1308.

25. Breuer, O.; Sundararaj, U. Big returns from small fibers: A Review of Polymer/Carbon Nanotube Composites. *Polym. Compos.* **2004,** *25,* 630–645.

26. Wise, K. E.; Park, C.; Siochi, E. J.; Harrison, J. S. Stable Dispersion of Single Wall Carbon Nanotubes in Polyimide: the Role of Noncovalent Interactions. *Chem. Phys. Lett.* **2004,** *391,* 207–211.

27. Ajayan, P. M.; Schadler, L. S.; Giannaris, C.; Rubio, A. Single-Walled Carbon Nanotube-Polymer Composites: strength and Weakness. *Adv. Mater.* **2000,** *12,* 750–753.

28. Shaffer, M.; Kinloch, I. A. Prospects for Nanotube and Nanofibre Composites. *Compos. Sci. Technol.* **2004,** *64,* 2281–2282.

29. Al-Hartomy, O.; Al-Ghamdi, A.; Said, S. F. A.; Dishovsky, N.; Mihaylov, M.; Ivanov, M.; Zaimova, D. Comparison of the Dielectric Thermal Properties and Dynamic Mechanical Thermal Properties of Natural Rubber-Based Composites Comprising Multiwall Carbon Nanotubes and Graphene Nanoplatelets. *Fullerenes, Nanotubes and Carbon Nanostruct.* **2015,** *23,* 1001–1007.

30. Salavagione, H. J.; Martínez, G.; Ellis, G. Graphene-Based Polymer Nanocomposites. In *Physics and Applications of Graphene—Experiments*; Mikhailov, S., Ed.; InTech: Rijeka, Croatia, 2011; pp 169–192.

31. Novoselov, K. S.; Geim, A. K.; Morozov, S. V.; Jiang, D.; Zhang, Y.; Dubonos, S. V.; Grigorieva, I. V.; Firsov, A. A. Electric Field Effect in Atomically Thin Carbon Films. *Science* **2004,** *306,* 666–669.

32. Todorova, Z.; Dishovsky, N.; Dimitrov, R.; El-Tantawy, F.; Abdel Aal, N.; Al-Hajry, A.; Bououdina, M. Natural Rubber Filled SiC and B4C Ceramic Composites as a New NTC Thermistors and Piezoresistive Sensor Materials. *Polym. Compos.* **2008,** *29,* 109–118.

33. Jeong, D.-Y.; Ryu, J.; Lim, Y.-S.; Dong, S.; Park, D.-S. Piezoresistive TiB2/Silicone Rubber Composites for Circuit Breakers. *Sens. Actuators A: Phys.* **2009,** *149,* 246–250.

34. Ismail, A. M.; Mahmoud, K. R.; Abd-El Salam, M. H. Electrical Conductivity and Positron Annihilation Characteristics of Ternary Silicone Rubber/Carbon Black/TiB2 Nanocomposites. *Polym. Test.* **2015,** *48,* 37–43.

35. Barba, A. A.; Lamberti, G.; d'Amore, M.; Acierno, D. Carbon Black/Silicone Rubber Blends as Absorbing Materials to Reduce Electromagnetic Interferences (EMI). *Polym. Bull.* **2006,** *57,* 587–593.

36. Joshi, A.; Datar, S. Carbon Nanostructure Composite for Electromagnetic Interference Shielding. *Pramana* **2015,** *84,* 1099–1116.

37. Krishnan, Y.; Chandran, S.; Usman, N.; Smitha, T. R.; Parameswaran, P. S.; Prema, K. H. Processability, Mechanical and Magnetic Studies on Natural Rubber-Ferrite Composites. *Int. J. Chem. Stud.* **2015,** *3,* 15–22.

38. Ismail, H.; Sam, S. T.; Mohd Noor, A. F.; Bakar, A. A. Properties of Ferrite-Filled Natural Rubber Composites. *Polym.-Plast. Technol. Eng.* **2007,** *46,* 641–650.

39. Kong, I.; Hj Ahmad, S.; Hj Abdullah, M.; Hui, D.; Nazlim Yusoff, A.; Puryanti, D. Magnetic and Microwave Absorbing Properties of Magnetite–Thermoplastic Natural Rubber Nanocomposites. *J. Magn. Magn. Mater.* **2010,** *322,* 3401–3409.

40. Urogiova, E.; Hudec, I.; Bellusova, D. Magnetic and Mechanical Properties of Strontium Ferrite–Rubber Composites. *KGK-Kautschuk Gummi Kunststoffe* **2006**, *6*, 224–228.

41. Easterbrook, E. K.; Allen, R. D. Ethylene-Propylene Rubber. In *Rubber Technology*; Morton, M., Ed.; Springer Netherlands: Dordrecht, 1999; pp 260–283.

42. Graff, R. S.; Baseden, G. A. Neoprene and Hypalon. In *Rubber Technology*; Morton, M., Ed.; Springer Netherlands: Dordrecht, 1999; pp 339–374.

43. Schroeder, H. Fluorocarbon Elastomers. In *Rubber Technology*; Morton, M., Ed.; Springer Netherlands: Dordrecht, 1999; pp 410–437.

44. Stephens, H. L. The Compounding and Vulcanization of Rubber. In Rubber Technology; Morton, M., Ed.; Springer Netherlands: Dordrecht, 1999; pp 20–58.

45. Wypych, G. *Handbook of Plasticizers*, 2nd ed.; ChemTec Publishing: Toronto, Canada, 2012.

46. Rodgers, B.; Waddell, W. The Science of Rubber Compounding. In *The Science and Technology of Rubber*, 4th ed.; Mark, J. E., Erman, B., Roland, C. M., Eds.; Academic Press: Boston, 2013; pp 417–471.

47. NPCS Board of Consultants and Engineers. *The Complete book on Rubber Chemicals*; Asia Pacific Business Press Inc., 2009.

48. Gelling, I. R.; Porter, M. Chapter 10. In *Natural Rubber Science and Technology*; Robert, A. D., Ed.; Oxford University Press: Oxford, 1988.

49. Chapman, A. V. *Natural Rubber and nr-Based Polymers: Renewable Materials with Unique Properties*; 24th International H F Mark Symposium on Advances in the Field of Elastomers & Thermoplastic Elastomers, Vienna, Austria, 2007.

50. Bac, N. V.; Huu, C. C. Synthesis and Application of Epoxidized Natural Rubber. *J. Macromol. Sci., Part A* **1996**, *33*, 1949–1955.

51. Bac, N. V.; Terlemezyan, L.; Mihailov, M. Epoxidation of Natural Rubber in Latex in the Presence of a Reducing Agent. *J. Appl. Polym. Sci.* **1993**, *50*, 845–849.

52. Heping, Y.; Sidong, L.; Zheng, P. Preparation and Study of Epoxidized Natural Rubber. *J. Therm. Anal. Calorim.* **1999**, *58*, 293–299.

53. Rasutis, D.; Soratana, K.; McMahan, C.; Landis, A. E. A Sustainability Review of Domestic Rubber from the Guayule Plant. *Ind. Crops Prod.* **2015**, *70*, 383–394.

54. de Rodríguez, D. J.; Angulo-Sánchez, J. L.; Rodríguez-García, R. Mexican High Rubber Producing Guayule Shrubs: A Potential Source for Commercial Development. *J. Polym. Environ.* **2006**, *14*, 37–47.

55. Barrera, C. S.; Cornish, K. High Performance Waste-Derived Filler/Carbon Black Reinforced Guayule Natural Rubber Composites. *Ind. Crops Prod.* **2016**, *86*, 132–142.

56. Begum, S. P. M.; Dominic, M. C. D.; Joseph, R.; Aiswrya, E. P.; Kumar, P. Synthesis Characterization and Application of Rice Husk Nanosilica in Natural Rubber. *Int. J. Sci., Environ. Technol.* **2013**, *2*, 1027–1035.

57. Liou, T.-H. Preparation and Characterization of Nano-Structured Silica from Rice Husk. *Mater. Sci. Eng.: A* **2004**, *364*, 313–323.

58. Real, C.; Alcalá, M. D.; Criado, J. M. Preparation of Silica from Rice Husks. *J. Am. Ceram. Soc.* **1996**, *79*, 2012–2016.

59. Zurina, M.; Ismail, H.; Bakar, A. A. Partial Replacement of Silica by Rice Husk Powder in Polystyrene–Styrene Butadiene Rubber Blends. *J. Reinf. Plast. Compos.* **2004**, *23*, 1397–1408.

60. Setua, D. K., Shukla, M. K., Nigam, V., Singh, H., and Mathur, G. N. Lignin Reinforced Rubber Composites. *Polym. Compos.* **2000**, *21*, 988–995.

61. Kumaran, M. G.; De, S. K. Utilization of Lignins in Rubber Compounding. *J. App. Polym. Sci.* **1978,** *22,* 1885–1893.
62. Le Corre, D.; Bras, J.; Dufresne, A. Starch Nanoparticles: A Review. *Biomacromolecules* **2010,** *11,* 1139–1153.
63. Liu, C.; Shao, Y.; Jia, D. Chemically Modified Starch Reinforced Natural Rubber Composites. *Polymer* **2008,** *49,* 2176–2181.
64. Wang, Z.-F.; Peng, Z.; Li, S.-D.; Lin, H.; Zhang, K.-X.; She, X.-D.; Fu, X. The Impact of Esterification on the Properties of Starch/Natural Rubber Composite. *Compos. Sci. Technol.* **2009,** *69,* 1797–1803.
65. Bai, W. New Application of Crystalline Cellulose in Rubber Composites. Ph.D. Thesis, Oregon State University, 2009.
66. Xu, S. H.; Gu, J.; Luo, Y. F.; Jia, D. M. Effects of Partial Replacement of Silica with Surface Modified Nanocrystalline Cellulose on Properties of Natural Rubber Nanocomposites. *eXPRESS Polym. Lett.* **2012,** *6,* 14–25.
67. Karadeniz, K.; Ergüler, N. Investigation of Plasticizer Effect of Hazelnut Oil and its Epoxidized Derivative on Chloroprene and Nitrile Rubbers. *KGK-Kautschuk Gummi Kunststoffe* **2012,** *65,* 49–54.
68. Bergmann, C. Replacement of Phthalates by Vegetable Oil Derived Plasticizers in NBR Compounds. *Rubber World* **2014,** *250,* 42–46.

CHAPTER 3

FILLERS FOR ELASTOMER-BASED COMPOSITE MATERIALS: SYNTHESIS AND CHARACTERIZATION

CONTENTS

ABSTRACT

Practical application of all elastomers is almost impossible without rein-forcing them by fillers. Nowadays, the usage of fillers in the prepara-tion of elastomer composites not only means to change and improve their mechanical and dynamic properties and lower the final price but also to develop composites possessing novel and more attractive properties. This chapter describes the synthesis methods, options for modification of fillers, and some possibilities for their characterization. Utilization of such fillers contributes to the complexity of elastomer properties and also to widening the range of their applications.

3.1   INTRODUCTION

Elastomers are usually reinforced with active fillers to achieve substantial improvement in the strength and stiffness of the resultant products. It is widely accepted that the extent of improvement of these properties depends on the size of filler particles, specific surface area, structurality (their ability to form aggregates and agglomerates), chemistry of their surface, degree of dispersion, and interaction with the polymer matrix.[1]

Carbon black (CB) and silica are the two most commonly used fillers in rubber industry. However, both of them have advantages and disadvantages. Hence, in the past years a considerable number of studies has been devoted to the so-called hybrid fillers, which combine the advantages of two fillers by overcoming their drawbacks.[2-7] Most publications in this field deal with hybrid CB/silica fillers designed for production of tire treads.[8-10] Many works have also been devoted to hybrid fillers designed for microwave absorbers and highly effective electromagnetic shields.[11-17] The reasons for partic-ular interest in these fillers is the fact that most of the present-day micro-wave absorbing composites are produced from a dielectric rubber matrix and specific functional fillers. These fillers—CB, graphite, active carbon, short carbon or metal fibers, carbonyl iron, micro- and nano-sized metal powders—possess high values of imaginary part of the complex permittivity or permeability and absorb high frequency energy. In the past few years, the composites containing both dielectric and magnetic fillers have been a well-considered topic in the field of EMI shielding and radar absorbing materials.[18,19] As an electromagnetic wave has both a dielectric and magnetic component, obviously, both dielectric and magnetic materials are effective for the absorption of microwave radiation. Therefore, it has been of great interest to combine components of high dielectric and magnetic losses into one hybrid filler, which opens new opportunities of preparing modern microwave absorbers with specific properties.[20–22]

The aim of the chapter is to present briefly the synthesis methods and possibilities to modify the reinforcing fillers. The discussion covers some characteristics of the fillers. Using such fillers contributes to improving the complexity of properties possessed by elastomer composites as well as to widen the range of their applications.

3.2 METHODS FOR SYNTHESIS OF FILLERS

3.2.1 METHODS FOR SYNTHESIS OF CARBON BLACK

CB is a material composed mainly of elemental carbon in the form of quasi-spherical particles bonded in aggregates and agglomerates of different shape and size. Although there are various methods for obtaining it: oil furnace process, gas furnace process, thermal process, acetylene black process, lampblack process, and channel process, more than 90% of the world's production is by using furnace CB.[23] A special furnace type reactor is used to run the oil furnace process for obtaining CB. The reactor is the core of the production line and in most cases, has been developed and patented by the respective manufacturer. The feedstock oil used to obtain CB is injected into the flame of natural gas, where its partial dehydrogenation leads to the formation of CB particles. In general, the production reactor may be divided into different sections that can be described as fuel, furnace, cooling, and a section for collecting the final product. The feedstock is introduced into the combustion section under atomized conditions and ignited by burning natural gas. CB particles are formed in the combustion zone. Parameters such as oxygen concentration, temperature, time of passing through the reaction zone, form of the combustion section, various additives, and many other factors are strictly controlled. The time spent in the reaction chamber is of the order of a few milliseconds and is one of the basic parameters determining the different types of furnace CB produced, that is, the size and shape of the particles and aggregates are under control. After the reaction zone, the CB particles undergo drastic cooling and the formation process is considered to be complete.[23]

3.2.2   METHODS FOR SYNTHESIS OF SILICA

Silica is a highly dispersible amorphous filler with more or less bound water (chemically, co-ordinationally, or by adsorption). It disperses in the rubber compound worse than carbon black, because it contains hydroxyl and silanol groups, as a result of which the filler particles are more prone to aggregation and agglomeration. Therefore, higher shear stress is required for the preparation of rubber compounds containing silica. Silica is modified with bifunctional organosilanes, so that its interaction with elastomers and reinforcing ability are enhanced.[24]

Silica used as reinforcing filler is obtained mainly by precipitation. The method consists in precipitation of amorphously hydrated silica $(SiO_2)_m \cdot (H_2O)_n$ at 70–90°C from solutions of sodium silicates with acids in the presence of salts of metals belonging to the second and third group of the Periodic Table.

Depending on precipitation conditions and on the nature of coagulant, an acidic, neutral, or alkaline product is obtained, which is successively filtered, washed, dried, and ground.[25] Maurice Abou Rida[26] et al. have prepared amorphous silica by precipitation of sodium silicate with hydrochloric acid by using Stöber method. The particle size of the resulting silica is controlled by a number of reactions in the presence of surfactants as dispersing agents. These surfactants are cetyltrimethylammonium bromide (CTAB) and dodecyltrimethylammonium bromide (DTAB). The authors state that the particle size of the resulting product can be controlled within 148–212 nm.

Another possibility to synthesize silica is to implement the sol-gel method. Jafarzadeh[27] and co-workers obtained nanosized silica modified by the sol-gel technique using different modes of reactants mixing and drying techniques. The method is based on dissolving tetraethyl orthosilicate (TEOS) in ethanol in a low-frequency ultrasonic bath, followed by addition of a small amount of distilled water in order to hydrolyze TEOS. Ammonia was used as a catalyst. The resulting gel was washed with ethanol and distilled water and then dried at a temperature of about 70°C. Finally, the dried product is calcined at temperatures of up to about 600°C. The authors state the particle size of silica prepared by that method to be in the 10–15 nm scale.[27]

Another method for the preparation of silica is the so-called fumed process. The preparation of the fumed silica consists of hydrolyzing silicon tetrachloride or tetrafluoride at 1000–1100°C by means of water vapor, that is, pyrolysis of tetrachloride is carried out in superheated vapor. The filler thus obtained is of a very stable morphology, highly dispersible, and with reactive ability.[25,28]

Regardless of the process used for the preparation of silica, a strict control of the conditions under which it is obtained is essential for the properties of the final product, namely, particle size, chemical activity of the surface, and so forth.

3.2.3 METHODS FOR SYNTHESIS OF HYBRID FILLERS

In a wider sense, the term a "hybrid material" refers to materials comprising two phases—organic and inorganic—with/without chemical interactions between them. The more proper referent of the term, however, is a material composed of an organic phase and an inorganic one at a molecular level. The interaction of the two phases could be either weak, for example, either by Van der Waals forces or hydrogen bonds; or by strong covalent bonds.[29]

Hybrid materials, finding application in preparation of elastomer composites, in particular, could be obtained by several methods: in situ formation of both phases, in situ formation of the inorganic phase, or a sol-gel process.[29]

In the recent years, Cabot Corporation has launched a dual phase filler of carbon-silica type (CSDPF 2000) in the market. The manufacturing method involves unique co-fuming process based on pyrolysis of feedstock of oils and silica-comprising ingredients.[8,30–32] With this material, the filler–filler interaction is weakened substantially due to surface modification, while the polymer–filler interaction is enhanced by the increased surface energy of the carbon domain of the filler and by the chemical bonds formed via a coupling reaction between polymer chains and silanols over the silica domain.[33]

We consider the process of obtaining carbon–silica fillers via pyrolysis-cum-water vapor of worn out waste tires to be a method alternative to the co-fuming process. Generally, the pyrolysis of tires is described as a tire recycling method.[34] We assumed that a dual phase filler could be obtained if the tires subjected to pyrolysis were not standard ones (filled only with CB), but the so called green tires comprising a high quantity of silica. We expected carbon to be formed due to the pyrolysis itself, as a result of the destruction of elastomers. The method is attractive, mostly because an essential environmental problem might be solved. On the other hand, the dual phase filler obtained has the advantages described above.

Impregnation is one of the most widely used and popular methods for applying different phases on carriers (first of all in the production of catalysts) or for the purpose of surface and texture modification of adsorbents.[35] By this impregnation method, the second phase is deposited from the solution of the so-called precursor (most often salts or soluble complexes). The impregnated substances are subjected to chemical or physical influence that may be considered to act as a thermal activation. The method of applying the second phase via impregnation allows controlled deposition of the second phase (silica) over the CB surface. A part of the latter may also be retained within the pores of carbon black particles and in the spaces between the particles. The method is simple and easy to accomplish. Its advantage is that it facilitates easy control over the quantity ratio between the two phases in the final hybrid product. Moreover, the impregnation method is suitable not only for producing hybrid fillers of CB-silica type but also for other types of fillers containing completely different inorganic components, such as magnetite.

3.3 METHODS FOR MODIFICATION OF FILLERS

Modification of fillers is performed in order to expand the range of their properties and the possibilities for application.

Nowadays, fillers are being modified by two basic methods. The first is carried out with the process of their preparation and consists of adding various modifying agents to the starting materials. The second, which corresponds more aptly to the concept of modification, is related to the direct effect on the very fillers upon completion of their synthesis. The effect can be achieved by chemical and physical agents or by a combined treatment, which most often involves the filler surface.

One possibility to prepare modified CB at the time of its production is placing the nozzle into the reaction zone through which metal salts[36,37] or other substances[38] are introduced. Under the influence of these additives, the produced carbon black acquires a more ordered structure or electric and magnetic properties.

The opportunities to modify already-prepared carbon black are not few. For instance, being modified in an aqueous solution of NaOCl and HNO_3 carbon black acquires high hydrophilicity[38] that makes it more compatible with polar elastomers and favors its dispersion in the matrix.

The thermal treatment of carbon black over 900°C is related to an enhancement in its conductivity. Incandescence of carbon black, concomitant with reduction of oxygen and hydrogen content is followed by a high degree of graphitization. This leads to a lower resistivity to an increase in the electrical conductivity of the modified CB, respectively. The lowest resistivity values have been obtained for CB treated in the 1000–1500°C temperature range. At temperatures above 1500°C, the parameters of the primary crystallites change. This worsens the filler–filler interaction, decreasing the conductivity of the treated CB.[39]

Treatment of silica surface with various organosilanes can also be regarded as a modification method. Silanization of silica surface by addition of organosilanes is an effective method for surface modification of the filler. This type of modification is carried out during the filler compounding with the elastomer matrix. Hydrolysis and condensation occur as a result of the high temperature (up to 150°C) during compounding. This gives rise to an interaction between the alkoxy groups of the bifunctional organosilane with the silanol groups of silica. Alcohol is evolved at the end of the reaction. This approach to silica modification results into "blocking" of hydroxyl groups over the filler surface, that is, up to its hydrophobization, which makes it considerably more compatible with the elastomeric matrix.[40–43] According to Theppradit[44] et al., silanization of the silica is more effective when the process is carried out while synthesizing the filler. In addition to the improved mechanical and dynamic properties of the elastomeric composites, the risk of premature vulcanization (scorching) during the mixing process is practically zero, since it is unnecessary to add extra organosilanes (which in most cases contain high amounts of sulfur in their molecules) to the pre-silanized silica.

Studies have shown that plasma is a suitable medium to produce modified fillers whose qualities and properties may be altered within very wide limits. This is due to the easy manipulation of the conditions yielding plasma. Plasma is partially or fully ionized gas (or vapor) wherein the positive and negative electrical charges are practically at an equal concentration. Plasma is regarded as the fourth state of matter characterized by partial or total ionization of the medium (atoms, molecules, radicals) resulting from the inelastic collision between the particles (electrons, atoms, molecules, ions, radicals). As a part of their kinetic energy is transformed into internal energy of the particles—they excite and ionize.

Plasma as a medium for obtaining materials with different qualities and variable properties is not a novelty in terms of production conditions. Recently, however, it has been a tool for modifying surfaces and volumes of materials of different nature. Plasma is used to change the surface characteristics of various types of polymer films and fibers. The processing of PET fibers with cold plasma at atmospheric pressure leads to enriching the surface of a material in nitrogen-containing groups; to an increase in the number and size of pores on the surface thereof, and to improvement in their adhesion to polymers—commensurate with that from standard chemical processing of these fibers.[45] The same results have also been achieved by

Hudec[46] et. al. Increasing the hydrophilicity of polymeric film surfaces is also attained by plasma treatment. These films find practical application in medicine (for treatment of catheters and implants).[47]

According to Pishinkov[48] et al., modification with cold plasma discloses the possibility, in terms of combined treatment with oxygen and ammonia plasma, for preparation of carbon fillers offering increased modules of the composites based on non-polar elastomers. The practical application of this effect could be found in materials requiring stronger carcass and especially in the various methods for processing elastomers, for example, a combination of extrusion and continuous vulcanization of thick-walled profiles requiring completely retained shape of the cross section until the vulcanization proceeds. Treatment with acrylic acid plasma offers a product having the ability to provide reduced heat build-up in non-polar elastomer composites, making them suitable for dynamic applications—with regards to the rubber industry.

Plasma treatment of active silica proves to be a quite suitable method for improving the degree of filler dispersion in the elastomeric matrix. This, together with improved mechanical properties has been found by the number of researchers.[49,50] The authors have also established that curing rate of the treated fillers is considerably higher than that of the untreated one. This effect is explained by the authors with the lower ability of the active filler to adsorb the vulcanization ingredients. A different strategy for surface treatment of sedimentary silica has been offered by Jesionowski and co-workers.[51] They initially treated the filler surface with three types of commercial silanes: (3-mercaptopropyl) trimethoxy silane, (aminoethyl)-3-aminopropyl) triethoxy silane, and 3-metacryloxypropyltrimethoxy silane, and subjected it to atmospheric plasma and explored its properties in a matrix of polybutylene terephthalate. The authors found by means of scanning tunneling microscope that plasma treatment of silanized filler particles results in much better filler dispersion in the polymer matrix—in a much smaller number of aggregates compared with the reference—untreated plasma samples, respectively.

3.4 FILLERS CHARACTERIZATION

The main characteristics of the fillers used in the rubber industry are essential for their reinforcing effect. These characteristics are particle size and shape, specific surface area, density, presence of functional groups on their surface, their ability to form secondary structures—aggregates and agglomerates, and so forth. There are standardized conventional methods for the determination of these fillers properties. For example, the size and form of the particles can be determined using a number of microscopic methods: scanning electron microscope (SEM), transmission electron microscope (TEM), scanning transmission electron microscope (STEM), and so forth. The specific surface area of the fillers is estimated by determining the nitrogen adsorption using BET method. CTAB adsorption is a relatively new method, which is associated mainly with measurement of the external surface area, and with the filler surface accessible by the polymer. The presence of functional groups on the surface of filler particles can be examined by FTIR or NMR spectroscopy. The ability of filler particles to form secondary structures (aggregates and agglomerates) is determined by measuring their oil absorption number (OAN), in other words, the ability of the filler to absorb dibutylphthalate (DBP).[52,53]

The iodine number (or iodine adsorption value) is the mass of iodine in grams that is consumed by 100 g of filler (e.g., CB). Iodine numbers are often used to determine the amount of unsaturation on filler surface. This unsaturation is in the form of double bonds, which react with iodine compounds. The higher the iodine number, the more $C = C$ bonds are present on the filler surface, and more active the filler is. Therefore, iodine number is an important parameter when characterizing a filler.

Besides the abovementioned characteristics, when hybrid fillers are in question, one should find an answer to several questions, namely what the distribution of the two phases (organic and inorganic) in the hybrid filler is, and to what extent the phases interpenetrate.

Without a doubt, the most appropriate analytical method for the purpose is energy dispersive X-ray (EDX) spectroscopy in a STEM, or STEM–EDX. High angle annular dark field (HAADF) imaging in the STEM displays compositional contrast that results from the different atomic numbers of the elements and their distribution. EDX allows identifying the particular elements and their relative proportions. Initial EDX analysis usually involves the generation of an X-ray spectrum from the entire scan area of the STEM image. The Y-axis shows the counts (number of X-rays received and processed by the detector) and the X-axis shows the energy of those X-rays. In addition, EDX software can:

- Keep the electron beam stationary on a spot or series of spots and generate spectra that provides more localized elemental information.
- Have the electron beam follow a line drawn on the sample image and generate a plot of the relative proportions of different elements along that line (one-dimensional line scanning).

- Map the distribution and relative proportion (intensity) of different elements over the scanned area (two-dimensional mapping)

Modern software now collects the entire cumulative spectra from each point, so element choices can be made post acquisition.

A great part of the described methods for synthesis, modification, and characterization of fillers which find application in manufacturing elastomer composites have been used for our investigations. The following chapters of this book present in detail the effect of the obtained fillers upon the complexity of properties, mainly those of composites based on natural rubber.

KEYWORDS

- **elastomer-based composites**
- **reinforcing fillers**
- **carbon black**
- **silica**
- **hybrid fillers**
- **modification of fillers**
- **characterization of fillers**

REFERENCES

1. Donnet, J.-B.; Custodero, E. Reinforcement of Elastomers by Particulate Fillers. In *The Science and Technology of Rubber* 4th ed.; Mark, J. E., Erman, B., Roland, C. M., Eds.; Academic Press: Boston, 2013; pp 383–416.
2. Sun, D.; Li, X.; Zhang, Y.; Li, Y. Effect of Modified Nano-silica on the Reinforcement of Styrene Butadiene Rubber Composites. *J. Macromol. Sci., Part B* **2011**, *50*, 1810–1821.
3. Zafarmehrabian, R.; Gangali, S. T.; Ghoreishy, M. H. R.; Davallu, M. The Effects of Silica/Carbon Black Ratio on the Dynamic Properties of the Tread Compounds in Truck Tires. *E-J. Chem.* **2012**, *9*.
4. Wolff, S.; Wang, M.-J. Filler—Elastomer Interactions. Part IV. The Effect of the Surface Energies of Fillers on Elastomer Reinforcement. *Rubber Chem. Technol.* **1992**, *65*, 329–342.
5. Seyvet, O.; Navard, P. Collision-Induced Dispersion of Agglomerate Suspensions in a Shear Flow. *J. Appl. Polym. Sci.* **2000**, *78*, 1130–1133.

6. Donnet, J.-B. Black and White Fillers and Tire Compound. *Rubber Chem. Technol.* **1998,** *71,* 323–341.

7. Sengloyluan, K.; Sahakaro, K.; Dierkes, W. K.; Noordermeer, J. W. M. Silica-reinforced Tire Tread Compounds Compatibilized by Using Epoxidized Natural Rubber. *Eur. Polym. J.* **2014,** *51,* 69–79.

8. Wang, M. J.; Kutsovsky, Y.; Zhang, P.; Mehos, G.; Murphy, L. J.; Mahmud, K. Using Carbon-silica Dual Phase Filler Improve Global Compromise Between Rolling Resistance, Wear Resistance and Wet Skid Resistance for Tires. KGK-*Kautsch. Gummi Kunstst.* **2002,** *55,* 33–40.

9. Kawazura, T. Process for Production of Modified Carbon Black for Rubber Reinforcement and Rubber Composition Containing Modified Carbon Black. U.S. Patent 20,020,169,242, A1.

10. Sone, K.; Ishida, M.; Mizushima, K. Process for Preparing Surface-treated Carbon Black and Rubber Composition. U.S. Patent 6,248,808, B1.

11. Shtarkova, R.; Dishovsky, N. Elastomer-based Microwave Absorbing Materials. *J. Elastomers Plast.* **2009,** *41,* 163–174.

12. Dishovsky, N.; Grigorova, M. On the Correlation Between Electromagnetic Waves Absorption and Electrical Conductivity of Carbon Black Filled Polyethylenes. *Mater. Res. Bull.* **2000,** *35,* 403–409.

13. Dishovsky, N. Rubber Based Composites with Active Behaviour to Microwaves (Review). *J. Univ. Chem. Technol. Metall.* **2009,** *44,* 115–122.

14. Zhang, C.-S.; Ni, Q.-Q.; Fu, S.-Y.; Kurashiki, K. Electromagnetic Interference Shielding Effect of Nanocomposites with Carbon Nanotube and Shape Memory Polymer. *Compos. Sci. Technol.* **2007,** *67,* 2973–2980.

15. Duan, Y.; Li, G.; Liu, L.; Liu, S. Electromagnetic Properties of Carbonyl Iron and Their Microwave Absorbing Characterization as Filler in Silicone Rubber. *Bull. Mater. Sci.* **2010,** *33,* 633–636.

16. Ling, Q.; Sun, J.; Zhao, Q.; Zhou, Q. Effects of Carbon Black Content on Microwave Absorbing and Mechanical Properties of Linear Low Density Polyethylene/ethylene-octene Copolymer/calcium Carbonate Composites. *Polym.-Plast. Technol. Eng.* **2011,** *50,* 89–94.

17. Das, T. K.; Prusty, S. Review on Conducting Polymers and Their Applications. *Polym.-Plast. Technol. Eng.* **2012,** *51,* 1487–1500.

18. Kim, J.-H.; Kim, S.-S. Microwave Absorbing Properties of Ag-coated Ni–Zn Ferrite Microspheres Prepared by Electroless Plating. *J. Alloys Compd.* **2011,** *509,* 4399–4403.

19. Shi, G. M.; Zhang, J. B.; Yu, D. W.; Chen, L. S. Synthesis and Microwave-absorbing Properties of Al_2O_3 Coated Polyhedral Fe Nanocapsules Prepared by Arc-discharge Method. *Adv. Mate. Res.* **2011,** *299–300,* 739–742.

20. Liu, X. G.; Li, B.; Geng, D. Y.; Cui, W. B.; Yang, F.; Xie, Z. G.; Kang, D. J.; Zhang, Z. D. (Fe, Ni)/C Nanocapsules for Electromagnetic-Wave-Absorber in the Whole Ku-band. *Carbon* **2009,** *47,* 470–474.

21. Meng, F.; Zhao, R.; Zhan, Y.; Lei, Y.; Zhong, J.; Liu, X. Preparation and Microwave Absorption Properties of Fe-phthalocyanine Oligomer/Fe_3O_4 Hybrid Microspheres. *Appl. Surf. Sci.* **2011,** *257,* 5000–5006.

22. Che, R. C.; Zhi, C. Y.; Liang, C. Y.; Zhou, X. G. Fabrication and Microwave Absorption of Carbon Nanotubes/$CoFe_2O_4$ Spinel Nanocomposite. *Appl. Phys. Lett.* **2006,** *88,* 033105.

23. Drogin, I. Carbon Black. *J. Air Pollut. Control Assoc.* **1968,** *18,* 216–228.
24. Dannenberg, E. M. Filler Choices in the Rubber Industry the Incumbents and Some New Candidates. *Elastomerics* **1981,** *113,* 30–50.
25. Donnet, J.-B.; Custodero, E. Reinforcement of Elastomers by Particulate Fillers. In *The Science and Technology of Rubber,* 4th ed.; Mark, J. E., Erman, B., Roland, C. M., Eds.; Academic Press: Boston, 2013; pp 383–416.
26. Abou Rida, M.; Harb, F. Synthesis and Characterization of Amorphous Silica Nanoparticles from Aqueous Silicates Using Cationic Surfactants. *J. Met., Mater. Miner.* **2014,** *24,* 37–42.
27. Jafarzadeh, M.; Rahman, I. A.; Sipaut, C. S. Synthesis of Silica Nanoparticles by Modified Sol–gel Process: The Effect of Mixing Modes of the Reactants and Drying Techniques. *J. Sol-Gel Sci. Technol.* **2009,** *50,* 328–336.
28. Khavryutchenko, V.; Khavryutchenko, A.; Barthel, H. Fumed Silica Synthesis: Influence of Small Molecules on the Particle Formation Process. *Macromol. Symp.* **2001,** *169,* 1–6.
29. Kickelbick, G. Introduction to Hybrid Materials. In *Hybrid materials: Synthesis, characterization, and applications;* Kickelbick, G., Ed.; WILEY-VCH Verlag GmbH and Co.: Weinheim, 2007; pp 1–48.
30. Mahmud, K.; Wang, M. J.; Francis, R. A.; Belmont, J. A. Elastomeric Compounds Incorporating Silicon-treated Carbon Blacks. U.S. Patent 7,199,176, B2.
31. Wang, M.-J.; Mahmud, K.; Murphy, L. J.; Patterson, W. J. Carbon-silica Dual Phase Filler, a New Generation Reinforcing Agent for Rubber: Part I. Characterization. *KGK-Kautsch. Gummi Kunstst.* **1998,** *51,* 348–360.
32. Tarantili, P. A. Reinforced Elastomers: Interphase Modification and Compatibilization in Rubber-based Nanocomposites. In *Advances in Elastomers II Composites and Nanocomposites;* Visakh, P. M., Thomas, S., Chandra, A. K., P., M. A., Eds.; Springer-Verlag Berlin Heidelberg: Berlin, Germany, 2013; pp 108–154.
33. Fröhlich, J.; Niedermeier, W.; Luginsland, H. D. The Effect of Filler–filler and Filler–elastomer Interaction on Rubber Reinforcement. *Composites, Part A* **2005,** *36,* 449–460.
34. Isayev, A. I. Recycling of Rubbers. In *The Science and Technology of Rubber* 4th ed.; Erman, B., Mark, J. E., Roland, C. M., Eds.; Academic Press: Boston, 2013; pp 697–764.
35. Marsh, H.; Heintz, E.; Rodrigues-Reinoso, F. *Introduction to Carbon Technologies Alicante;* University of Alicante: Spain, 1997.
36. Burbine, W. G.; Cole, H. M.; Merrill, J. E.; Patterson, D. L. Carbon Black Process. U.S. Patent 3,088,806, A.
37. Burbine, W. G.; Cole, H. M.; Merrill, J. E.; Patterson, D. L. Process for Producing Magnetic Carbon Black Compositions. U.S. Patent 3,213,026, A.
38. Parker, L. Process of Producing Hydrophilic Carbon Black. U.S. Patent 3,347,632, A.
39. Schaeffer, W. D.; Smith, W. R. Effect of Heat Treatment on Reinforcing Properties of Carbon Black. *Rubber Chem. Technol.* **1956,** *29,* 286–295.
40. Wolff, S. Silanes in Tire Compounding After Ten Years—a Review. *Tire Sci. Technol.* **1987,** *15,* 276–294.
41. Hunsche, A.; Görl, U.; Müller, A.; Knaack, M.; Göbel, T. Investigations Concerning the Reaction Silica/organosilane and Organosilane/polymer—part 1: Reaction Mechanism and Reaction Model for Silica/organosilane. *KGK-Kautsc. Gummi Kunstst.* **1997,** *50,* 881–889.
42. Hunsche, A.; Görl, U.; Koban, H. G.; Lehmann, T. Investigations on the Reaction Silica/organosilane and Organosilane/polymer—part 2. Kinetic Aspects of the Silica-organosilane Reaction. *KGK-Kautsch. Gummi Kunstst.* **1998,** *51,* 525–533.

43. Görl, U.; Parkhouse, A. Investigations on the Reaction Silica/organosilane and Organosilane/polymer Part 3: Investigations Using Rubber Compounds. *KGK-Kautsch. Gummi Kunstst.* **1999;** 493–500.

44. Theppradit, T.; Prasassarakich, P.; Poompradub, S. Surface Modification of Silica Particles and its Effects on Cure and Mechanical Properties of the Natural Rubber Composites. *Mater. Chem. Physics* **2014,** *148,* 940–948.

45. Krump, H.; Hudec, J.; Jasso, M. M.; Dayss, E.; Crimmann, P. Surface Modification of PET by Two Different Types of Plasma Reactors. *KGK-Kautsch. Gummi Kunstst.* **2004,** *57,* 662–667.

46. Hudec, I.; Jaššo, M.; Černák, M.; Krump, H.; Dayss, E.; Šuriová, V. Plasma Treatment and Polymerization-method for Adhesion Strength Improvement of Textile Cord to Rubber. *KGK-Kautsch. Gummi Kunstst.* **2005,** *58,* 525–528.

47. Vladkova, T. G.; Keranov, I. L.; Dineff, P. D.; Youroukov, S. Y.; Avramova, I. A.; Krasteva, N.; Altankov, G. P. Plasma Based Ar+ Beam Assisted Poly(dimethylsiloxane) Surface Modification. *Nuclear Instruments and Methods in Physics Research Section B: Beam Interactions with Materials and Atoms* **2005,** *236,* 552–562.

48. Pishinkov, D.; Dishovsky, N.; Borros, S. New Possibilities for Cold Plasma Modification of Furnace and Acetylene Carbon Black. *KGK-Kautsch. Gummi Kunstst.* **2010,** *63,* 548–553.

49. Nah, C.; Huh, M.-Y.; Rhee, J. M.; Yoon, T.-H. Plasma Surface Modification of Silica and its Effect on Properties of Styrene–butadiene Rubber Compound. *Polym. Int.* **2002,** *51,* 510–518.

50. Mathew, G.; Huh, M. Y.; Rhee, J. M.; Lee, M. H.; Nah, C. Improvement of Properties of Silica-filled Styrene-butadiene Rubber Composites Through Plasma Surface Modification of Silica. *Polym. Adv. Technol.* **2004,** *15,* 400–408.

51. Jesionowski, T.; Bula, K.; Janiszewski, J.; Jurga, J. The Influence of Filler Modification on its Aggregation and Dispersion Behaviour in Silica/pbt Composite. *Compos. Interfaces* **2003,** *10,* 225–242.

52. Melsom, J. A. Testing Carbon Black. In *Basic Rubber Testing: Selecting Methods for a Rubber Test Program;* Dick, J. S., Ed.; ASTM International: USA, 2003; pp 89–104.

53. Melsom, J. A. Testing Silica and Organosilanes. In *Basic Rubber Testing: Selecting Methods for a Rubber Test Program;* Dick, J. S., Ed.; ASTM International: USA, 2003; pp 105–110.

CHAPTER 4

ELASTOMER-BASED COMPOSITE MATERIALS COMPRISING HYBRID FILLERS OBTAINED BY DIFFERENT TECHNIQUES

CONTENTS

ABSTRACT

Carbon black and silica are the two widely used fillers in the rubber industry. Although both of them have their pros and cons, researchers have been exploring the possibilities to synthesize various hybrid fillers combining valuable properties of the each of the two phases to meet the constantly growing requirements of producers. This chapter aims at demonstrating the capabilities of two radically different methods—impregnation and pyrolysis-cum-water vapor—to produce various types of hybrid fillers, such as furnace carbon black–silica, conductive carbon black (CCB)–silica, CCB–magnetite, pyrolysis carbon–silica filler and It also investigates their impacts upon the curing, mechanical, dynamic, dielectric, and microwave properties of composites based on different types of elastomeric matrices. The fillers have been characterized by a number of modern methods. Attention has been paid to the effect of the method used for obtaining the fillers on the distribution of the two phases and their interpenetration, which is essential to the final performance properties of articles containing these fillers. The impact of fillers on vulcanization, mechanical, dynamic, electric, and microwave properties of elastomer composites has been studied. It has been found that the ratio between the two phases, their distribution, and interpenetration are the determining factors for the properties discussed.

4.1   INTRODUCTION

Unique properties of elastomer composites have been the subject of many research and development investigations. They have been in use in various fields, such as tire manufacturing, car industry, sports, defense, space research, medicine, and so forth due to their durability, high strength, lightweight, design, and process flexibility.

Particulate fillers or reinforcements, such as carbon black, silica, clays, metal powders, and ceramics are added to rubber formulations to meet the requirements for targeted material properties, such as tensile strength, wear resistance, thermal, dielectric, microwave properties, and so forth. Therefore, the appropriate choice of filler is as important as the choice of rubber for desired formulation, so that a product could realize its performance specifications.[1-4]

Carbon black and silica are the two widely used fillers in rubber industry. However, both of them have their advantages and disadvantages. The presence of carbon black in rubber compounds reduces heat aging resistance and increases heat build-up at dynamic deformations in produced vulcanizates. Rubber compounds comprising carbon black are black and cannot be used for color articles and their compounding processes pollute the workshop. On the other hand, synthetic silica is a filler that is equivalent to carbon black and is capable of producing good mechanical and dynamical properties, good heat aging resistance, and low heat build-up in vulcanizates. Silica particles are more prone to agglomeration than those of carbon black owing to the formation of hydrogen bonds between silanol groups over its surface. When no coupling agent is used, those disadvantages of silica cause its poor dispersion in the rubber matrix and the polymer–filler interaction is less pronounced. Hence, the vulcanizates produced have poor mechanical and dynamic properties.[5–10]

In order to meet the constantly growing demand of producers, Cabot Corporation has launched a dual-phase filler of the carbon–silica type, produced by a co-fuming process. Thus, the new dual-phase (hybrid) product acquires the valuable properties of both carbon black and silica.[11,12] Such fillers facilitate lower filler–filler interactions and stronger polymer–filler interactions. As a result, the rubber composites are of better mechanical properties. These fillers are commonly used in the manufacturing of passenger and truck tires. The tires produced with these fillers possess better wet skid and rolling resistance.[11] Altering the quantitative ratio between the two phases in the hybrid product and changing their chemical nature could expand the applications of the hybrid fillers, hence those of the elastomer composites to a great extent.[12–19]

The aim of this chapter is to demonstrate the possibilities to use two cardinally different methods—impregnation and pyrolysis-cum-water-vapor—for obtaining hybrid fillers of the types: furnace carbon black–silica, conductive carbon black (CCB)–silica, CCB–magnetite, and pyrolysis carbon—silica, as well as to investigate the effect that the fillers have on the curing, mechanical, dynamic, dielectric, and microwave properties of composites based on various elastomer matrices.

4.2   EXPERIMENT

4.2.1   MATERIALS

Two types of elastomers, differing in their chemical nature and properties, have been used as polymer matrix. These are natural rubber (NR)—SVR

10 (supplied by the Hong Thanh Rubber Pty. Ltd.) and epoxidized natural rubber (ENR)—Epoxyprene 25 (purchased from San-Thap International Co., Ltd).

Four different types of commercial fillers were used as substrate for the preparation of hybrid fillers: Industrial furnace carbon black type PM-15 (produced in Russia with characteristics close to those of carbon black type N 776), carbon black type PM-75 (produced in Russia with characteristics close to those of carbon black type N 330), CCB Printex XE-2B (by Orion Engineered Carbons GmbH), and Corax N 330 carbon black (by Orion Engineered Carbons GmbH). Silicasol (containing 40% of silica; pH: 9; density: 1.3 g/cm^3) and magnetite (containing 95–100% of Fe$_3$O$_4$ and 0–5% of silica, purchased from Inoxia, UK) were chosen as impregnating agents. Corax N 220 carbon black (by Orion Engineered Carbons GmbH), Ultrasil 7000 GR (Evonik Industries), and Vulkasil®S (by Lanxess) were also used as standard fillers. Two organosilanes were used as coupling agents—bis(triethoxysilylpropyl) tetrasulfide (TESPT, Si-69) and bis(triethoxysilylpropyl) disulfide (TESPD, Si-266), both produced by Evonik Industries. Other ingredients, such as zinc oxide (ZnO), stearic acid (SA), N-tert-butyl-2-benzothiazolesulfenamide (TBBS), sulfur (S), and so forth, were commercial grades and used without further purification.

4.2.2 PREPARATION OF HYBRID FILLERS

4.2.2.1 PREPARATION OF FURNACE CARBON BLACK–SILICA HYBRID FILLERS

Industrial furnace carbon black, type PM-15 and carbon black, type PM-75 were used as substrates. Silicasol was chosen as an impregnating agent so as to obtain carbon–silica dual-phase fillers. The choice of the two fillers was determined by the great difference in their main characteristics: specific surface area, particle size, oil absorption number (OAN), iodine adsorption (IA), and so on. Both fillers were impregnated with silicasol, equivalent to 3 and 7% of silica. The obtained hybrid fillers were denoted as ICSF-15-3, ICSF-15-7, ICSF-75-3, and ICSF-75-7. The letters denote hybrid fillers having two phases (carbon black and silica) obtained by an impregnation technology. The two digits following the letters denote the specific surface area of the carbon black used as a hybrid filler substrate, and the last digit denotes the percentage of silica used.

The hybrid fillers were prepared via impregnation according to the following procedure:

1. The needed amount of silicasol (estimated such that the impregnation solution contained 3 or 7% silica) was diluted in 400 ml of distilled water. The carbon black (100 g) was loaded into a ball mill, and the silicasol solution was poured over it. The impregnation and homogenization were run for 2 h.
2. After the impregnation, the filler was dried, first at 50°C for 30 min and then at 200°C for 2 h. After that, it was ground in a ball mill for 2 h.
3. The dry ground filler was transferred into a customized reactor (Figure 4.1), where it was thermally treated at 440°C under a 10^{-2}-mm Hg vacuum for 2 h. The reactor was designed such that the thermal activation could be run in vacuo at higher temperatures without the carbon material being spoiled.
4. After being removed from the reactor, the sample was again ground in the ball mill for 2 h.

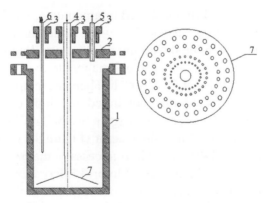

FIGURE 4.1   Design of the reactor used: (1) reactor vessel, (2) top (sealing flange), (3) sealing nuts, (4) gas inlet, (5) gas outlet, (6) thermocouple, and (7) Buchner funnel. (Reprinted from Al-Ghamdi, A. A.; Al-Hartomy, O. A.; Al-Solamy, F. R.; Dishovsky, N.; Malinova, P.; Lakov, L. Characterization of Hybrid Fillers Based on Carbon Black of Different Types Obtained by Impregnation. *Proceedings of the Institution of Mechanical Engineers, Part L: Journal of Materials Design and Applications*, © 2015 with permission from SAGE Publications.)

4.2.2.2 PREPARATION OF CONDUCTIVE CARBON BLACK–SILICA HYBRID FILLERS

The hybrid fillers studied were prepared by impregnation of conductive Printex XE-2B carbon black (by Orion Engineered Carbons GmbH) with silicasol. A mixture of 100 g of Printex XE-2B carbon black and the needed amount of silicasol, corresponding to 3 or 7% of silica, was placed into a ball mill and 1.6 l of distilled water was poured over it. The mixture was impregnated for 2 h. The product was placed into a drying chamber at 150°C till its complete drying. Then, it was ground in a ball mill for 2 h. Subsequently, it was thermally treated at 440°C under 10^{-2} mm Hg vacuum for 2 h in a reactor designed especially for the purpose (Figure 4.1). Thus, the thermal activation could be run at higher temperature without a negative effect upon the carbon black. After being removed from the reactor, the product was ground again in ball mill and was ready for further investigations. Two hybrid fillers based on Printex XE-2B conductive black and 3 or 7% of silica were prepared according to the above method. The fillers were denoted as Pr/Si 3 and Pr/Si 7.

4.2.2.3   PREPARATION OF CONDUCTIVE CARBON BLACK–MAGNETITE HYBRID FILLERS

Conductive carbon black Printex XE-2B in amounts of 90, 70, and 50 g and magnetite in amounts of 10, 30, and 50 g were loaded into a ball mill, 1500 ml of ethyl alcohol was poured, and the mixture was impregnated for 2 h. Ethyl alcohol was used to avoid eventual magnetite oxidation. The suspension thus obtained was dried at 50°C for 2 h. Then, the temperature was raised to 150°C and the drying was further continued for 2 h till complete drying of the compound. The yield was grounded again in a ball mill for 2 h. After that the ground yield was loaded into the reactor (Figure 4.1) and heated under 10^{-2} mm Hg vacuum at 440°C for 2 h. If necessary, the product was ground once more in a ball mill. Three types of hybrid fillers were synthesized and denoted according to their carbon–magnetite phase ratios. CCB/M-10—carbon:magnetite phase ratio 90:10; CCB/M-30—carbon : magnetite phase ratio 70:30; CCB/M-50—carbon:magnetite phase ratio 50:50.

4.2.2.4   PREPARATION OF PYROLYSIS CARBON BLACK–SILICA HYBRID FILLERS

Carbon–silica filler (marked as PCSF) was obtained via pyrolysis-cum-water vapor[20,21] of waste green tire tread (Michelin Energy 195/65 R15) under the following conditions: temperature 500°C, water vapor concentration—40%.

According to the method, the tires were subjected to pyrolysis process where the raw materials are gradually heated between 300 and 1000°C by constantly purging some of the following gases: air, smoke gases, carbon dioxide, nitrogen, and/or addition of vapor from 0 to 100%. It is also possible to run pyrolysis and activation simultaneously using ammonium as well as mixtures of the former gases. The final product was evacuated for a period of 1 min to 48 h, and then cooled in an airless atmosphere.

4.2.3 MEASUREMENTS

4.2.3.1 CHARACTERIZATION OF THE STUDIED HYBRID FILLERS

The hybrid fillers studied were characterized according to their:

- Ash content—ISO 1125:1999;
- IA—ISO 1304:2006;
- OAN—ISO 4656:2012;
- Specific surface area (BET—Brunauer–Emmett–Teller adsorption method)—ISO 4652:2012.

The texture characteristics of the studied fillers were determined by low-temperature (77.4 K) nitrogen adsorption on a Quantachrome Instruments NOVA 1200e (USA) apparatus. The nitrogen adsorption–desorption isotherms were analyzed to evaluate the following parameters: the specific surface area (S_{BET}) that was determined on the basis of the BET equation, the total pore volume (V_t) was estimated in accordance with the Gurvich rule at a relative pressure close to 0.99. The volume of the micropores (V_{MI}) and the specific surface area related to micropores (S_{MI}), as well as the external specific surface area (S_{EXT}) were evaluated according to Vt-method. Additionally, the pore-size distributions were calculated by density functional theory (DFT) method using nonlocal density function theory (NLDFT) equilibrium model. All samples were outgassed for 16 h under vacuum at 120°C before the measurements. The average diameter of the mesopores and their size distribution were determined by Barett–Joyner–Halenda (BJH) method.[22]

Powder X-ray diffraction (XRD) patterns were collected at room temperature on a Bruker D8 Advance diffractometer using CuKα radiation and a Lynx Eye position sensing detector (PSD) within the 5.3–80° 2θ range, step 0.04° 2θ and 0.1 sec/strip (total of 17.5 sec/step) counting time. To improve the statistics, sample rotating speed of 30 rpm was used. The identification of crystalline phases was performed by Diffrac Plus EVA 12 software and ICDD PDF-2 (2009) database.

For ascertaining the distribution of both phases, the hybrid products were also investigated by transmission electron microscopy (TEM). The TEM investigations were performed on a TEM JEOL 2100 instrument at accelerating voltage of 200 kV. The specimens were prepared by grinding and dispersing them in ethanol by ultrasonic treatment for 6 min. The suspension was dripped on standard holey carbon/Cu grids. The measurements of lattice-fringe spacing recorded in high-resolution TEM (HRTEM) micrographs were made using digital image analysis of reciprocal space parameters. The analysis was carried out by the Digital Micrograph software; TEM JEOL 2100; XEDS: Oxford Instruments, X-MAXN 80 T; CCD Camera Orius 1000, 11 Mp, Gatan.

Thermo-gravimetric analysis (TGA) and differential thermal analysis (DTA) were performed on STAPT1600 TG-DTA/DSC [simultaneous thermal analysis (STA)] apparatus of LINSEIS Messgeräte GmbH, Germany, from ambient temperature to1000°C in airflow with a 1 dm³/min debit and heating rate of 10°C/min.

Infrared spectroscopy with Fourier transformation (VARIAN 660-FTIR Spectrometer, KBr pellet) was also used for characterization of the studied fillers.

4.2.3.2 CHARACTERIZATION OF THE STUDIED COMPOSITES

The vulcanization characteristics of the rubber composites were determined at 150°C on a moving die rheometer (MDR 2000 Alpha Technologies) according to ISO 3417:2008. The mechanical properties of the composites studied were determined according to ISO 37:2011. Shore A hardness of the composites studied was determined in agreement with ISO 7619-1:2010, while the wear resistance was determined according to ISO 4649:2010.

The swelling was carried out according to ISO 1817:2015 at room temperature in toluene. The cross-link (n_c=1/2.M_c) density was estimated using Flory–Rehner[23] equation for calculating the molecular mass of the segment between two cross-links (M_c):

$$M_c = \frac{-\rho_r V_s \left(V_r\right)^{1/3}}{\ln\left(1-V_r\right)+V_r + \chi V_r^2} \tag{4.1}$$

where:

M_c is average molecular weight of the rubber segments between two crosslinks; ρ_r is rubber density; V_s is molar volume of the solvent; χ is the parameter characterizing rubber–solvent interaction; and V_r is volume fraction of the swollen network.

The complex dynamic modulus and the heat build-up were determined on a Goodrich flexometer at 850 min^{-1} deformation rate. Dynamic properties (Dynamic storage modulus (E) and mechanical loss angle tangent (tan δ) of the studied composites were investigated using a Dynamic Mechanical Thermal Analyzer Mk III system (Rheometric Scientific). The data were obtained at a frequency of 5 Hz and strain amplitude of 64 μm in the temperature range of -100 to 100°C using a heating rate of 3°C/min under single cantilever bending mode. The dimensions of the investigated samples were as follows—width 10 mm, length 25 mm, and the thickness measured using a micrometer varied between 1 and 2 mm.

Volume resistivity (ρ_v, Ω.m) of the studied composites was measured using two electrodes (2-terminal method) and calculated by the equation:

$$\rho_v = R_v \frac{S}{h} \tag{4.2}$$

where:

R_v is ohmic resistance between the electrodes; h is sample thickness between the electrodes in meter; S is cross-sectional area of the measuring electrode in sq. meter.

A Wheatstone bridge was used to measure the resistance.

The surface resistivity of the samples was calculated using the equation:

$$\rho_s = 100R_s \tag{4.3}$$

where R_s is the resistance measured in Ω.

For estimating the correlation between the contact point distance and the ohmic resistance to verify the composite homogeneity, approximately 100 × 12 mm large strips of the cured nanocomposite were cut; exact width and thickness of the strips were measured using a Mitutoyo micrometer (0.001 mm). Resistance (R_S) was measured using a DT 9208A digital multimeter at ambient temperature.[24]

Correlation coefficient is a measure of the degree of dependence between two variables.

Pearson correlation coefficient (R_P) is most used one and is calculated according to the formula:[25]

$$R_P = \frac{\sum_{i=1}^{n}\left(Y_i - \overline{Y}\right)\left(X_i - \overline{X}\right)}{\sqrt{\sum_{i=1}^{n}\left(Y_i - \overline{Y}\right)^2 \sum_{i=1}^{n}\left(X_i - \overline{X}\right)^2}} \tag{4.4}$$

Where X_i, Y_i are experimental (current) values of two variables whose correlation is searched;

$\bar{X}, \bar{Y}$ are average values of the same variables.

The degree of dependence between two variables can be between +1 and −1, inclusive:

$0 < R_p < 0.3$—low degree of correlation;

$0.3 < R_p < 0.5$—medium degree of correlation;

$0.5 < R_p < 0.7$—significant degree of correlation;

$0.7 < R_p < 0.9$—very high degree of correlation; and

$0.9 < R_p < 1.0$—perfect degree of correlation.

Direction of the correlation:

+ positives: one variable increases, the other variable also increases.

− negatives: one variable increases, the other variable decreases.

The electromagnetic parameters of the composite materials were measured by the resonant perturbation method. According to this method, the tested sample was introduced into a resonator, and the electromagnetic parameters of the sample were deduced from the change in the resonant frequency and quality factor of the resonator.[26] For the rectangular cavity, the TE_{10n} modes were used for the complex permittivity and permeability measurements. The sample was placed at the position of maximum intensity of electric field, where $n = odd$ was always adopted, because the sample position could be located easily, as the geometric center of the cavity was the one of the maximum positions.[27] The formulas for the real and imaginary parts of the relative permittivity were as follows:[28]

$$\varepsilon_r' = \left(\frac{f_c - f_s}{2f_s}\right) \cdot \left(\frac{V_c}{V_s}\right) + 1 \tag{4.5}$$

$$\varepsilon_r'' = \left(\frac{V_c}{4V_s}\right) \cdot \left(\frac{1}{Q_s} - \frac{1}{Q_c}\right) \tag{4.6}$$

where f_c and Q_c are resonance frequency and Q-factor of the cavity without an inserted sample, f_s and Q_s, are with an inserted sample, respectively; V_c is the volume of the cavity; V_s is the volume of the sample.

For permeability measurements, the sample was placed at the position of maximum magnetic field, where $n = even$ was always adopted. The formulas for the real and imaginary parts of the relative permeability were as follows:[29]

$$\mu'_r = \left(\frac{f_c - f_s}{f_s}\right) \cdot \left(\frac{V_c}{V_s}\right) \cdot \left(\frac{\lambda_g^2 + 4a^2}{8a^2}\right) + 1 \qquad (4.7)$$

$$\mu''_r = \left(\frac{V_c}{V_s}\right) \cdot \left(\frac{1}{Q_s} - \frac{1}{Q_c}\right) \cdot \left(\frac{\lambda_g^2 + 4a^2}{16a^2}\right) \qquad (4.8)$$

where $\lambda_g = 2d/L$, is the guided wavelength and $L = 1, 2, 3,...$; d is cavity length and a is cavity width.

Therefore, the odd resonant modes provided for the determination of permittivity and conductivity, while the even modes provided for the permeability determination.

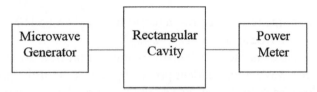

FIGURE 4.2 Block diagram of the measurement setup for evaluating electromagnetic parameters. (Reprinted with permission from Ahmed A. Al-Ghamdi,Omar A. Al-Hartomy, Falleh R. Al-Solamy, Nikolay Dishovsky et al. Conductive carbon black/magnetite hybrid fillers in microwave absorbing composites based on natural rubber, Composites Part B: Engineering, © 2016 Elsevier.)

The measurement setup is illustrated in Figure 4.2. Three different cavities, designed to resonate at approximately 3.128 GHz (TE$_{103}$), 4.061 GHz (TE$_{105}$), 6.148 GHz (TE$_{105}$), 7.480 GHz (TE$_{107}$), and 8.527 GHz (TE$_{105}$), were used to determine the electrical parameters, and three other different cavities, designed to resonate at approximately 3.566 GHz (TE$_{104}$), 6.814 GHz (TE$_{106}$), and 9.284 GHz (TE$_{106}$), were used for the determination of the magnetic parameters. The cavities were fabricated by standard brass waveguides. Conducting plates with small apertures on either ends of the cavity provided inductive coupling to the cavity. The power transmitted through the cavity was measured by a power meter, and the resonant frequency and quality factor were evaluated. The measurements were performed at room temperature varying from 19 to 24°C within the frequency range of 3.0–9.5 GHz, at incident power 5 mW.

Measured values of ε_r' and ε_r'' were used to determine the conductivity of the composite samples. For a dielectric material having nonzero conductivity, we apply Maxwell's curl equation:

$$\nabla \times H = \left(\sigma_V + j\omega\bar{\varepsilon}\right).E = \left(\sigma_V + \omega\varepsilon_r''\right).E + j\omega\varepsilon_r'E \qquad (4.9)$$

where E and H are the respective electric and magnetic fields in the medium, σ_V is the electrical conductivity of the medium, ω is the angular frequency, $\bar{\varepsilon}$ is the complex permittivity of the medium, $\varepsilon_r{}'$ and $\varepsilon_r{}''$ are the real and imaginary part of the complex permittivity, respectively.

The loss tangent:

$$\tan \delta = \frac{\sigma_V + \omega \varepsilon_r^{'}}{\omega \varepsilon_r^{'}} \qquad (4.10),$$

where $\sigma_V + \omega \varepsilon_r{}'' = \sigma_e$ which is the effective conductivity of the medium. When σ_V is very small, the effective conductivity is reduced to:

$$\sigma_e = \omega \varepsilon_r^{'} = 2\pi f \varepsilon_0 \varepsilon_r^{'} \qquad (4.11)$$

where f is the frequency of electromagnetic field, $\varepsilon_0 = 8.85410^{-12}$ F/m is the permittivity of the free space, and $\varepsilon_r{}'' = \varepsilon''/\varepsilon_0.{}^{30}$

Figure 4.3 presents the mechanism of interaction of a composite to an electromagnetic wave with incident power P_I. A portion of the power of the wave P_R reflects back by the surface of the material. Another portion passes through the material while being absorbed by it and gets converted into heat P_A, whereas remainder gets transmitted P_T.

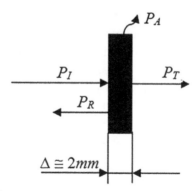

FIGURE 4.3 Schematic presentation of a shielding material interacting with an electromagnetic wave of incident power (P_I). (Reprinted from Al-Ghamdi, A. A.; Al-Hartomy, O. A.; Al-Solamy, F. R.; Dishovsky, N.; Mihaylov, M.; Malinova, P.; Atanasov, N. Dielectric and Microwave Properties of Elastomer Composites Loaded with Carbon–silica Hybrid Fillers. *J. Appl. Polym. Sci.* **2016**, *133*, 42978(1)–42978(9). © 2016 with permission from John Wiley.)

The total shielding effectiveness (SE_T) was defined as the ratio between the incident power on the sample P_I and the transmitted power P_T in accordance with Eq. (4.12):[31–33]:

$$SE_T = 10 \log \frac{P_I}{P_T} \qquad (4.12)$$

The SE_T, in dB and the reflective shielding effectiveness of the sample surface (SE_R, in dB) were determined by Eqs. (4.13) and (4.14):[34–37]:

$$SE_T = -10 \lg T, \qquad (4.13)$$

$$T = \left| P_T / P_I \right| = \left| S_{21} \right|^2$$

$$SE_R = -10 \lg (1 - R), \qquad (4.14)$$

$$R = \left| P_R / P_I \right| = \left| S_{11} \right|^2$$

S_{11} and S_{21} are complex scattering parameters or S-parameters (S_{11} corresponds to the reflection coefficient and S_{21} to the transmission coefficient).

The absorptive shielding effectiveness (SE_A) was calculated as the difference between (4.13) and (4.14), as shown in Eq. (4.15):

$$SE_A = SE_T - SE_R \qquad (4.15)$$

The attenuation coefficient (α, dB/cm) was determined using Eq. (4.15). Substituting SE_T and SE_R in Eq. (4.15) with Eqs. (4.13) and (4.14), Eq. (4.16) is obtained.

$$SE_A = 10 \lg \left(\frac{P_I \left(1 - \left| S_{11} \right|^2 \right)}{P_T} \right) \qquad (4.16)$$

For determining the attenuation coefficient, Eq. (4.16) was divided by the thickness of the sample d in centimeters, as shown in Eq. (4.17).

$$\alpha = \frac{SE_A}{d} = 10 \lg \left(\frac{P_T}{P_I \left(1 - \left| S_{11} \right|^2 \right)} \right) \qquad (4.17)$$

Figure 4.4 presents the schematic diagram of the measurement system for the electromagnetic interference shielding effectiveness in a broadband frequency range. The experimental setup consists of a coaxial reflectometer (coaxial directional couplers Narda model 4222-16 and detectors Narda FSCM 998999 model 4503A, separating the incident power from the reflected power in the transmission line); a ratio meter (HP Model 416A) that calculates and detects the amplitude of the reflection coefficient; a series of radiofrequency generators (G4-37A, G4-79 to G4-82 and HP 68A in the frequency range from 1 to 12 GHz); a signal generator (BM492 releasing a modulating signal at 1 kHz directed toward the radiofrequency generator); a coaxial transmission line (Orion type E2M for frequencies from 1 to 5 GHz); a coaxial measuring line (APC-7 mm for frequencies from 6 to 12 GHz); and a power meter (HP 432A). The cutoff frequencies for the coaxial measuring lines were determined by the formulas presented in literature.[26,38] The sample holder, reflectometer setup, and radiofrequency generator were connected with a rugged phase stable cable (N9910X-810 Agilent) and through the connectors without interference from other components.

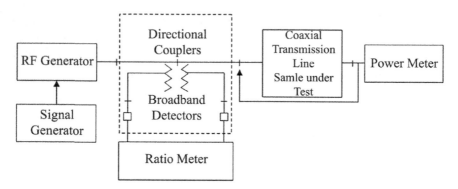

FIGURE 4.4 Schematic diagram of the system for measuring EMI SE. (Reprinted from Al-Ghamdi, A. A.; Al-Hartomy, O. A.; Al-Solamy, F. R.; Dishovsky, N.; Mihaylov, M.; Malinova, P.; Atanasov, N. Dielectric and Microwave Properties of Elastomer Composites Loaded with Carbon–silica Hybrid Fillers. *J. Appl. Polym. Sci.* **2016,** *133,* 42978(1)–42978(9). © 2016 with permission from John Wiley.)

The ratio meter was calibrated prior to carrying out the actual measurements. A calibrating "Open-Short-Load" procedure was applied with Agilent N9330 and Agilent 1250 calibration kits.

Measurements of the incident power, transmitted power, and reflection coefficient were performed using the following procedure:

a) Preparation of the measuring system.
b) Calibration of the system with calibration kits in order to eliminate systematic errors.
c) Preparation of the samples from the obtained vulcanized materials by cutting out pieces having following dimensions:

- External diameter of 20 mm and internal diameter of 7 mm in case of using coaxial transmission line Orion E2M
- External diameter of 7 mm and internal diameter of 3 mm in case of using coaxial measuring line APC-7 mm

d. Carrying out measurements for determining the module of the reflection coefficient using:

- Standard load of the type Agilent 1250, connected in the position of the coaxial line
- Blank coaxial line with a standard load at the end (Agilent N9330 calibrating kit of tools)

e. Carrying out measurement of incident power P_I.

f. Careful placement of the cut-out sample inside the coaxial line and measurement of the reflection coefficient and transmitted power P_T.

g. The cut-out standard material (polytetrafluoroethylene—PTFE load of thickness 1 mm) was placed into the coaxial line and the reflection coefficient and transmitted power P_T measured for confirming the correctness of the measurements done.

The measurements were carried out at room temperature varying from 19 to 24°C and incident power P_I at the inlet of the coaxial measuring line varying from 800 to 1300 µW within the frequency range of 1–12 GHz.

4.3 RESULTS AND DISCUSSION

4.3.1 FURNACE CARBON BLACK–SILICA HYBRID FILLERS OBTAINED VIA IMPREGNATION

4.3.1.1 CHARACTERIZATION OF THE STUDIED HYBRID FILLERS

It is known that the main characteristics determining the properties and effectiveness of the fillers are: particle size, specific surface area, OAN, IA, ash content, and their surface chemical activity.[1]

The main properties of the impregnated hybrid carbon–silica fillers are summarized in Table 4.1. The characteristics of the conventional carbon black used as a substrate for the hybrid fillers are given as references.

As seen from Table 4.1, the ash content of the two types of conventional carbon black (PM-15 and PM-75) increases upon the impregnation with silicasol owing to the introduction of a second inorganic silica phase.

The higher silica amount in the fillers inevitably leads to an increase in their ash content. Assuming the ash content of the initial carbon black to be about 0.5%, and introducing 3 and 7% of silica, respectively, via the impregnation, the hybrid fillers obtained should have theoretical ash content about 3.5% (those comprising 3% of silica) and about 7.5% (for those with 7% of silica). As the Table 4.1 shows, the ash content is slightly higher, probably due to other inorganic by-products resulting from the impregnation.

TABLE 4.1 Main Properties of the Studied Hybrid Fillers.

	Ash Content (%)	S_{BET} (m²/g)	OAN (ml/100 g)	IA (mg/g)	V_t (cm³ (STP)/g)	D_{av} (Nm)
PM-15	0.50	15	85	19	–	–
PM-75	0.30	75	96	84	–	–
ICSF-15-3	4.12	31	75	21	0.16	2.1
ICSF-15-7	9.80	33	70	20	0.24	2.9
ICSF-75-3	3.43	81	89	90	0.68	3.3
ICSF-75-7	8.30	85	86	87	0.11	1.2

(Reprinted from Al-Ghamdi, A. A.; Al-Hartomy, O. A.; Al-Solamy, F. R.; Dishovsky, N.; Malinova, P.; Lakov, L. Characterization of Hybrid Fillers Based on Carbon Black of Different Types Obtained by Impregnation. *Proceedings of the Institution of Mechanical Engineers, Part L: Journal of Materials Design and Applications,* © 2015 with permission from SAGE Publications.)

According to Table 4.1, the specific surface area of the obtained hybrid fillers is higher than that of the conventional carbon black used as a substrate. The effect is more pronounced in the case of dual-phase fillers ICSF-15-3 and ICSF-15-7, obtained by the impregnation of the low active conventional carbon black PM-15. It is also worth noting that the higher silica amounts in the dual-phase fillers increase their specific surface area negligibly. The higher specific surface area could facilitate stronger interactions between the filler and the elastomer macromolecules. This proves the effectiveness of the impregnation method used herein. The latter also benefits the structurality of the dual-phase fillers thus obtained. As seen from Table 4.1, the OAN of the hybrid fillers obtained is lower than that of the conventional carbon black used as a substrate. The lower OAN values of the hybrid fillers suggest that they are less prone to agglomeration leading to their better dispersion in the rubber matrix. The better dispersion of the filler particles in the rubber matrix and the better pronounced polymer–filler interaction are crucial to obtain vulcanizates of good mechanical and dynamic properties.

The specific surface area of the hybrid fillers obtained by impregnating carbon black with silicasol is higher than that of the substrate carbon black due to the penetration of silica phase into carbon black agglomerates and their destruction into smaller ones. The smaller the filler particles, its aggregates, and agglomerates, the higher is the specific surface area of filler. Moreover, when the silica phase penetrates into carbon black agglomerates in the course of impregnation, it not only destructs them but also prevents their further agglomeration. Therefore, the hybrid fillers obtained have lower OAN values.

The IA of fillers reveals their adsorption activity. Table 4.1 evidences that although being slightly higher, the IA of the hybrid fillers obtained is commeasurable with that of the conventional carbon black used as a substrate. This is due to the more active surface of silica (comprising silanol groups) which has been introduced as a second phase into the filler obtained by impregnating carbon black with silicasol. The above discussion about the specific surface area, OAN, and IA of the hybrid fillers obtained is proven by their texture characteristics presented in Figures 4.5–4.8. In the case of PM-15 carbon black (Figures 4.5 and 4.6) the addition of silicasol enhances volume expansion of intra-aggregate pores as revealed by the comparison of the adsorption isotherms of the samples comprising various silica amounts and by the comparison of their hysteresis loops in particular. At higher silica amounts, the loop grows. This corresponds to the larger volume of intra-aggregate pores and to the occurrence of inter-aggregate pores. The types of hysteresis loops, however, do not differ. All this is confirmed by the comparison of the mesopore size distribution curves. Most probably silica is distributed into the inter-aggregate pores. The inter-aggregate pores when filled with silica, although partially, start to form secondary aggregates comprising silica, especially when silica amount in the filler increases.

In the case of PM-75 carbon black (Figures 4.7 and 4.8), besides the effects described above, one observes how the larger specific surface area of that type of carbon black affects the aggregation and formation of secondary structures. At lower concentrations of silica, the secondary aggregates are dominant, while at higher concentrations the aggregates become dense, the structure is consolidated and bulky, therefore the pore volume shrinks. There is also a change in the type of isotherms for the fillers with low and high silica content.

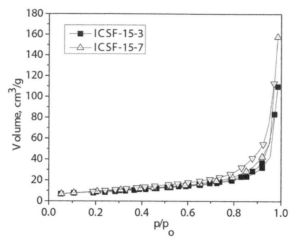

FIGURE 4.5 Volume of adsorbed gas dependence (cm³/g) on relative pressure (p/p₀) of the hybrid fillers ICSF-15-3 and ICSF-15-7. (Reprinted from Al-Ghamdi, A. A.; Al-Hartomy, O. A.; Al-Solamy, F. R.; Dishovsky, N.; Malinova, P.; Lakov, L. Characterization of Hybrid Fillers Based on Carbon Black of Different Types Obtained by Impregnation. *Proceedings of the Institution of Mechanical Engineers, Part L: Journal of Materials Design and Applications,*© 2015 with permission from SAGE Publications.)

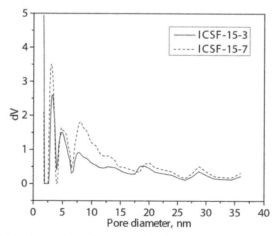

FIGURE 4.6 Derivative volume (dV) as a function of pore diameter (nm) of the hybrid fillers ICSF-15-3 and ICSF-15-7. (Reprinted from Al-Ghamdi, A. A.; Al-Hartomy, O. A.; Al-Solamy, F. R.; Dishovsky, N.; Malinova, P.; Lakov, L. Characterization of Hybrid Fillers Based on Carbon Black of Different Types Obtained by Impregnation. *Proceedings of the Institution of Mechanical Engineers, Part L: Journal of Materials Design and Applications,*© 2015 with permission from SAGE Publications.)

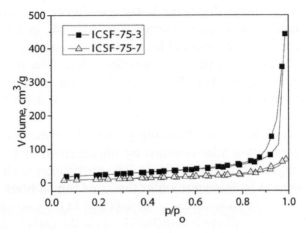

FIGURE 4.7 Volume of adsorbed gas dependence (cm³/g) on relative pressure (p/p₀) of the hybrid fillers ICSF-75-3 and ICSF-75-7. (Reprinted from Al-Ghamdi, A. A.; Al-Hartomy, O. A.; Al-Solamy, F. R.; Dishovsky, N.; Malinova, P.; Lakov, L. Characterization of Hybrid Fillers Based on Carbon Black of Different Types Obtained by Impregnation. *Proceedings of the Institution of Mechanical Engineers, Part L: Journal of Materials Design and Applications,* © 2015 with permission from SAGE Publications.)

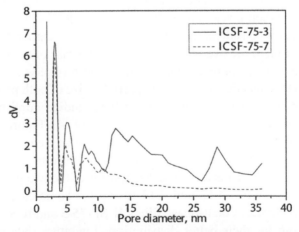

FIGURE 4.8 Derivative volume (dV) as a function of pore diameter (nm) of the hybrid fillers ICSF-75-3 and ICSF-75-7. (Reprinted from Al-Ghamdi, A. A.; Al-Hartomy, O. A.; Al-Solamy, F. R.; Dishovsky, N.; Malinova, P.; Lakov, L. Characterization of Hybrid Fillers Based on Carbon Black of Different Types Obtained by Impregnation. *Proceedings of the Institution of Mechanical Engineers, Part L: Journal of Materials Design and Applications,* © 2015 with permission from SAGE Publications.)

Without doubt, energy dispersive X-ray spectroscopy (EDS), used with the scanning transmission electron microscopy (STEM), or STEM-EDS, is the most appropriate modern method for answering questions about the distribution of carbon and silica phases within the hybrid filler obtained, about their interpenetration, and about the effect of the silica amount as a factor determining the interpenetration. Figures 4.9–4.11 present the EDS layered images and the compositional maps of the impregnated carbon–silica dual-phase fillers. The figures show that both phases (the carbon and silica) are evenly distributed in the filler obtained by impregnation of PM-75 carbon black comprising 3% silica (ICSF-75-3). The distribution is slightly worse in the filler obtained by impregnation of the same carbon black comprising 7% silica (ICSF-75-7). Noteworthy is the fact that in the case of ICSF-75-3 (Figure 4.10), the silica phase is distributed within the carbon black, that is, the two phases interpenetrate quite well. However, in the case of CSF-75-7 (Figure 4.11), the silica phase is flocculated in certain domains forming distinct aggregates, the interpenetration of the two phases is worse than in the case of ICSF-75-3. Indubitably, the particle size of the two phases is important for their distribution. The particle size depends on the phase quantity. When the difference among the particle size of both phases is not much, a better effect is achieved. This is confirmed by the images in Figure 4.9 which show that the worse distribution of the two phases is that of the filler obtained by impregnation of PM-15 carbon black comprising 3% silica (ICSF-15-3). In this case, silica aggregates of considerable size are in contact with carbon black ones but no interpenetration of the two phases, as seen in the case of the filler based on PM-75 carbon black comprising 3% silica, has been observed. The phenomenon is due to the great difference in the particles size of both phases.

The bright field TEM images of the fillers studied confirm the above discussion and are presented in Figures 4.12–4.14. As Figure 4.12 shows, the particles of PM-15 substrate carbon black, their aggregates and agglomerates are of considerably larger size and distinguish from the impregnating silica particles. This hinders the good distribution of the two phases and their interpenetration. Figures 4.13 and 4.14 reveal the commeasurable particle size of the two phases of dual-phase fillers ICSF-75-3 and ICSF-75-7 which is a prerequisite for their better distribution. Literature data also confirm PM-15 carbon black to have considerably larger particles than carbon black PM-75.[39]

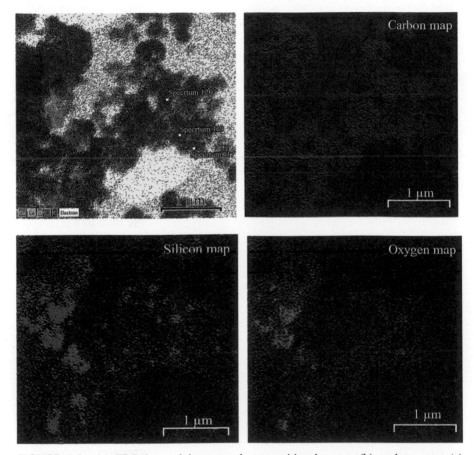

FIGURE 4.9 (a) EDS layered image and compositional maps, (b) carbon map, (c) silicon map, (d) oxygen map of the hybrid filler ICSF-15-3. EDS energy dispersive X-ray spectroscopy. (Reprinted from Al-Ghamdi, A. A.; Al-Hartomy, O. A.; Al-Solamy, F. R.; Dishovsky, N.; Malinova, P.; Lakov, L. Characterization of Hybrid Fillers Based on Carbon Black of Different Types Obtained by Impregnation. *Proceedings of the Institution of Mechanical Engineers, Part L: Journal of Materials Design and Applications,*© 2015 with permission from SAGE Publications.)

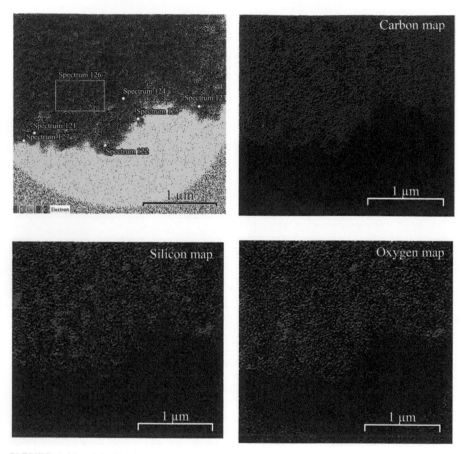

FIGURE 4.10 (a) EDS layered image and compositional maps, (b) carbon map, (c) silicon map, (d) oxygen map of the hybrid filler ICSF-75-3. (Reprinted from Al-Ghamdi, A. A.; Al-Hartomy, O. A.; Al-Solamy, F. R.; Dishovsky, N.; Malinova, P.; Lakov, L. Characterization of Hybrid Fillers Based on Carbon Black of Different Types Obtained by Impregnation. *Proceedings of the Institution of Mechanical Engineers, Part L: Journal of Materials Design and Applications,* © 2015 with permission from SAGE Publications.)

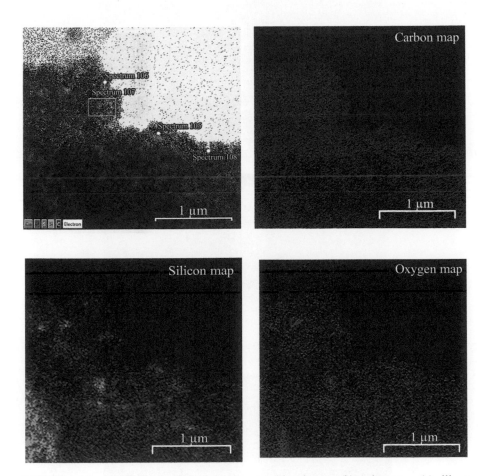

FIGURE 4.11   (a) EDS layered image and compositional maps, (b) carbon map, (c) silicon map, (d) oxygen map of the hybrid filler ICSF-75-7. (Reprinted from Al-Ghamdi, A. A.; Al-Hartomy, O. A.; Al-Solamy, F. R.; Dishovsky, N.; Malinova, P.; Lakov, L. Characterization of Hybrid Fillers Based on Carbon Black of Different Types Obtained by Impregnation. *Proceedings of the Institution of Mechanical Engineers, Part L: Journal of Materials Design and Applications,*© 2015 with permission from SAGE Publications.)

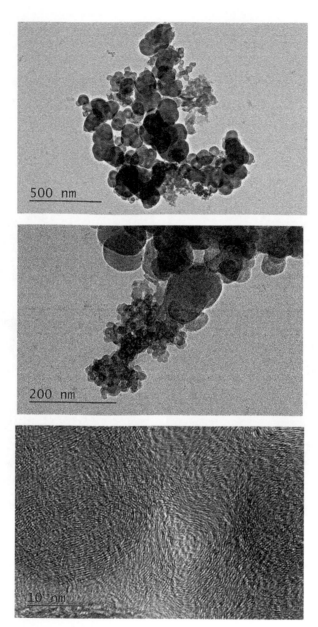

FIGURE 4.12 Bright field TEM images of ICSF-15-3 at different magnifications. TEM transmission electron microscopy. (Reprinted from Al-Ghamdi, A. A.; Al-Hartomy, O. A.; Al-Solamy, F. R.; Dishovsky, N.; Malinova, P.; Lakov, L. Characterization of Hybrid Fillers Based on Carbon Black of Different Types Obtained by Impregnation. *Proceedings of the Institution of Mechanical Engineers, Part L: Journal of Materials Design and Applications,*© 2015 with permission from SAGE Publications.)

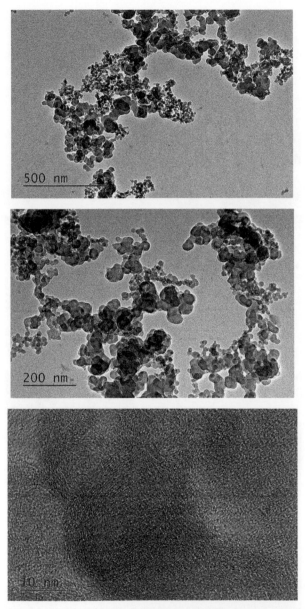

FIGURE 4.13   Bright field TEM images of ICSF-75-3 at different magnifications. (Reprinted from Al-Ghamdi, A. A.; Al-Hartomy, O. A.; Al-Solamy, F. R.; Dishovsky, N.; Malinova, P.; Lakov, L. Characterization of Hybrid Fillers Based on Carbon Black of Different Types Obtained by Impregnation. *Proceedings of the Institution of Mechanical Engineers, Part L: Journal of Materials Design and Applications,* © 2015 with permission from SAGE Publications.)

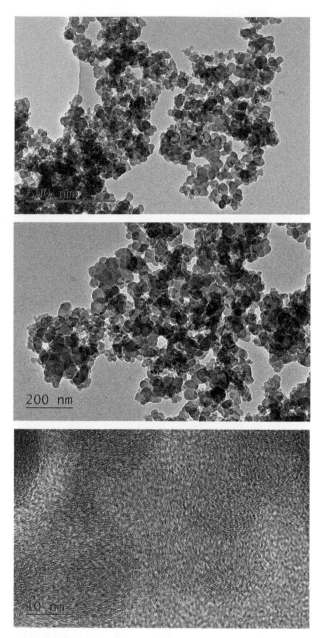

FIGURE 4.14 Bright field TEM images of ICSF-75-7 at different magnifications. (Reprinted from Al-Ghamdi, A. A.; Al-Hartomy, O. A.; Al-Solamy, F. R.; Dishovsky, N.; Malinova, P.; Lakov, L. Characterization of Hybrid Fillers Based on Carbon Black of Different Types Obtained by Impregnation. *Proceedings of the Institution of Mechanical Engineers, Part L: Journal of Materials Design and Applications,* © 2015 with permission from SAGE Publications.)

4.3.1.2 PREPARATION OF RUBBER COMPOSITES

The rubber compounds studied were prepared on a two-roll laboratory mill (rolls length/diameter 320 × 160 mm). Table 4.2 presents the compositions of the rubber compounds studied.

TABLE 4.2 Compositions of the Investigated Rubber Compounds (phr).

	NR 1	NR 2	NR 3	NR 4	NR 5	NR 6
Natural rubber—SVR 10	100.0	100.0	100.0	100.0	100.0	100.0
Zinc oxide	3.0	3.0	3.0	3.0	3.0	3.0
Stearic acid	2.0	2.0	2.0	2.0	2.0	2.0
Carbon black PM 15	70.0	–	–	–	–	–
Carbon black PM 75	–	70.0	–	–	–	–
ICSF-15-3	–	–	70.0	–	–	–
ICSF-15-7	–	–	–	70.0	–	–
ICSF-75-3	–	–	–	–	70.0	–
ICSF-75-7	–	–	–	–	–	70.0
TBBS	1.5	1.5	1.5	1.5	1.5	1.5
Sulfur	2.0	2.0	2.0	2.0	2.0	2.0

(Reprinted from Al-Ghamdi, A. A.; Al-Hartomy, O. A.; Al-Solamy, F. R.; Dishovsky, N.; Malinova, P.; Lakov, L. Characterization of Hybrid Fillers Based on Carbon Black of Different Types Obtained by Impregnation. *Proceedings of the Institution of Mechanical Engineers, Part L: Journal of Materials Design and Applications,*© 2015 with permission from SAGE Publications.)

The vulcanization of the NR-based compounds was carried out on an electrically heated hydraulic press using a special homemade mold at 150°C and 10 MPa. Each rubber compound was vulcanized for a certain cure time (T_{90}), determined by its cure curves (Figure 4.15).

4.3.1.3 CURING CHARACTERISTICS OF NATURAL RUBBER-BASED COMPOUNDS COMPRISING THE STUDIED HYBRID FILLERS

The hybrid products obtained were used as reinforcing fillers of NR-based compounds and vulcanizates. Table 4.3 summarizes the main curing properties of the studied rubber compounds determined by their cure curves plotted in Figure 4.15.

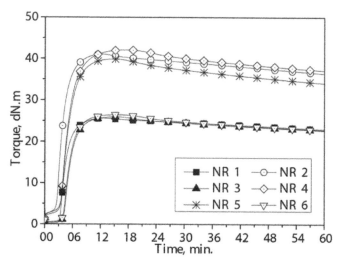

FIGURE 4.15 Cure curves of the investigated rubber compounds. (Reprinted from Al-Ghamdi, A. A.; Al-Hartomy, O. A.; Al-Solamy, F. R.; Dishovsky, N.; Malinova, P.; Lakov, L. Characterization of Hybrid Fillers Based on Carbon Black of Different Types Obtained by Impregnation. *Proceedings of the Institution of Mechanical Engineers, Part L: Journal of Materials Design and Applications,* © 2015 with permission from SAGE Publications.)

TABLE 4.3 Curing Properties of the Investigated Rubber Compounds.

	NR 1	NR 2	NR 3	NR 4	NR 5	NR 6
M_L (dN.m)	0.17	2.23	0.44	0.54	2.38	2.13
M_H (dN.m)	25.62	41.05	25.92	26.43	42.07	39.85
$\Delta M = M_H - M_L$ (dN.m)	25.45	38.82	25.48	25.89	39.69	37.72
T_{s2} (min/s)	3:14	1:46	3:57	3:48	2:22	2:41
T_{90} (min/s)	6:50	6:19	7:58	7:52	8:14	7:51
Cure rate, V (%/min)	27.8	22.0	24.9	20.9	17.1	19.3

(Reprinted from Al-Ghamdi, A. A.; Al-Hartomy, O. A.; Al-Solamy, F. R.; Dishovsky, N.; Malinova, P.; Lakov, L. Characterization of Hybrid Fillers Based on Carbon Black of Different Types Obtained by Impregnation. *Proceedings of the Institution of Mechanical Engineers, Part L: Journal of Materials Design and Applications,* © 2015 with permission from SAGE Publications.)

Figure 4.15 shows that the cure curves for all compounds have a similar pattern—curves with reversion. Reversion is a phenomenon related to reactions leading to desulfurization of sulfur cross-links as well as changes in the vulcanizate structure. As a result, the longer curing of such rubber compounds yields vulcanizates of poor mechanical properties. The reversion is typical of NR-based compounds and occurs due to the large number of double bonds in the rubber macromolecule.[40-42]

There are two distinct areas in the curve patterns corresponding to the two types of carbon black. The curves for the compounds comprising PM-15 (NR 1) carbon black and the dual-phase fillers (NR 3 and NR 4) based on it are located in the lower part of the plot, above them are the curves for the compounds comprising PM-75 (NR 2) and the dual-phase fillers (NR 5 and NR 6) based on NR 2 carbon black.

As shown in Table 4.3, the minimum torque (M_L) values and the viscosity of the rubber compounds comprising conventional PM-15 (NR 1) carbon black and the dual-phase fillers (NR 3 and NR 4) based on it are significantly lower than those of the compounds comprising PM-75 (NR 2) and the dual-phase fillers (NR 5 and NR 6) based on that type of carbon black. This is due to the larger particles of PM-15 carbon black and its low specific surface area, which lead to weaker polymer–filler interactions. Therefore, the viscosity of the rubber compounds remains relatively low. The minimum torque (M_L) values of the compounds comprising the dual-phase fillers (NR 3 and NR 4) obtained by impregnation of low active carbon black PM-15 are higher than those of the compound comprising the conventional filler (NR 1). This tendency is not observed in case of the compounds comprising carbon black PM-75 (NR 2) and the dual-phase fillers (NR 5 and NR 6) based on it. This result is quite obvious as the increase in the specific surface area of the dual-phase fillers following the impregnation with silicasol is more pronounced in the case of those based on the low active carbon black PM-15.

As seen from Table 4.3, the maximum torque (M_H) values and ΔM of the compounds comprising the dual-phase fillers (NR 3 and NR 6) do not change significantly, as compared to those of the respective compounds based on conventional carbon black (NR 1 and NR 2). Hence, the impregnation of both types of conventional carbon black with silica does not affect the cross-link density of the vulcanizates of the rubber compounds they comprise.

Due to impregnation of both types of conventional carbon black with silicasol, the rubber compounds thus filled have longer scorch time and cure time than that of the respective compounds filled with the conventional carbon black (NR 1 and NR 2), while the curing process runs at a lower rate. This is due to silica which is present as a second phase in the hybrid filler.

It is known that silanol groups distributed over the silica surface are able to adsorb the vulcanization accelerators, thus lowering the vulcanization rate.

4.3.1.4 MECHANICAL PROPERTIES OF NATURAL RUBBER-BASED COMPOSITES COMPRISING THE STUDIED HYBRID FILLERS

Table 4.4 presents the mechanical properties of the vulcanizates of the studied rubber compounds prior to the accelerated heat aging as well as their aging coefficients in terms of tensile strength (K_σ) and elongation at break (K_ε).

The mechanical properties of the composites filled with the hybrid fillers obtained by impregnation (modulus100, modulus300, tensile strength, elongation at break, Shore A hardness and abrasion) do not change significantly in comparison to the respective vulcanizates comprising the conventional carbon black used as a substrate, although the values are slightly lower. This is due to the fact that no silane coupling agent ensuring the chemical interaction between the silica particles and elastomer macromolecules was used since our aim was to study the reinforcing effect of the pure dual-phase fillers we had obtained.

TABLE 4.4 Mechanical Properties of the Investigated Rubber Composites.

	NR 1	NR 2	NR 3	NR 4	NR 5	NR 6
M_{100} (MPa)	3.2	6.6	3.0	2.9	5.6	5.5
M_{300} (MPa)	15.3	–	14.9	11.0	–	–
σ (MPa)	17.2	19.4	17.4	15.6	17.8	16.3
ε_{rel} (%)	360	255	360	340	270	270
ε_{res} (%)	10	10	10	10	10	10
Shore A hardness	66	79	67	67	80	78
Abrasion (mm³)	107	87	101	108	91	100
K_σ (%)	− 16.5	− 31.5	− 16.6	− 7.8	− 24.7	− 25.5
K_ε (%)	− 22.7	− 29.9	− 24.3	− 12.3	− 24.0	− 28.0

It is obvious from the aging coefficients presented in Table 4.4 that the dual-phase fillers obtained facilitate the studied vulcanizates with a better aging resistance than the carbon black used as a substrate to obtain them. This is due to the second phase consisting of silica which has higher thermo-stability than carbon black.

The results show that the fillers thus prepared do not significantly change the curing and mechanical characteristics of the vulcanizates comprising them, but improve their thermal aging resistance. In our opinion, this result is satisfactory since our aim is to improve the microwave properties of the final composites without impacting their mechanical properties. The isolation of the carbon black aggregates by the silica phase and the interpenetration of the two phases is a prerequisite for improving the microwave properties of composites comprising those fillers.

4.3.1.5   DIELECTRIC PROPERTIES OF NATURAL RUBBER-BASED COMPOSITES COMPRISING THE STUDIED HYBRID FILLERS

Figure 4.16 presents the frequency dependence of the real part of permittivity (ε_r'—dielectric constant) of the studied NR-based composites comprising the obtained hybrid fillers. As seen from Figure 4.16, the dependence is of a well pronounced resonance character. The highest values of the real part of permittivity for all composites studied are observed at about 3 GHz. In the range between 5 and 12 GHz, there are no significant changes in ε_r' values depending on frequency. Composite NR 2 filled with PM-75 carbon black possesses the highest values throughout the entire frequency range studied. Composite NR 1 filled with PM-15 carbon black has the lowest values of the dielectric constant. This is due to the difference in particle size of the two fillers and in their ability to form chain structures (conductive pathways).[43] The smaller size of PM-75 carbon black particles and their aptitude to form chain structures favor the creation of electrical (conductive) pathways. Hence, the interfacial polarization resulting from inhomogeneity introduced by the filler particles becomes easier and consequently the dielectric constant gets higher. The higher values of the dielectric constant evidence the easily accomplished molecular polarization.[44]

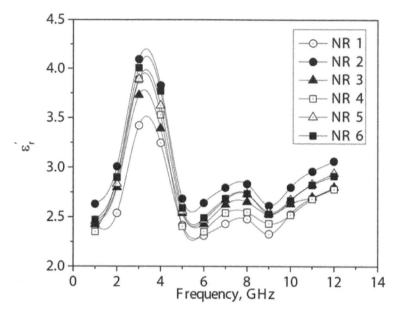

FIGURE 4.16 Frequency dependence of the real part of permittivity for the studied composites. (Reprinted from Al-Ghamdi, A. A.; Al-Hartomy, O. A.; Al-Solamy, F. R.; Dishovsky, N.; Mihaylov, M.; Malinova, P.; Atanasov, N. Dielectric and Microwave Properties of Elastomer Composites Loaded with Carbon–silica Hybrid Fillers. *J. Appl. Polym. Sci.* **2016**, *133*, 42978(1)–42978(9). © 2016 with permission from John Wiley.)

The dielectric constant values of composites NR 5 and NR 6 comprising hybrid fillers ICSF-75-3 and ICSF-75-7, prepared by impregnation of conventional PM-75 carbon black with silicasol, are slightly lower than those of composite NR 2 filled with the substrate PM-75 carbon black. The penetration of the silica phase among the chain structures of carbon black and their isolation hinders the creation of conductive pathways. This impedes the molecular polarization; hence, the dielectric constant of the composites obtained is lower. Composite NR 1 filled with conventional PM-15 carbon black and composites NR 3 and NR 4 comprising the hybrid fillers based on conventional PM-15 carbon black do not share the above-mentioned tendency, since the interpenetration of the two phases is less pronounced.[43]

As seen from Figure 4.16, in all cases the real part of permittivity values are generally rather low. This is due to the fact that the carbon black used is of very low conductivity and the polymer matrix is nonpolar and a crystallizing one that hinders the molecular polarization.[45,46]

The frequency dependence of the imaginary part of permittivity (ε_r'') of the composites studied is presented in Figure 4.17.

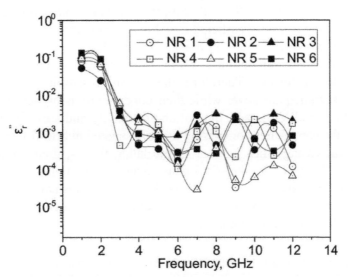

FIGURE 4.17 Frequency dependence of the imaginary part of permittivity for the studied composites. (Reprinted from Al-Ghamdi, A. A.; Al-Hartomy, O. A.; Al-Solamy, F. R.; Dishovsky, N.; Mihaylov, M.; Malinova, P.; Atanasov, N. Dielectric and Microwave Properties of Elastomer Composites Loaded with Carbon–silica Hybrid Fillers. *J. Appl. Polym. Sci.* **2016**, *133*, 42978(1)–42978(9). © 2016 with permission from John Wiley.)

Figure 4.17 shows ε_r'' to be a parameter very sensitive to frequency changes as well as to the type of filler used. In the frequency range of 1–5 GHz, ε_r'' decreases with the increasing frequency, while in the range of 5–12 GHz the parameter is of a well pronounced resonance character.

4.3.1.6 ELECTRICAL PROPERTIES OF NATURAL RUBBER-BASED COMPOSITES COMPRISING THE STUDIED HYBRID FILLERS

Table 4.5 presents the values of volume resistivity (ρ_v) and surface resistivity (ρ_s) for the composites studied. The results obtained are in full agreement with the dielectric properties data. As Table 4.5 shows, the volume resistivity values of composite NR 1—filled with conventional PM-15 carbon black as well as of composites NR 3 and NR 4 comprising PM-15 based hybrid fillers, vary in the range from 3.0×10^{12} to 1.2×10^{13} Ω.m. In that case, the composites studied are attributed as dielectrics. The volume resistivity values of composite NR 2 filled with PM-75 carbon black and NR 5 and NR 6 comprising PM-75 based hybrid fillers are much lower and vary in the range from 2.2×10^3 to 3.8×10^3 Ω.m, that is, those composites are semi-conductors. As mentioned earlier, this is due to the difference in the particles size of the two types of carbon black. The smaller size of PM-75 carbon black particles and their aptitude to form chain structures favor the creation of conductive pathways. Therefore, the volume resistivity values of the composites studied are lower while their conductivity is higher. According to the table, in all the cases, the volume resistivity and surface resistivity values of the composites comprising the investigated hybrid fillers increase with the increasing silica amount in the latter, that is, their conductivity decreases. This is due to the silica phase penetration among the chain-like structures of carbon black. Thus, the dielectric phase insulates the conductive one and the formation of conductive pathways is hindered resulting in a lower conductivity of the composite.

TABLE 4.5 Volume Resistivity and Surface Resistivity of the Studied Composites.

	ρ_v (Ω.m)	ρ_s (Ω)
NR 1	3.0×10^{12}	5.1×10^{14}
NR 2	2.2×10^3	2.3×10^6
NR 3	4.2×10^{12}	5.9×10^{14}
NR 4	1.2×10^{13}	6.5×10^{14}
NR 5	3.0×10^3	3.4×10^6
NR 6	3.8×10^3	6.0×10^6

(Reprinted from Al-Ghamdi, A. A.; Al-Hartomy, O. A.; Al-Solamy, F. R.; Dishovsky, N.; Mihaylov, M.; Malinova, P.; Atanasov, N. Dielectric and Microwave Properties of Elastomer Composites Loaded with Carbon–silica Hybrid Fillers. J. Appl. Polym. Sci. 2016, 133, 42978(1)–42978(9). © 2016 with permission from John Wiley.)

4.3.1.7 MICROWAVE PROPERTIES OF NATURAL RUBBER-BASED COMPOSITES COMPRISING THE STUDIED HYBRID FILLERS

Figures 4.18–4.20 present the frequency dependencies of the SE_T, SE_R, and SE_A for the composites studied and for the control sample PTFE.

As seen from Figure 4.18, elastomer composite NR 2 filled with PM-75 carbon black and composites NR 5 and NR 6 comprising hybrid fillers based on PM-75 carbon black have the highest values of SE_T. Composite

NR 1 filled with PM-15 carbon black and composites NR 3 and NR 4 whose hybrid fillers are based on PM-15 possess SE_T values lower than those of the composites comprising PM-75. Obviously, there is a tendency of increasing SE_T values with the increase in silica amount in the hybrid fillers. This is explained by the interpenetration of the two phases which causes the changes occurring in SE_R and SE_A (Figures 4.19 and 4.20). Frequency changes also affect the SE_T values of the composites studied. As Figure 4.18 shows, at frequencies of 1 and 2 GHz all samples are transparent to electromagnetic waves and their SE_T values tend to zero. In the frequency range 2–12 GHz, the dependency is of a resonance character. Throughout the entire frequency range, the control sample (PTFE) does not exhibit shielding properties, that is, its SE_T values tend to zero. This justifies the correctness of the experiment carried out.

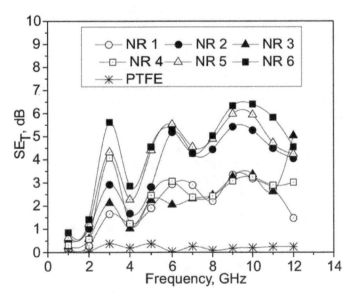

FIGURE 4.18 Frequency dependence of the total shielding effectiveness (SE_T) of the studied composites. (Reprinted from Al-Ghamdi, A. A.; Al-Hartomy, O. A.; Al-Solamy, F. R.; Dishovsky, N.; Mihaylov, M.; Malinova, P.; Atanasov, N. Dielectric and Microwave Properties of Elastomer Composites Loaded with Carbon–silica Hybrid Fillers. *J. Appl. Polym. Sci.* **2016**, *133*, 42978(1)–42978(9). © 2016 with permission from John Wiley.)

As shown in Figure 4.19, elastomer composite NR 2 filled with conventional PM-75 carbon black has the highest SE_R values. Composites NR 5 and NR 6 comprising hybrid fillers possess lower SE_R values. Moreover, there is a tendency of decreasing SE_R values of the composites with increasing the

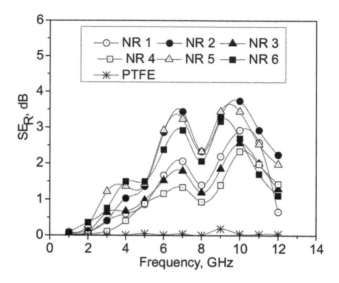

FIGURE 4.19 Frequency dependence of the reflective shielding effectiveness (SE$_R$) of the studied composites. (Reprinted from Al-Ghamdi, A. A.; Al-Hartomy, O. A.; Al-Solamy, F. R.; Dishovsky, N.; Mihaylov, M.; Malinova, P.; Atanasov, N. Dielectric and Microwave Properties of Elastomer Composites Loaded with Carbon–silica Hybrid Fillers. *J. Appl. Polym. Sci.* **2016**, *133*, 42978(1)–42978(9). © 2016 with permission from John Wiley.)

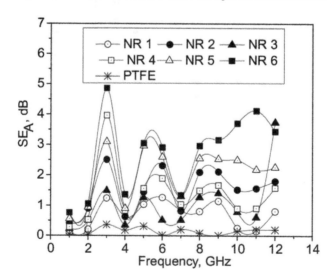

FIGURE 4.20 Frequency dependence of the absorptive shielding effectiveness (SE$_A$) of the studied composites. (Reprinted from Al-Ghamdi, A. A.; Al-Hartomy, O. A.; Al-Solamy, F. R.; Dishovsky, N.; Mihaylov, M.; Malinova, P.; Atanasov, N. Dielectric and Microwave Properties of Elastomer Composites Loaded with Carbon–silica Hybrid Fillers. *J. Appl. Polym. Sci.* **2016**, *133*, 42978(1)–42978(9). © 2016 with permission from John Wiley.)

amount of the filler's second phase (silica). This tendency is also observed in the case of composite NR 1 filled with conventional PM-15 carbon black, as well as in the case of composites NR 3 and NR 4 comprising hybrid fillers having PM-15 carbon black as a substrate. However, SE_R values in those cases are lower than SE_R values of the composites comprising PM-75 carbon black and PM-75 based hybrid fillers. The figure shows that the frequency dependence of SE_R is of resonance character, while the lower values of PTFE (absolutely transparent to microwaves) tending to zero justify the correctness of the experiment carried out.

As shown in Figure 4.20, the SE_A of the composites studied is reverse to the one of the SE_R. The composites comprising hybrid fillers possess the highest SE_A values which decrease with lowering the silica amount in the fillers. Here, NR 2 composite comprising PM-75 carbon black and composites NR 5 and NR 6 comprising hybrid fillers with PM-75 carbon black as a substrate also exhibit the best properties.

The results obtained have shown that the use of dual-phase fillers of different ratios of the conductive and dielectric phase would furnish the easy control and tailoring the microwave properties of electromagnetic composites. On the other hand, the impregnation technology is an unsophisticated method for controlling the quantitative ratio of the two phases in such fillers.

4.3.2 CONDUCTIVE CARBON BLACK–SILICA HYBRID FILLERS OBTAINED VIA IMPREGNATION

4.3.2.1 CHARACTERIZATION OF THE STUDIED HYBRID FILLERS

Printex XE-2B carbon black (from Orion Engineered Carbons GmbH) has been used as a substrate for the hybrid fillers studied due to its high conductivity, specific surface area, IA, OAN, and so forth. Table 4.6 summarizes the main properties of the hybrid fillers obtained. The properties of the fillers we have modified are compared to those of Printex XE-2B carbon black as given in the data sheet of the product provided by the producers.

The OAN or dibutylphthalate absorption (DBPA) of the fillers reveals their ability to form secondary structures, that is, aggregates and agglomerates. Therefore, OAN is a key parameter for a great part of the fillers used in rubber industry.[1,47] As seen from Table 4.6, the non-modified carbon black has OAN of 420 ml/100 g. The values of OAN for both carbon black/silica hybrid fillers (Pr/Si 3 and Pr/Si 7) obtained by impregnation of Printex XE-2B carbon black with different silicasol amounts are 460 and 480 ml/100 g, respectively. Hence, the hybrid fillers we have obtained are more prone to forming different secondary structures than non-modified Printex XE-2B carbon black. This is due to the silica available in the fillers. The effect is more pronounced at higher silica amounts in the total mass of the hybrid fillers. It is known that silica is hydrophilic and has hydroxyl (silanol) groups over its surface. Silanol groups located over the silica particles interact with each other via hydrogen bonds which ensure their greater aptitude to agglomeration than that of carbon black.

TABLE 4.6   Main Properties of the Studied Fillers.

	OAN (ml/100 g)	IA (mg/g)	S_{BET} (m²/g)	S_{MI} (m²/g)	S_{EXT} (m²/g)	V_t (cm³/g)	V_{MI} (cm³/g)	D_{AV} (nm)
Printex XE-2B	420	1125	1000	–	–	–	–	–
Pr/Si 3	460	695	912	82	830	1.73	0.04	7.6
Pr/Si 7	480	750	899	76	823	1.70	0.03	7.6

(Reprinted from Al-Ghamdi, A. A.; Al-Hartomy, O. A.; Al-Solamy, F. R.; Dishovsky, N.; Mihaylov, M.; Atanasov, N.; Atanasova, G.; Nihtianova, D. Microwave Properties of Natural Rubber Based Composites Comprising Conductive Carbon Black/silica Hybrid Fillers. *J. Polym. Res.* **2016,** *23,* 180. © 2016 With permission of Springer.)

The IA of fillers reveals their adsorption activity. The method is applicable only to the characterization of carbon black.[47] For this reason IA values for the prepared hybrid fillers comprising certain amounts of silica presented in Table 4.6 are subjective.

As seen from Table 4.6, the specific surface area (S_{BET}) of non-modified Printex XE-2B carbon black is about 1000 m²/g. Its impregnation with different silicasol amounts (equivalent to 3 and 7% of silica, respectively) leads to gradual decrease of the specific surface area of the hybrid fillers. This decrease is caused by the lower specific surface area of silica phase.

Obviously, the difference in silicasol amounts used for the modification is not great enough to cause drastic differences in the values for the adsorption-texture parameters of the hybrid fillers obtained. Nevertheless, Table 4.6 shows the synchronic (though insignificant) decrease of the total volume of Pr/Si 7 pores, as compared to that of Pr/Si 3. On the contrary, the average diameter of the pores is the same for both fillers which prove the identical process of forming the carbon/silica aggregates.

Figures 4.21–4.24 present the volume of adsorbed gas dependence (cm³/g) on relative pressure (p/p₀) and the derivative volume (dV) as a function of pore diameter (nm) of the studied fillers.

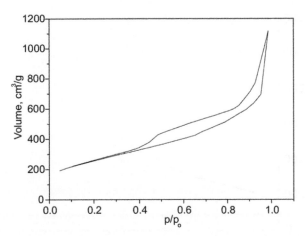

FIGURE 4.21 Volume of adsorbed gas dependence (cm³/g) on relative pressure (p/p0) of hybrid filler Pr/Si 3. (Reprinted from Al-Ghamdi, A. A.; Al-Hartomy, O. A.; Al-Solamy, F. R.; Dishovsky, N.; Mihaylov, M.; Atanasov, N.; Atanasova, G.; Nihtianova, D. Microwave Properties of Natural Rubber Based Composites Comprising Conductive Carbon Black/silica Hybrid Fillers. *J. Polym. Res.* **2016,** *23,* 180. © 2016 With permission of Springer.)

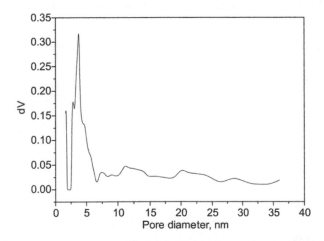

FIGURE 4.22 Derivative volume (dV) as a function of pore diameter (nm) of hybrid filler Pr/Si 3. (Reprinted from Al-Ghamdi, A. A.; Al-Hartomy, O. A.; Al-Solamy, F. R.; Dishovsky, N.; Mihaylov, M.; Atanasov, N.; Atanasova, G.; Nihtianova, D. Microwave Properties of Natural Rubber Based Composites Comprising Conductive Carbon Black/silica Hybrid Fillers. *J. Polym. Res.* **2016,** *23,* 180. © 2016 With permission of Springer.)

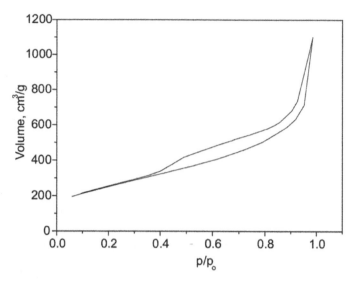

FIGURE 4.23   Volume of adsorbed gas dependence (cm³/g) on relative pressure (p/p₀) of hybrid filler Pr/Si 7. (Reprinted from Al-Ghamdi, A. A.; Al-Hartomy, O. A.; Al-Solamy, F. R.; Dishovsky, N.; Mihaylov, M.; Atanasov, N.; Atanasova, G.; Nihtianova, D. Microwave Properties of Natural Rubber Based Composites Comprising Conductive Carbon Black/silica Hybrid Fillers. *J. Polym. Res.* **2016,** *23,* 180. © 2016 With permission of Springer.)

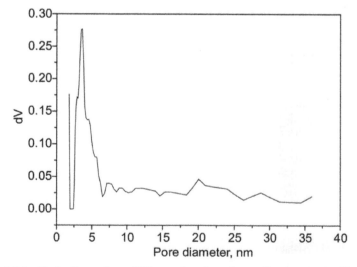

FIGURE 4.24   Derivative volume (dV) as a function of pore diameter (nm) of hybrid filler Pr/Si 7. (Reprinted from Al-Ghamdi, A. A.; Al-Hartomy, O. A.; Al-Solamy, F. R.; Dishovsky, N.; Mihaylov, M.; Atanasov, N.; Atanasova, G.; Nihtianova, D. Microwave Properties of Natural Rubber Based Composites Comprising Conductive Carbon Black/silica Hybrid Fillers. *J. Polym. Res.* **2016,** *23,* 180. © 2016 With permission of Springer.)

As seen from Figures 4.21–4.24, the curves for the pores size distribution for both fillers are more or less of quite a similar pattern. Though, it could be assumed that the major peaks at D_p: 25-65-70 nm for both fillers are due to inter-aggregate pores dominating in their structure. The availability of intra-aggregate pores is more pronounced in the case of Pr/Si 3, while in the case of Pr/Si 7 those are practically massively blocked by silicasol. This means silica is located over the surface of carbon aggregates in Pr/Si 3, while in Pr/Si 7 the silica phase is distributed both over and inside the carbon black aggregates. Obviously, in the case of Pr/Si 7 aggregates clustering is more pronounced than in the case of Pr/Si 3. The occurrence of visible ill-resolved peaks for Pr/Si 7 (distribution curve) is related to the intra-cluster pores (in the clustered aggregates which are of large size than those of filler Pr/Si 3).

Figures 4.25 and 4.26 present the bright-field STEM images and the compositional maps of the hybrid fillers studied, answering the questions about the distribution of carbon-silica phases within the hybrid fillers obtained.

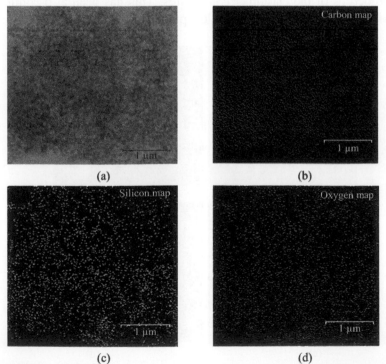

(a)

(b)

(c)

(d)

FIGURE 4.25 (a) Bright-field STEM image and compositional maps, (b) Carbon map, (c) Silicon map, (d) Oxygen map of hybrid filler Pr/Si 3. (Reprinted from Al-Ghamdi, A. A.; Al-Hartomy, O. A.; Al-Solamy, F. R.; Dishovsky, N.; Mihaylov, M.; Atanasov, N.; Atanasova, G.; Nihtianova, D. Microwave Properties of Natural Rubber Based Composites Comprising Conductive Carbon Black/silica Hybrid Fillers. *J. Polym. Res.* **2016,** *23,* 180. © 2016 With permission of Springer.)

Figures 4.25 and 4.26 confirm the successful modification of conductive Printex XE-2B carbon black with different silicasol amounts which resulted into carbon black/silica hybrid fillers with a different quantitative ratio between the conductive (carbon) and dielectric (silica) phases. The compositional maps of the fillers obtained show that the silica phase is evenly distributed in that of Printex XE-2B carbon black in filler Pr/Si 3 (comprising 3% of silica). It is mainly distributed over the surface of carbon aggregates. There are larger lumps of the inorganic phase (clusters, agglomerates) and domains where silica is less in filler Pr/Si 7 (comprising 7% of silica). This reveals the silica phase to be distributed over the surface of carbon black aggregates as well as inside them. STEM analysis is in agreement with the texture characteristics of the hybrid fillers investigated and evidences that the distribution of silica phase in the carbon one in both fillers hinders the interaction between the conductive particles to a certain extent (i.e., the formation of conductive pathways). On the other hand, this could provide an easy control and tailoring of the microwave properties of the elastomer composites. This has been also confirmed by TEM micrographs of the hybrid fillers presented in Figures 4.27 and 4.28.

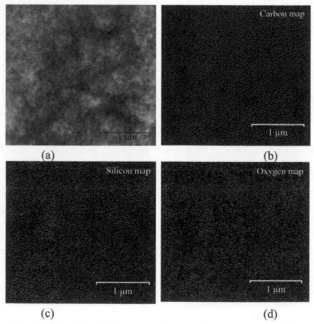

FIGURE 4.26   (a) Bright-field STEM image and compositional maps, (b) Carbon map, (c) Silicon map, (d) Oxygen map of hybrid filler Pr/Si 7. (Reprinted from Al-Ghamdi, A. A.; Al-Hartomy, O. A.; Al-Solamy, F. R.; Dishovsky, N.; Mihaylov, M.; Atanasov, N.; Atanasova, G.; Nihtianova, D. Microwave Properties of Natural Rubber Based Composites Comprising Conductive Carbon Black/silica Hybrid Fillers. *J. Polym. Res.* **2016,** *23,* 180. © 2016 With permission of Springer.)

As seen from TEM micrographs of higher magnifications (400,000×), presented in Figures 4.27(b) and 4.28(b), in both hybrid fillers the darker and more ordered structures of silica are located over the surface of carbon black (the brighter structures). Low-magnification TEM micrographs of (Figures 4.27(a) and 4.28(a)), prove the above statement about the distribution of the two phases in the hybrid fillers obtained which is based on their texture characteristics and STEM analysis.

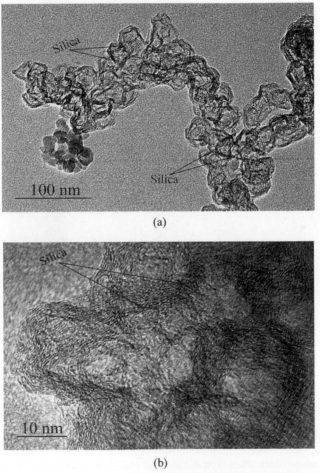

(a)

(b)

FIGURE 4.27 Bright-field TEM micrographs of hybrid filler Pr/Si 3 at different magnifications: (a) 60,000×, (b) 400,000×. (Reprinted from Al-Ghamdi, A. A.; Al-Hartomy, O. A.; Al-Solamy, F. R.; Dishovsky, N.; Mihaylov, M.; Atanasov, N.; Atanasova, G.; Nihtianova, D. Microwave Properties of Natural Rubber Based Composites Comprising Conductive Carbon Black/silica Hybrid Fillers. *J. Polym. Res.* **2016,** *23,* 180. © 2016 With permission of Springer.)

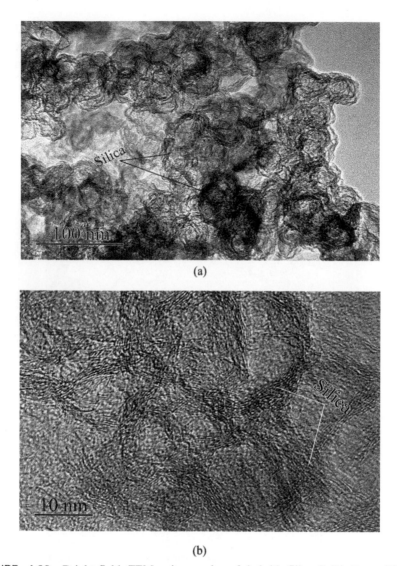

(a)

(b)

FIGURE 4.28 Bright-field TEM micrographs of hybrid filler Pr/Si 7 at different magnifications: (a) 60,000×, (b) 400,000×. (Reprinted from Al-Ghamdi, A. A.; Al-Hartomy, O. A.; Al-Solamy, F. R.; Dishovsky, N.; Mihaylov, M.; Atanasov, N.; Atanasova, G.; Nihtianova, D. Microwave Properties of Natural Rubber Based Composites Comprising Conductive Carbon Black/silica Hybrid Fillers. *J. Polym. Res.* **2016,** *23,* 180. © 2016 With permission of Springer.)

Figures 4.29 and 4.30 present the energy dispersive X-ray spectra of the hybrid fillers studied. The figures show that the fillers comprise traces of small amounts of vanadium, nickel, iron, and so forth but carbon and silica. Those results are expected, as the metals are present in Printex XE-2B CCB used as a substrate for the fillers preparation.

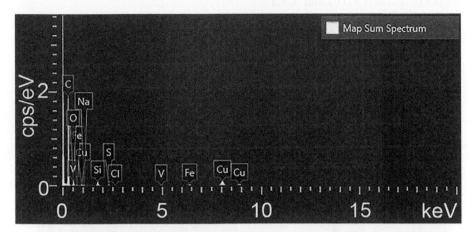

FIGURE 4.29 Energy dispersive X-ray spectrum of hybrid filler Pr/Si 3. (Reprinted from Al-Ghamdi, A. A.; Al-Hartomy, O. A.; Al-Solamy, F. R.; Dishovsky, N.; Mihaylov, M.; Atanasov, N.; Atanasova, G.; Nihtianova, D. Microwave Properties of Natural Rubber Based Composites Comprising Conductive Carbon Black/silica Hybrid Fillers. *J. Polym. Res.* **2016,** *23,* 180. © 2016 With permission of Springer.)

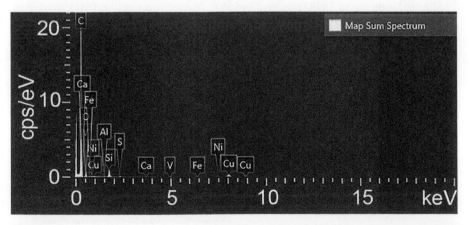

FIGURE 4.30 Energy dispersive X-ray spectrum of hybrid filler Pr/Si 7. (Reprinted from Al-Ghamdi, A. A.; Al-Hartomy, O. A.; Al-Solamy, F. R.; Dishovsky, N.; Mihaylov, M.; Atanasov, N.; Atanasova, G.; Nihtianova, D. Microwave Properties of Natural Rubber Based Composites Comprising Conductive Carbon Black/silica Hybrid Fillers. *J. Polym. Res.* **2016,** *23,* 180. © 2016 With permission of Springer.)

Figures 4.31(a) and 4.32(a) present the bright-field micrographs which show the morphologies of the particles, available in the hybrid fillers studied, involved in selected area electron diffraction (SAED) measurements (Figures 4.31(b) and 4.32(b)).

SAED patterns shown in Figures 4.31(b) and 4.32(b) confirm the availability of FeS and Fe_2O_3 (powder diffraction file—PDF 76-0961, 84-0311) in the hybrid fillers studied. This evidences the existence of the abovementioned metals in the obtained products in the form of various sulfides and chiefly in the form of oxides. Regarding the aim of our investigations, the result is positive since some of those oxides find application in manufacturing elastomer-based EMI SM.[48]

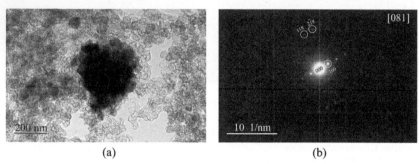

(a)   (b)

FIGURE 4.31   (a) Bright-field micrograph of FeS particle (in the middle) in hybrid filler Pr/Si 3, (b) Single crystal SAED pattern of FeS PDF 76-0961 in orientation [081]. (Reprinted from Al-Ghamdi, A. A.; Al-Hartomy, O. A.; Al-Solamy, F. R.; Dishovsky, N.; Mihaylov, M.; Atanasov, N.; Atanasova, G.; Nihtianova, D. Microwave Properties of Natural Rubber Based Composites Comprising Conductive Carbon Black/silica Hybrid Fillers. *J. Polym. Res.* **2016,** *23,* 180. © 2016 With permission of Springer.)

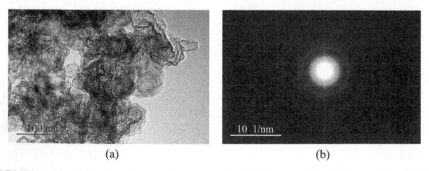

(a)   (b)

FIGURE 4.32   (a) Bright-field micrograph of hybrid filler Pr/Si 7, (b) polycrystalline SAED pattern of hybrid filler Pr/Si 7. (Reprinted from Al-Ghamdi, A. A.; Al-Hartomy, O. A.; Al-Solamy, F. R.; Dishovsky, N.; Mihaylov, M.; Atanasov, N.; Atanasova, G.; Nihtianova, D. Microwave Properties of Natural Rubber Based Composites Comprising Conductive Carbon Black/silica Hybrid Fillers. *J. Polym. Res.* **2016,** *23,* 180. © 2016 With permission of Springer.)

4.3.2.2 PREPARATION OF RUBBER COMPOSITES

The rubber compounds studied were prepared on a two-roll laboratory mill (rolls Length/Diameter 320 × 160 mm). Table 4.7 presents the compositions of the rubber compounds studied.

TABLE 4.7 Compositions of the Investigated Natural Rubber-Based Composites (phr).

	NR 1	NR 2	NR 3
Natural rubber—SVR 10	100.0	100.0	100.0
Zinc oxide	3.0	3.0	3.0
Stearic acid	2.0	2.0	2.0
Printex XE-2B	70.0	–	–
Pr/Si 3	–	70.0	–
Pr/Si 7	–	–	70.0
TBBS	1.5	1.5	1.5
Sulfur	2.0	2.0	2.0

(Reprinted from Al-Ghamdi, A. A.; Al-Hartomy, O. A.; Al-Solamy, F. R.; Dishovsky, N.; Mihaylov, M.; Atanasov, N.; Atanasova, G.; Nihtianova, D. Microwave Properties of Natural Rubber Based Composites Comprising Conductive Carbon Black/silica Hybrid Fillers. *J. Polym. Res.* **2016**, *23*, 180. © 2016 With permission of Springer.)

The vulcanization of the NR-based compounds was carried out on an electrically heated hydraulic press using a special homemade mold at 150°C and 10 MPa.

4.3.2.3 MICROWAVE PROPERTIES OF NATURAL RUBBER-BASED COMPOSITES COMPRISING THE STUDIED HYBRID FILLERS

Figures 4.33–4.35 show measured SE_T, SE_R, and SE_A of the studied composites over the frequency range of 1–12 GHz. The measurements were conducted on samples 2.71 mm (NR-1), 2.65 mm (NR 2) and 2.85 mm (NR 3) thickness.

As seen from Figures 4.33–4.35, sample NR 1 comprising non-modified Printex XE-2B has the lowest SE_T. The average SE_T value in the case is about 11.31 dB. In all cases, the SE_T of composites NR 2 and NR 3 comprising the studied hybrid fillers is higher than that of composite NR 1 filled with their substrate carbon black. Moreover, with the increase in the silica amount in the total mass of the hybrid fillers the SE_T of the composites studied improves. In the frequency range 1–12 GHz, the best SE_T was obtained for NR 3, with average value of 16.06 dB. Taking into

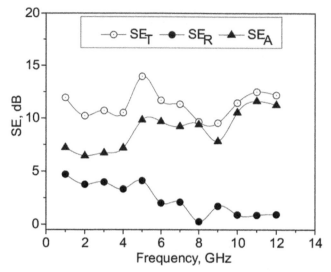

FIGURE 4.33 Frequency dependence of SE_T, SE_R and SE_A of sample NR 1. (Reprinted from Al-Ghamdi, A. A.; Al-Hartomy, O. A.; Al-Solamy, F. R.; Dishovsky, N.; Mihaylov, M.; Atanasov, N.; Atanasova, G.; Nihtianova, D. Microwave Properties of Natural Rubber Based Composites Comprising Conductive Carbon Black/silica Hybrid Fillers. *J. Polym. Res.* **2016,** *23,* 180. © 2016 With permission of Springer.)

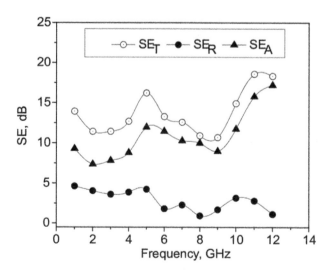

FIGURE 4.34 Frequency dependence of SE_T, SE_R and SE_A of sample NR 2. (Reprinted from Al-Ghamdi, A. A.; Al-Hartomy, O. A.; Al-Solamy, F. R.; Dishovsky, N.; Mihaylov, M.; Atanasov, N.; Atanasova, G.; Nihtianova, D. Microwave Properties of Natural Rubber Based Composites Comprising Conductive Carbon Black/silica Hybrid Fillers. *J. Polym. Res.* **2016,** *23,* 180. © 2016 With permission of Springer.)

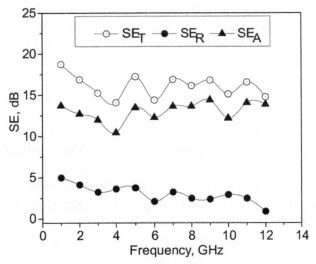

FIGURE 4.35   Frequency dependence of SE_T, SE_R and SE_A of sample NR 3. (Reprinted from Al-Ghamdi, A. A.; Al-Hartomy, O. A.; Al-Solamy, F. R.; Dishovsky, N.; Mihaylov, M.; Atanasov, N.; Atanasova, G.; Nihtianova, D. Microwave Properties of Natural Rubber Based Composites Comprising Conductive Carbon Black/silica Hybrid Fillers. *J. Polym. Res.* **2016,** *23,* 180. © 2016 With permission of Springer.)

account the fact that the SE_R of the composites studied does not change significantly depending on the type of filler, the SE_T has been determined by the changes in the SE_A. As the figures show, SE_A of the composites studied increases with the increasing silica amount in the hybrid fillers. This is because of the inorganic (dielectric) phase of the fillers distributed both over the surface and inside the carbon black aggregates (conductive phase). Thus, the interaction between the conductive particles (building of conductive pathways) is restricted, what in fact is one of the prerequisites for obtaining microwave absorbers. The higher silica amount in the fillers, the more pronounced the said effect is. Besides, with increasing silica amount in the total mass of the hybrid fillers probably the number of formed electric dipoles in the composites thus filled increases. This leads to a deeper penetration of the electromagnetic waves into the material and to an increase in the SE_A of the latter.

Figures 4.33–4.35 reveal the resonance character of SE_T, SE_R, and SE_A dependence on frequency. In all studied cases, absorption is a dominating mechanism. This could be explained by examining the values of the modulus of the reflection coefficient presented in Figure 4.36 in more detail. As seen, the modulus of the reflection coefficient for the composite filled with Pr/Si 3 (NR 2), for instance, at 1 GHz is 0.81. The amplitude of the reflected wave will be approximately 81% of the incident wave. This means the power of the reflected wave ($P_R = |\Gamma|^2 \cdot P_I$, Figure 4.3) will be 66% of the power of the incident wave (P_I). According to Eq. (4.14), the SE_R in that case is about 4.6 dB at 1 GHz. According to the law of conservation of energy, the wave penetrating the sample will have power equal to 34% of that of the incident wave. In accordance with Figure 4.34, the SE_T at 1 GHz is approximately 14 dB. As it is clear from Eq. (4.13), the transmitted power (P_T) is approximately 4% of the incident power (P_I). According to Eqs. (4.13) and (4.15), the low P_T value determines the high SE_A value and the dominating absorption character of the microwave shielding effectiveness.

The frequency dependence of the attenuation coefficient (α, dB/cm) of the studied composites is presented in Figure 4.37. Obviously, the attenuation coefficient (which is a direct function of the absorptive shielding effectiveness) of the studied composites increases with the increasing silica amount in the fillers they comprise. As the data in the figure shows, in the frequency interval 1–4 GHz the attenuation coefficient values for NR 3 decrease to 35 dB/cm approximately. The increase in the attenuation coefficient values for NR 2 upto 65 dB/cm is observed with the increasing frequency from 9 to 12 GHz exhibiting frequency selectivity in the X-band. The low, almost zero, values of the attenuation coefficient for the completely transparent for electromagnetic waves PTFE confirm the correctness of the performed measurements.

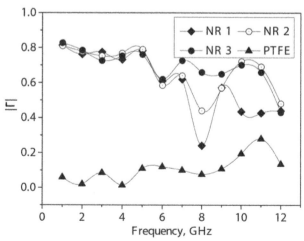

FIGURE 4.36 Frequency dependence of the modulus of reflection coefficient $|\Gamma|$ of the studied composites. (Reprinted from Al-Ghamdi, A. A.; Al-Hartomy, O. A.; Al-Solamy, F. R.; Dishovsky, N.; Mihaylov, M.; Atanasov, N.; Atanasova, G.; Nihtianova, D. Microwave Properties of Natural Rubber Based Composites Comprising Conductive Carbon Black/silica Hybrid Fillers. *J. Polym. Res.* **2016**, *23*, 180. © 2016 With permission of Springer.)

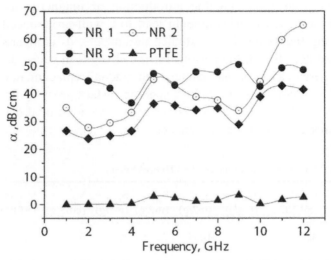

FIGURE 4.37 Frequency dependence of the attenuation coefficient (α) of the studied composites. (Reprinted from Al-Ghamdi, A. A.; Al-Hartomy, O. A.; Al-Solamy, F. R.; Dishovsky, N.; Mihaylov, M.; Atanasov, N.; Atanasova, G.; Nihtianova, D. Microwave Properties of Natural Rubber Based Composites Comprising Conductive Carbon Black/silica Hybrid Fillers. *J. Polym. Res.* **2016**, *23*, 180. © 2016 With permission of Springer.)

The high conductivity of the studied composites determined by their high concentration (above the percolation threshold) of fillers wherein predominates the conductive phase, is the reason for their shielding effectiveness. The conductive and dielectric properties of polymer materials depend on many factors and the filler/matrix ratio is among the most important ones. In the presence of small amounts of a conductive filler elastomer composites are good dielectric materials but above the critical filler concentration (percolation threshold) the conductivity may increase drastically.[49] Literature studies have established that the EMI SE of a composite is a function of its electrical conductivity. Moreover, EMI SE of composites increases with the filler concentration.[50,51]

4.3.3 CONDUCTIVE CARBON BLACK–MAGNETITE HYBRID FILLERS OBTAINED VIA IMPREGNATION

4.3.3.1 CHARACTERIZATION OF THE STUDIED HYBRID FILLERS

The main properties of the hybrid CCB/M fillers are summarized in Table 4.8. The characteristics of the CCB used as a substrate for the hybrid fillers are given as references.

As seen from Table 4.8, the introduction of magnetite as a second phase and the increase in its amount lead to a well-pronounced tendency of decreasing the values for all studied parameters, as compared to those of the substrate. Though the reasons for the changes in each parameter are different, there is a major one, namely, the gradual replacement of the very active carbon phase (possessing high specific surface area) by the magnetic one having much lower specific surface area. The additional densifying of the structures also has its certain impact.

TABLE 4.8 Main Properties of the Studied Hybrid Fillers.

	OAN (ml/100 g)	IA (mg/g)	S_{BET} (m2/g)	S_{MI} (m2/g)	S_{EXT} (m2/g)	V_{MI} (cm3/g)	V_t (cm3 (STP)/g)	D_{AV} (nm)
CCB	420	1125	1003	70	850	0.15	1.221	7.0
CCB/M-10	410	695	861	66	794	0.03	1.450	6.7
CCB/M-30	357	653	669	43	626	0.02	1.060	6.4
CCB/M-50	287	376	482	56	426	0.03	0.770	6.4

(Reprinted from Al-Ghamdi, A. A.; Al-Hartomy, O. A.; Al-Solamy, F. R.; Dishovsky, N.; Zaimova, D.; Malinova, P.; Nihtianova, D. Preparation and Characterization of Natural Rubber Composites Comprising Conductive Carbon Black/Magnetite Hybrid Fillers Obtained by Impregnation Technology. *Polym. Plast. Technol. Eng.* Published online: 01 Mar 2016. http://dx.doi.org/10.1080/03602559.2015.1132461 (accessed 01 Mar 2016) © 2016 with permission from Taylor & Francis.)

The texture analysis and texture characteristics of the obtained hybrid fillers are plotted in Figures 4.38 and 4.39. Regardless the varying of magnetite amount in a wide range (from 10 to 50%), the pattern of adsorption–desorption isotherms and hysteresis loops for the three samples does not change (Fig 4.38). However, the increase in magnetite amount from 10 to 50% the values of specific surface area decrease duly (with 22 and 44%, respectively) as those of the total pore volume of samples CCB/M-30 and CCB/M-50 (with 27 and 47%, respectively).

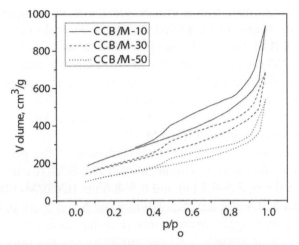

FIGURE 4.38 Dependence of adsorbed gas volume in cm^3/g on relative pressure p/p_0. (Reprinted from Al-Ghamdi, A. A.; Al-Hartomy, O. A.; Al-Solamy, F. R.; Dishovsky, N.; Zaimova, D.; Malinova, P.; Nihtianova, D. Preparation and Characterization of Natural Rubber Composites Comprising Conductive Carbon Black/Magnetite Hybrid Fillers Obtained by Impregnation Technology. *Polym. Plast. Technol. Eng.* Published online: 01 Mar 2016. http://dx.doi.org/10.1080/03602559.2015.1132461 (accessed 01 Mar 2016) © 2016 with permission from Taylor & Francis.)

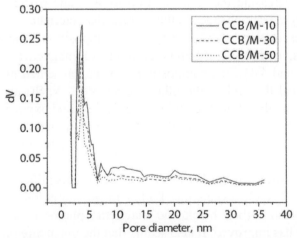

FIGURE 4.39 Dependence of derivative volume dV on pore size, nm. (Reprinted from Al-Ghamdi, A. A.; Al-Hartomy, O. A.; Al-Solamy, F. R.; Dishovsky, N.; Zaimova, D.; Malinova, P.; Nihtianova, D. Preparation and Characterization of Natural Rubber Composites Comprising Conductive Carbon Black/Magnetite Hybrid Fillers Obtained by Impregnation Technology. *Polym. Plast. Technol. Eng.* Published online: 01 Mar 2016. http://dx.doi.org/10.1080/03602559.2015.1132461 (accessed 01 Mar 2016) © 2016 with permission from Taylor & Francis.)

The pores size distribution curves for the three samples have quite similar patterns. It could be assumed that the main peaks having maxima at D_p: 4.2 nm (CCB/M-10), D_p: 4.0 nm (CCB/M-30) and D_p: 3.9 nm (CCB/M-50) (Figure 4.39) are due to the inter-aggregate pores dominating in those samples. The existence of intra-aggregate pores in the samples having D_p max: 2.7–3.1 is more pronounced for CCB/M-10, while in the other two samples in fact the intra-aggregate pores are more blocked by the magnetite phase, if compared to those of CCB/M-10.

The gradual minimization and transformation of the peaks at 4.3 nm, 7.1 nm and the ill-resolved peak at 5.6 nm (CCB/M-10) into ill-resolved peaks at 4.7–5.0 nm, 5.5–6.5 nm and 6.5–8.6 nm (CCB/M-30) and into ill-resolved peaks at 4.4–4.6 nm, 5.5–6.0 nm and 6.5–8.5 nm (CCB/M-50) is caused by the intra-agglomerate pores (existing in aggregates of carbon black and magnetite particles). Those agglomerates not only enlarge with the increasing magnetite amount but become denser additionally.

In this aspect, the shrunk average pore size of the samples from $D_{av} = 6.7$ nm for CCB/M-10 to $D_{av} = 6.4$ nm for CCB/M-30 and CCB/M-50 could be explained.

The association of the carbon black aggregates and magnetite particles into agglomerates is obvious at all magnetite concentrations. The process is enhanced at increasing the concentration to 30% and particularly to 50%; this causes significant changes in the texture characteristics of the fillers.

The representative HAADF image and compositional maps of CCB/M hybrid fillers, containing different amounts of magnetite are shown in Figures 4.40 and 4.41. The compositional maps, presented in Figures 4.40 and 4.41 reveal that unlike the other elements (C, V, Si) available in the fillers, which are distributed relatively evenly, iron and oxygen are localized as agglomerates—Figures 4.40(b), 4.40(d) and Figures 4.41(b), 4.41(d). Their size and polydispersity increase with the increasing magnetite amount in the synthesis course.

The comparison of TEM images at different magnification, presented in Figures 4.42–4.43, reveals that higher magnetite amounts decrease the contrast between carbon black and magnetite phases meaning the filler homogeneity has improved. It also seems that the magnetite phase interpenetrates the carbon black one. The bright fields on the images correspond to the carbon phase while the dark ones to the magnetite phase. EDS spectra (Figs. 4.40(a) and 4.41(a)) also confirm the improved homogeneity of the fillers. Table 4.9 presents the content of elements carbon, oxygen, silicon, and iron in weight percents by EDS data.

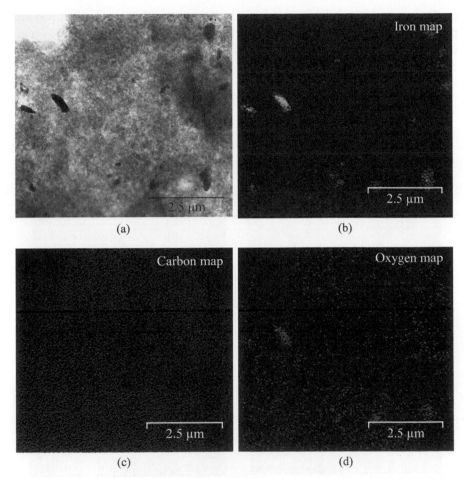

FIGURE 4.40 HAADF image (a) and compositional maps (b), (c), (d) of the CCB/M-10. (Reprinted from Al-Ghamdi, A. A.; Al-Hartomy, O. A.; Al-Solamy, F. R.; Dishovsky, N.; Zaimova, D.; Malinova, P.; Nihtianova, D. Preparation and Characterization of Natural Rubber Composites Comprising Conductive Carbon Black/Magnetite Hybrid Fillers Obtained by Impregnation Technology. *Polym. Plast. Technol. Eng.* Published online: 01 Mar 2016. http://dx.doi.org/10.1080/03602559.2015.1132461 (accessed 01 Mar 2016) © 2016 with permission from Taylor & Francis.)

According to the minimal and maximal percentage of each element in the spectra characterizing the certain fillers (CCB/M-10 and CCB/M-50), those differences in the filler comprising a higher magnetite amount are smaller, that is, that filler is a more homogeneous one (Table 4.10).

Taking into account the considerable number of elements (S, V, Ni, Fe) present in CCB, as well as silicon dioxide in magnetite, it has been worth analyzing the hybrid fillers in a diffraction regime. The results obtained are presented in Figures 4.44–4.45, showing the chemical compounds formed by the elements present on the compositional maps.

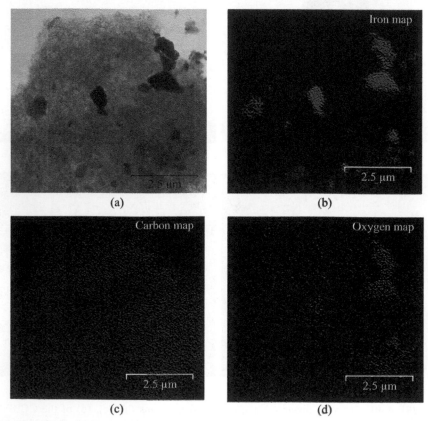

FIGURE 4.41   HAADF image (a) and compositional maps (b), (c), (d) of the CCB/M-50. (Reprinted from Al-Ghamdi, A. A.; Al-Hartomy, O. A.; Al-Solamy, F. R.; Dishovsky, N.; Zaimova, D.; Malinova, P.; Nihtianova, D. Preparation and Characterization of Natural Rubber Composites Comprising Conductive Carbon Black/Magnetite Hybrid Fillers Obtained by Impregnation Technology. *Polym. Plast. Technol. Eng.* Published online: 01 Mar 2016. http://dx.doi.org/10.1080/03602559.2015.1132461 (accessed 01 Mar 2016) © 2016 with permission from Taylor & Francis.)

The bright-field TEM images of the same samples are shown in Figures 4.42 and 4.43.

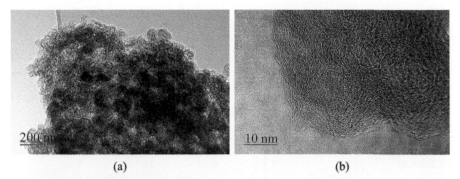

(a)   (b)

FIGURE 4.42 Bright-field TEM images at different magnifications of CCB/M-10: (a) 25,000×, (b) 400,000×. (Reprinted from Al-Ghamdi, A. A.; Al-Hartomy, O. A.; Al-Solamy, F. R.; Dishovsky, N.; Zaimova, D.; Malinova, P.; Nihtianova, D. Preparation and Characterization of Natural Rubber Composites Comprising Conductive Carbon Black/Magnetite Hybrid Fillers Obtained by Impregnation Technology. *Polym. Plast. Technol. Eng.* Published online: 01 Mar 2016. http://dx.doi.org/10.1080/03602559.2015.1132461 (accessed 01 Mar 2016) © 2016 with permission from Taylor & Francis.)

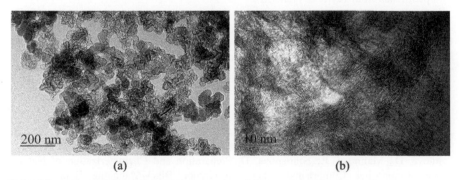

(a)   (b)

FIGURE 4.43 Bright -field TEM images at different magnifications of CCB/M-50: (a) 25,000, (b) 400,000×. (Reprinted from Al-Ghamdi, A. A.; Al-Hartomy, O. A.; Al-Solamy, F. R.; Dishovsky, N.; Zaimova, D.; Malinova, P.; Nihtianova, D. Preparation and Characterization of Natural Rubber Composites Comprising Conductive Carbon Black/Magnetite Hybrid Fillers Obtained by Impregnation Technology. *Polym. Plast. Technol. Eng.* Published online: 01 Mar 2016. http://dx.doi.org/10.1080/03602559.2015.1132461 (accessed 01 Mar 2016) © 2016 with permission from Taylor & Francis.)

TABLE 4.9 Content of the Elements Carbon, Oxygen, Silicon, and Iron in the Areas, Where Spectra No 113–122 were Scanned.

	Spectrum №	C (wt. %)	O (wt. %)	Fe (wt. %)	Si (wt. %)
CCB/M-10	113	8.38	17.24	74.21	0.18
CCB/M-10	114	40.78	25.78	14.29	12.72
CCB/M-10	115	10.90	11.86	76.82	0.43
CCB/M-10	116	56.60	4.98	37.81	0.57
CCB/M-50	118	35.77	3.18	1.51	0.60
CCB/M-50	119	67.30	6.71	24.31	1.06
CCB/M-50	120	40.69	10.01	48.15	0.95
CCB/M-50	121	72.68	0.93	20.09	2.56
CCB/M-50	122	50.87	2.51	45.07	0.95

(Reprinted from Al-Ghamdi, A. A.; Al-Hartomy, O. A.; Al-Solamy, F. R.; Dishovsky, N.; Zaimova, D.; Malinova, P.; Nihtianova, D. Preparation and Characterization of Natural Rubber Composites Comprising Conductive Carbon Black/Magnetite Hybrid Fillers Obtained by Impregnation Technology. *Polym. Plast. Technol. Eng.* Published online: 01 Mar 2016. http://dx.doi.org/10.1080/03602559.2015.1132461 (accessed 01 Mar 2016) © 2016 with permission from Taylor & Francis.)

TABLE 4.10 Differences Between the Minimal and Maximal Percentage (in wt. %) of the Elements Carbon, Iron, Oxygen and Silicon According to EDS Data Listed in Table 4.9.

	C (wt. %)	O (wt. %)	Fe (wt. %)	Si (wt. %)
CCB/M-10	48.2	20.8	62.5	12.5
CCB/M-50	36.9	9.8	46.6	1.9

(Reprinted from Al-Ghamdi, A. A.; Al-Hartomy, O. A.; Al-Solamy, F. R.; Dishovsky, N.; Zaimova, D.; Malinova, P.; Nihtianova, D. Preparation and Characterization of Natural Rubber Composites Comprising Conductive Carbon Black/Magnetite Hybrid Fillers Obtained by Impregnation Technology. *Polym. Plast. Technol. Eng.* Published online: 01 Mar 2016. http://dx.doi.org/10.1080/03602559.2015.1132461 (accessed 01 Mar 2016) © 2016 with permission from Taylor & Francis.)

Polycrystalline SAED data (Figure 4.44(b)) show that filler CCB/M-30 has a complicated phase composition: predominate magnetite (Fe_3O_4)-powder diffraction file (PDF) 82-1533, Fe_2O_3 PDF 84 0311, silica (coesite) PDF 83-1833, traces of FeO PDF 77-2355 and VO_2 PDF 82-1074. Polycrystalline SAED data (Figure 4.44(d)) show two phases: Fe_2O_3 PDF 84 0311 and FeS PDF 76-0965.

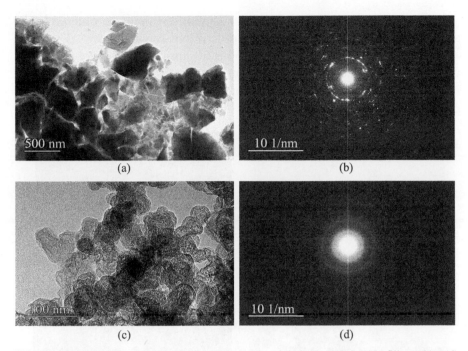

FIGURE 4.44 (a), (c) Bright-field micrograph of hybrid CCB/M-30 filler; (b), (d) polycrystalline SAED (selected area electron diffraction) patterns. (Reprinted from Al-Ghamdi, A. A.; Al-Hartomy, O. A.; Al-Solamy, F. R.; Dishovsky, N.; Zaimova, D.; Malinova, P.; Nihtianova, D. Preparation and Characterization of Natural Rubber Composites Comprising Conductive Carbon Black/Magnetite Hybrid Fillers Obtained by Impregnation Technology. *Polym. Plast. Technol. Eng.* Published online: 01 Mar 2016. http://dx.doi.org/10 .1080/03602559.2015.1132461 (accessed 01 Mar 2016) © 2016 with permission from Taylor & Francis.)

Polycrystalline SAED data show that the filler contains FeS_2 PDF 74-1051. Single crystalline SAED data are typical for α—quartz in orientation [11] PDF 46-1045.

Obviously, at the temperature at which the hybrid fillers thermal activation is carried out, regardless of vacuum atmosphere, do occur conditions for obtaining small amounts of various compounds, mainly because of the availability of other elements in the system (e.g., vanadium and sulfur up to 0.8%, according to[52]). The amounts of silicon dioxide (observed even in the monocrystalline form) present in magnetite also affect the phase composition (Figure 4.45).

The high-resolution TEM images also deserve interest (Figure 4.46—400,000×).

As seen from Figure 4.46, the filler structure is the most ordered in the absence of magnetite phase: the concentric microstructure is well developed and the spherulites are the largest (about 60 nm). The structure is densely packed and compact. The introduction of a magnetite phase, and with its increase, the spherulites shrink, the concentric microstructure is not so well pronounced.

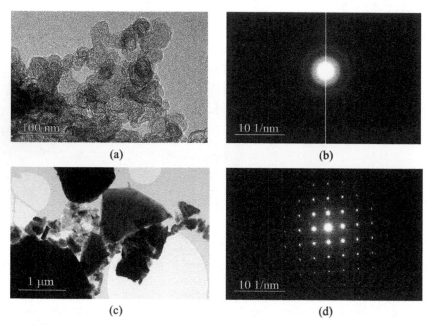

FIGURE 4.45   (a) Bright-field micrograph of hybrid filler CCB/M-50, (b) polycrystalline SAED pattern, (c) bright-field micrograph of α–quartz particle, (d) single crystalline SAED of α–quartz in orientation [111]. (Reprinted from Al-Ghamdi, A. A.; Al-Hartomy, O. A.; Al-Solamy, F. R.; Dishovsky, N.; Zaimova, D.; Malinova, P.; Nihtianova, D. Preparation and Characterization of Natural Rubber Composites Comprising Conductive Carbon Black/ Magnetite Hybrid Fillers Obtained by Impregnation Technology. *Polym. Plast. Technol. Eng.* Published online: 01 Mar 2016. http://dx.doi.org/10.1080/03602559.2015.1132461 (accessed 01 Mar 2016) © 2016 with permission from Taylor & Francis.)

In some cases, there is no order in the carbon phase. There are dark and bright zones, what allows the assumption that, at higher concentrations of the magnetite phase it penetrates into the carbon aggregates and insolates them. The phenomenon is not observed at lower magnetite concentrations. Our hypothesis is that the phenomenon mentioned is of great importance for improving the microwave properties of the composites comprising the particular hybrid fillers. The data have also been confirmed by the texture characteristics and by the characteristics summarized in Table 4.8.

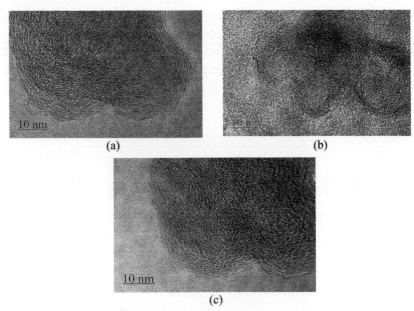

FIGURE 4.46 High-resolution TEM images of a filler comprising: (a) 0 wt. % magnetite, (b) 10 wt. % magnetite, (c) 50 wt. % magnetite. (Reprinted from Al-Ghamdi, A. A.; Al-Hartomy, O. A.; Al-Solamy, F. R.; Dishovsky, N.; Zaimova, D.; Malinova, P.; Nihtianova, D. Preparation and Characterization of Natural Rubber Composites Comprising Conductive Carbon Black/Magnetite Hybrid Fillers Obtained by Impregnation Technology. *Polym. Plast. Technol. Eng.* Published online: 01 Mar 2016. http://dx.doi.org/10.1080/03602559.2015.113 2461 (accessed 01 Mar 2016) © 2016 with permission from Taylor & Francis.)

The results from DTA and TGA characterization of the fillers are summarized in Table 4.11.

TABLE 4.11 DTA and TGA Characteristics of the Carbon Black.

	CCB	CCB/M-10	CCB/M-30	CCB/M-50
Commencement of the weight losses (°C)	330	355	379	410
Weight losses at 1000°C (%)	−98	−80	−64	−44
Maximum rate of oxidation (°C)	430	440	460	510
Temperature of complete combustion (°C)	550	560	584	599

(Reprinted from Al-Ghamdi, A. A.; Al-Hartomy, O. A.; Al-Solamy, F. R.; Dishovsky, N.; Zaimova, D.; Malinova, P.; Nihtianova, D. Preparation and Characterization of Natural Rubber Composites Comprising Conductive Carbon Black/Magnetite Hybrid Fillers Obtained by Impregnation Technology. *Polym. Plast. Technol. Eng.* Published online: 01 Mar 2016. http://dx.doi.org/10.1080/03602559.2015.1132461 (accessed 01 Mar 2016) © 2016 with permission from Taylor & Francis.)

With increasing the amount of magnetite phase the fillers become less reactive, therefore the maximum oxidation rate and complete combustion are reached at a higher temperature, while the weight losses decrease. The specific surface area (S_{BET}) dependence of the maximum oxidation rate temperature is well pronounced. It is presented in Figure 4.47 and confirms utterly the above conclusions.

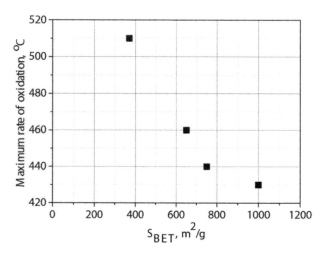

FIGURE 4.47 Dependence of the maximum oxidation rate temperature on the fillers specific surface area, on CCB/M phase ratio, respectively. (Reprinted from Al-Ghamdi, A. A.; Al-Hartomy, O. A.; Al-Solamy, F. R.; Dishovsky, N.; Zaimova, D.; Malinova, P.; Nihtianova, D. Preparation and Characterization of Natural Rubber Composites Comprising Conductive Carbon Black/Magnetite Hybrid Fillers Obtained by Impregnation Technology. *Polym. Plast. Technol. Eng.* Published online: 01 Mar 2016. http://dx.doi.org/10.1080/03602559.2015.113 2461 (accessed 01 Mar 2016) © 2016 with permission from Taylor & Francis.)

4.3.3.2 PREPARATION OF RUBBER COMPOSITES

The formulations of the NR-based compounds, containing a CCB/M hybrid fillers are shown in Table 4.12.

The elastomer compounds were prepared on an open laboratory two-roll mill with rolls dimensions L/D 320 × 160 and 1.27 friction. Test samples were vulcanized to plates with dimensions 6 × 8 cm by a steel press form on an electrically heated vulcanization hydraulic press at 10 MPa at the vulcanization time, determined according to their cure curves taken at 150°C.

TABLE 4.12   Formulations of the Studied NR-Based Compounds (phr).

	N 0	N 1	N 2	N 3
Natural rubber—SMR-10	100.0	100.0	100.0	100.0
Stearic acid	2.0	2.0	2.0	2.0
Zinc oxide	3.0	3.0	3.0	3.0
CCB	70.0	–	–	–
CCB/M-10	–	70.0	–	–
CCB/M-30	–	–	70.0	–
CCB/M-50	–	–	–	70.0
TBBS	1.5	1.5	1.5	1.5
Sulfur	2.0	2.0	2.0	2.0

(Reprinted from Al-Ghamdi, A. A.; Al-Hartomy, O. A.; Al-Solamy, F. R.; Dishovsky, N.; Zaimova, D.; Malinova, P.; Nihtianova, D. Preparation and Characterization of Natural Rubber Composites Comprising Conductive Carbon Black/Magnetite Hybrid Fillers Obtained by Impregnation Technology. *Polym. Plast. Technol. Eng.* Published online: 01 Mar 2016. http://dx.doi.org/10.1080/03602559.2015.1132461 (accessed 01 Mar 2016) © 2016 with permission from Taylor & Francis.)

4.3.3.3   CURING CHARACTERISTICS OF NATURAL RUBBER-BASED COMPOUNDS COMPRISING THE STUDIED HYBRID FILLERS

Table 4.13 summarizes the main curing characteristics of the studied rubber compounds determined according to their cure curves.

TABLE 4.13   Curing Characteristics of the Studied Rubber Compounds.

	N 1	N 2	N 3
M_L (dN.m)	47	25	10
M_H (dN.m)	75	56	39
$\Delta M = M_H - M_L$ (dN.m)	28	31	29
T_{90} (min)	15:42	12:11	10:51
T_{S2} (min)	1:10	4:06	3:52

(Reprinted from Al-Ghamdi, A. A.; Al-Hartomy, O. A.; Al-Solamy, F. R.; Dishovsky, N.; Zaimova, D.; Malinova, P.; Nihtianova, D. Preparation and Characterization of Natural Rubber Composites Comprising Conductive Carbon Black/Magnetite Hybrid Fillers Obtained by Impregnation Technology. *Polym. Plast. Technol. Eng.* Published online: 01 Mar 2016. http://dx.doi.org/10.1080/03602559.2015.1132461 (accessed 01 Mar 2016) © 2016 with permission from Taylor & Francis.)

Usually the minimum torque (M_L) is related to viscosity, while the maximum torque (M_H)—to resistivity and strain stress deformation. As seen from Table 4.13, the increase in the magnetite amount leads to a significant decrease of M_L and M_H values. Moreover the compound comprising 10% of magnetite practically has no thermoplastic interval. The observed effects are due to the high filler quantity (70 phr) and to the high specific surface area of Printex XE 2B carbon black used as a substrate for the hybrid fillers. The strong filler-filler and elastomer-filer interactions in the case of CCB/M-10 (N 1) is due to the prevailing amount of the active carbon phase and predetermines the high M_L and M_H values. A higher magnetite phase quantity leads to gradual weakening of the filler-filler and elastomer-filer interactions, hence to lowering M_L and M_H values and to expanding the thermoplastic interval. The optimum cure time (T_{90}) is longer for CCB/M-10 (N 1), while that for fillers CCB/M-30 (N 2) and CCB/M-50 (N 3) is shorter. The higher vulcanization rate for CCB/M-30 (N 2) and CCB/M-50 (N 3) is probably due to the higher amount of magnetite. The latter being a kind of ferrite enhances the vulcanization rate, as stated by some authors.[52]

4.3.3.4   EQUILIBRIUM SWELLING RATE OF NATURAL RUBBER-BASED COMPOSITES COMPRISING THE STUDIED HYBRID FILLERS

The molecular weight of the rubber segments between two cross-links (M_c) and the crosslink density (v) of the vulcanizates comprising the hybrid fillers are shown in Figures 4.48 and 4.49.

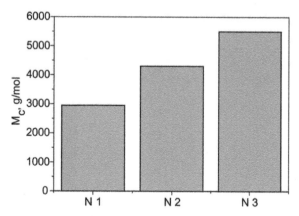

FIGURE 4.48   Molecular weights of the rubber segments between two cross-links (M_c) of the studied composites comprising CCB/M hybrid fillers. (Reprinted from Al-Ghamdi, A. A.; Al-Hartomy, O. A.; Al-Solamy, F. R.; Dishovsky, N.; Zaimova, D.; Malinova, P.; Nihtianova, D. Preparation and Characterization of Natural Rubber Composites Comprising Conductive Carbon Black/Magnetite Hybrid Fillers Obtained by Impregnation Technology. *Polym. Plast. Technol. Eng.* Published online: 01 Mar 2016. http://dx.doi.org/10.1080/03602559.2015.113 2461 (accessed 01 Mar 2016) © 2016 with permission from Taylor & Francis.)

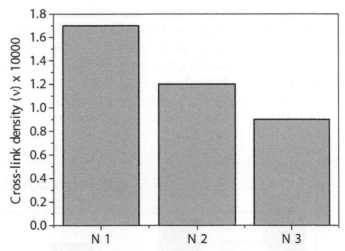

FIGURE 4.49 Cross-link density of the studied composites comprising CCB/M hybrid fillers. (Reprinted from Al-Ghamdi, A. A.; Al-Hartomy, O. A.; Al-Solamy, F. R.; Dishovsky, N.; Zaimova, D.; Malinova, P.; Nihtianova, D. Preparation and Characterization of Natural Rubber Composites Comprising Conductive Carbon Black/Magnetite Hybrid Fillers Obtained by Impregnation Technology. *Polym. Plast. Technol. Eng.* Published online: 01 Mar 2016. http://dx.doi.org/10.1080/03602559.2015.1132461 (accessed 01 Mar 2016) © 2016 with permission from Taylor & Francis.)

The results presented in Figures 4.48 and 4.49 confirm those obtained from the cure curves. The cross-link density of the vulcanizates comprising fillers with a small amount of magnetite as a second phase (N 1) is higher since the carbon black of high specific surface area yields a denser cross-link network. The filler with the lowest molecular weight of the rubber segments between two cross-links yields the densest crosslink network. Increasing the amount of the second phase leads to loosening of the cross-link network, while the molecular weight of the rubber segments between crosslinks increases, due to the much inerter character of magnetite.

4.3.3.5 MECHANICAL PROPERTIES OF NATURAL RUBBER-BASED COMPOSITES COMPRISING THE STUDIED HYBRID FILLERS BEFORE AND AFTER AGING

The dependence of the vulcanizates mechanical properties on the type of hybrid fillers studied is shown in Figures 4.50, 4.51. At higher magnetite amounts the carbon black content is lower what leads to lower tensile strength values (Figure 4.50) since there is a significant difference in the reinforcing activity of the two phases. The weaker elastomer–filler interaction also ensures greater mobility of the macromolecules, and increase in the elongation at break (Figure 4.51), that is the ratio of the two phases is of great importance for the mechanical properties of the vulcanizates.

As far as the thermal aging is concerned, the resistance of the vulcanizates decreases with the increase in the magnetite phase amount.

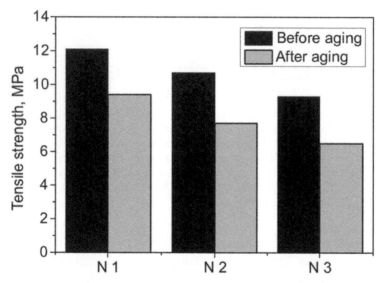

FIGURE 4.50 Tensile strength of the studied composites comprising CCB/M hybrid fillers. (Reprinted from Al-Ghamdi, A. A.; Al-Hartomy, O. A.; Al-Solamy, F. R.; Dishovsky, N.; Zaimova, D.; Malinova, P.; Nihtianova, D. Preparation and Characterization of Natural Rubber Composites Comprising Conductive Carbon Black/Magnetite Hybrid Fillers Obtained by Impregnation Technology. *Polym. Plast. Technol. Eng.* Published online: 01 Mar 2016. http://dx.doi.org/10.1080/03602559.2015.1132461 (accessed 01 Mar 2016) © 2016 with permission from Taylor & Francis.)

The Shore A hardness after the aging is plotted in Figure 4.52. Obviously, the Shore A hardness of the composites comprising the filler with 10% of magnetite is higher than that of the composites whose fillers have 30 and 50% of magnetite, respectively. The result is logical since the viscosity and the crosslink density of the vulcanizate comprising CCB/M-10 (N 1) are higher than those of the vulcanizates comprising CCB/M-30 (N 2) and CCB/M-50 (N 3).

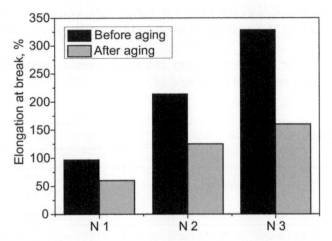

FIGURE 4.51 Elongation at break of the studied composites comprising CCB/M hybrid fillers. (Reprinted from Al-Ghamdi, A. A.; Al-Hartomy, O. A.; Al-Solamy, F. R.; Dishovsky, N.; Zaimova, D.; Malinova, P.; Nihtianova, D. Preparation and Characterization of Natural Rubber Composites Comprising Conductive Carbon Black/Magnetite Hybrid Fillers Obtained by Impregnation Technology. *Polym. Plast. Technol. Eng.* Published online: 01 Mar 2016. http://dx.doi.org/10.1080/03602559.2015.1132461 (accessed 01 Mar 2016) © 2016 with permission from Taylor & Francis.)

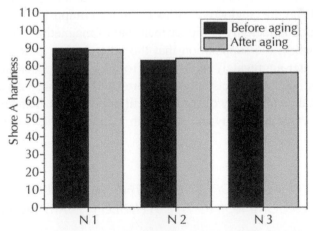

FIGURE 4.52 Shore hardness of the studied composites comprising CCB/M hybrid fillers. (Reprinted from Al-Ghamdi, A. A.; Al-Hartomy, O. A.; Al-Solamy, F. R.; Dishovsky, N.; Zaimova, D.; Malinova, P.; Nihtianova, D. Preparation and Characterization of Natural Rubber Composites Comprising Conductive Carbon Black/Magnetite Hybrid Fillers Obtained by Impregnation Technology. *Polym. Plast. Technol. Eng.* Published online: 01 Mar 2016. http://dx.doi.org/10.1080/03602559.2015.1132461 (accessed 01 Mar 2016) © 2016 with permission from Taylor & Francis.)

4.3.3.6 ELECTRIC PROPERTIES OF NATURAL RUBBER-BASED COMPOSITES COMPRISING THE STUDIED HYBRID FILLERS

The values for the volume resistivity and surface resistivity of the vulcanizates studied are presented in Table 4.14.

TABLE 4.14 Volume Resistivity and Surface Resistivity of Composites Comprising the Studied CCB/M Hybrid Fillers.

	ρ_v (Ω.m)	ρ_s (Ω)
N 1	3.7	5.3×10^2
N 2	5.6	8.1×10^2
N 3	7.4	3.0×10^3

(Reprinted from Al-Ghamdi, A. A.; Al-Hartomy, O. A.; Al-Solamy, F. R.; Dishovsky, N.; Zaimova, D.; Malinova, P.; Nihtianova, D. Preparation and Characterization of Natural Rubber Composites Comprising Conductive Carbon Black/Magnetite Hybrid Fillers Obtained by Impregnation Technology. *Polym. Plast. Technol. Eng.* Published online: 01 Mar 2016. http://dx.doi.org/10.1080/03602559.2015.1132461 (accessed 01 Mar 2016) © 2016 with permission from Taylor & Francis.)

As seen from the Table 4.14, the resistivity values are very low and rather close. This is due to the fact that the percolation threshold had already been passed and the conductive pathways in the composite have been built. Carbon black conductivity is higher than that of magnetite[53] and the data in the table confirm the supposition that the latter is distributed intra-aggregately and inter-aggregately with regard to the carbon phase in the hybrid fillers and limits the contacts between more CCB structures. As a result, the total electrical conductivity of the hybrid filler decreases. The volume resistivity and surface resistivity increase with increasing the magnetite phase, respectively.

4.3.3.7 DYNAMIC PROPERTIES OF THE NATURAL RUBBER-BASED COMPOSITES COMPRISING THE STUDIED HYBRID FILLERS

The temperature dependence of the dynamic storage modulus of the vulcanizates studied is presented in Figure 4.53.

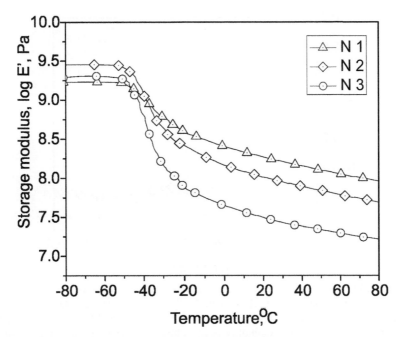

FIGURE 4.53 Temperature dependence of the dynamic storage modulus (E') of the studied composites comprising CCB/M hybrid fillers. (Reprinted from Al-Ghamdi, A. A.; Al-Hartomy, O. A.; Al-Solamy, F. R.; Dishovsky, N.; Zaimova, D.; Malinova, P.; Nihtianova, D. Preparation and Characterization of Natural Rubber Composites Comprising Conductive Carbon Black/Magnetite Hybrid Fillers Obtained by Impregnation Technology. *Polym. Plast. Technol. Eng.* Published online: 01 Mar 2016. http://dx.doi.org/10.1080/03602559.2015.113 2461 (accessed 01 Mar 2016) © 2016 with permission from Taylor & Francis.)

As Figure 4.53 shows, E' values are the highest in the interval from −40 to 80°C for the composite comprising 10% of magnetite. That is due to the great carbon black amount and to the rubber–filler interaction thus predetermined, since high E' values result from the restricted mobility of the vulcanizate macromolecules. Following the logic of the previous statement, the vulcanizates comprising 30 and 50% of magnetite have lower E' values because of the lower carbon black amount, hence their macromolecules are more mobile.

The mechanical loss angle tangent (tan δ) is the ratio between the dynamic loss modulus (E″) and the dynamic storage modulus (E)—(tan δ = E″/E). Figure 4.54 presents the temperature dependence of the mechanical loss angle tangent (tan δ) of the vulcanizates studied.

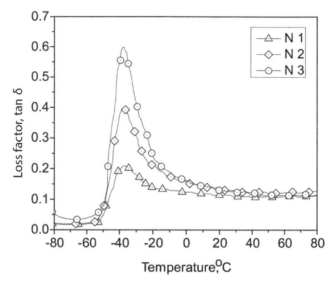

FIGURE 4.54   Temperature dependence of mechanical loss angle tangent (tan δ) of the studied composites comprising CCB/M hybrid fillers. (Reprinted from Al-Ghamdi, A. A.; Al-Hartomy, O. A.; Al-Solamy, F. R.; Dishovsky, N.; Zaimova, D.; Malinova, P.; Nihtianova, D. Preparation and Characterization of Natural Rubber Composites Comprising Conductive Carbon Black/Magnetite Hybrid Fillers Obtained by Impregnation Technology. *Polym. Plast. Technol. Eng.* Published online: 01 Mar 2016. http://dx.doi.org/10.1080/03602559.2015.113 2461 (accessed 01 Mar 2016) © 2016 with permission from Taylor & Francis.)

As known, tan δ peak corresponds to the glass transition temperature (T$_g$) of the studied composites. As seen from Figure 4.54, T$_g$ is not quite dependent on the magnetite content in the dual-phase fillers. The filler of lower magnetite content has T$_g$ = − 37°C. An increase in the amount of magnetite phase slightly increases T$_g$ (about 2–3°C). As far as the intensity of tan δ peak is concerned, the results are in accordance with those obtained for E' and are due to the ratio between the active and inert phase of the filler.

4.3.3.8   COMPOSITES HOMOGENEITY

A key issue in the manufacture of nanocomposites is ensuring their homogeneity. Herein, the homogeneity of the nanocomposites has been addressed from a macroscopic standpoint. Since the electrical properties are very sensitive to the microstructure, which is determined by processing, electrical measurements are expected to provide important feedback on the manufacturing process. Electrical measurements have several advantages over other methods used to characterize the microstructure of materials. Those include the relatively low cost of the electrical equipment used, easy sample preparation, and the fact that the information can be obtained in a non-destructive way.[54] Homogeneity has been assessed estimating variations of macro-electrical resistance by varying the distance between contact points. The dependence of resistance on the distance between contact points for different magnetite phase concentrations is shown in Figure 4.55. The coefficients of correlation between the resistance and the distance shown in Table 4.15, are calculated using Eq. 4.4

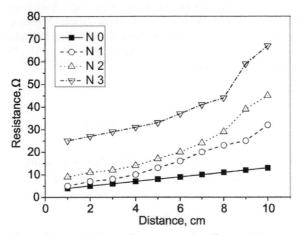

FIGURE 4.55 Dependence of the resistance on the distance between contact points for composites, containing CCB/M hybrid fillers. (Reprinted from Al-Ghamdi, A. A.; Al-Hartomy, O. A.; Al-Solamy, F. R.; Dishovsky, N.; Zaimova, D.; Malinova, P.; Nihtianova, D. Preparation and Characterization of Natural Rubber Composites Comprising Conductive Carbon Black/Magnetite Hybrid Fillers Obtained by Impregnation Technology. *Polym. Plast. Technol. Eng.* Published online: 01 Mar 2016. http://dx.doi.org/10.1080/03602559.2015.113 2461 (accessed 01 Mar 2016) © 2016 with permission from Taylor & Francis.)

As Figure 4.55 and Table 4.15 show, higher magnetite amounts in the dual-phase filler worsen the homogeneity of the composites, since CCB which disperses easily and evenly is of lower amount, while magnetite forms agglomerates which are harder to disperse. That is seen from TEM studies and from the analysis of magnetite distribution in the hybrid fillers by Fe and O-mapping shown in Figures 4.40–4.41. As seen, the coefficient of correlation varies considerably depending both on the type of substrate and magnetite amount. In the case of neat CCB, a perfect degree of correlation ($0.9 < R_p < 1.0$) has been observed, while in the case of the fillers modified with magnetite the degree of correlation is very high ($0.7 < R_p < 0.9$). However, for all composites comprising CCB/M fillers the homogeneity remains relatively higher.

TABLE 4.15   Coefficients of Correlation Between Electrical Resistance (Ω) and the Distance Between the Contact Points (cm) for Composites, Containing Hybrid Fillers.

	Amount of magnetite (%)	**Coefficient of correlation**
N 0	0	0.931
N 1	10	0.890
N 2	30	0.810
N 3	50	0.790

(Reprinted from Al-Ghamdi, A. A.; Al-Hartomy, O. A.; Al-Solamy, F. R.; Dishovsky, N.; Zaimova, D.; Malinova, P.; Nihtianova, D. Preparation and Characterization of Natural Rubber Composites Comprising Conductive Carbon Black/Magnetite Hybrid Fillers Obtained by Impregnation Technology. *Polym. Plast. Technol. Eng.* Published online: 01 Mar 2016. http://dx.doi.org/10.1080/03602559.2015.1132461 (accessed 01 Mar 2016) © 2016 with permission from Taylor & Francis.)

4.3.3.9   MICROWAVE PROPERTIES OF NATURAL RUBBER-BASED COMPOSITES COMPRISING THE STUDIED HYBRID FILLERS

The total EMI SE of the investigated composites depending on frequency is shown in Figure 4.56. The reflective (SE_R) and absorptive (SE_A) shielding effectiveness dependences are presented separately (Figures 4.57–4.58), so that the results related to EMI SE could be better comprehended and interpreted.

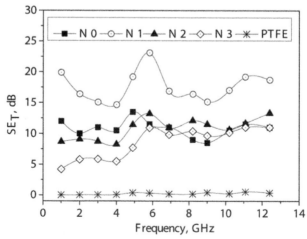

FIGURE 4.56   Frequency dependence of total shielding effectiveness of composites containing CCB/M hybrid fillers. (Reprinted from Al-Ghamdi, A. A.; Al-Hartomy, O. A.; Al-Solamy, F. R.; Dishovsky, N.; Malinova, P.; Atanasova, G.; Atanasov, N. Conductive Carbon Black/magnetite Hybrid Fillers in Microwave Absorbing Composites Based on Natural Rubber. *Compos. Part B: Eng.* **2016,** *96,* 231–241. © 2016 with permission from Elsevier.)

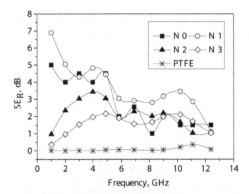

FIGURE 4.57 Frequency dependence of reflective shielding effectiveness of composites containing CCB/M hybrid fillers. (Reprinted from Al-Ghamdi, A. A.; Al-Hartomy, O. A.; Al-Solamy, F. R.; Dishovsky, N.; Malinova, P.; Atanasova, G.; Atanasov, N. Conductive Carbon Black/magnetite Hybrid Fillers in Microwave Absorbing Composites Based on Natural Rubber. *Compos. Part B: Eng.* **2016**, *96*, 231–241. © 2016 with permission from Elsevier.)

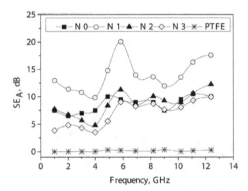

FIGURE 4.58 Frequency dependence of absorptive shielding effectiveness of composites containing CCB/M hybrid fillers. (Reprinted from Al-Ghamdi, A. A.; Al-Hartomy, O. A.; Al-Solamy, F. R.; Dishovsky, N.; Malinova, P.; Atanasova, G.; Atanasov, N. Conductive Carbon Black/magnetite Hybrid Fillers in Microwave Absorbing Composites Based on Natural Rubber. *Compos. Part B: Eng.* **2016**, *96*, 231–241. © 2016 with permission from Elsevier.)

As seen, composite N 1 exhibits good shielding properties in the entire frequency range. Its SE_T is within the 14.7–23.1 dB range, what means at the thickness of the material mentioned above, the incident power P_I will be attenuated 30–200 times. The Figures show that, at the same thickness of the material composite N 1 surpasses significantly in SE control sample N 0. Composite N 3 possesses the lowest SE_T which changes in the 5–10 dB range, probably due to its being thinner than the rest samples. The supposition has been proven by the results presented in Figures 4.59 and 4.60, which show the reflection and attenuation coefficient of composites N 2 and N 3 to be close at the same thickness of the samples. The shielding effectiveness (SE) of PTFE tends to zero in the entire frequency range.

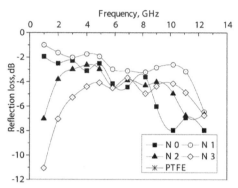

FIGURE 4.59 The frequency dependence of reflection coefficients |Γ| of composites containing CCB/M hybrid fillers. (Reprinted from Al-Ghamdi, A. A.; Al-Hartomy, O. A.; Al-Solamy, F. R.; Dishovsky, N.; Malinova, P.; Atanasova, G.; Atanasov, N. Conductive Carbon Black/magnetite Hybrid Fillers in Microwave Absorbing Composites Based on Natural Rubber. *Compos. Part B: Eng.* **2016,** *96,* 231–241. © 2016 with permission from Elsevier.)

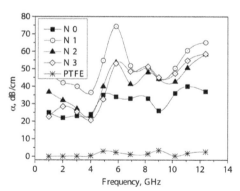

FIGURE 4.60 The frequency dependence of attenuation coefficient α, dB/cm of composites containing CCB/M hybrid fillers. (Reprinted from Al-Ghamdi, A. A.; Al-Hartomy, O. A.; Al-Solamy, F. R.; Dishovsky, N.; Malinova, P.; Atanasova, G.; Atanasov, N. Conductive Carbon Black/magnetite Hybrid Fillers in Microwave Absorbing Composites Based on Natural Rubber. *Compos. Part B: Eng.* **2016,** *96,* 231–241. © 2016 with permission from Elsevier.)

As Figures 4.57 and 4.58 show, the SE_T is due mainly to the absorption of EM power by the composite materials. According to Figure 4.57, SE_R of composites N 0 and N 1 decreases in the frequency range 1–6 GHz, while that of composites N 2 and N 3 increases at higher frequency. In the 6–12 GHz range SE_R of the studied conductive fillers is less sensitive to the changes in frequency. The results presented allow the conclusion that higher amounts of magnetite in the composites lower SE_R and widen the broadband of the shielding materials.

Figures 4.56 and 4.58 reveal similar behavior of SE_T and SE_A values for composites N 2 and N 3 in the 2–10 GHz range and resonance properties with a well pronounced peak at about 6 GHz. The analysis of the results for composite N 1 brings on the conclusion that its SE_T and SE_A have the best pronounced resonance properties.

The EM parameters of the materials and the results obtained are presented in Figures 4.61–4.67 in order to give a better explanation of SE_T and SE_A resonance behavior for composites N 1, N 2 and N 3 in the 3.0–9.5 GHz range.

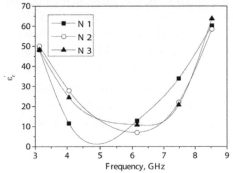

FIGURE 4.61   The frequency dependence of the real part of relative permittivity for the composites containing CCB/M hybrid fillers. (Reprinted from Al-Ghamdi, A. A.; Al-Hartomy, O. A.; Al-Solamy, F. R.; Dishovsky, N.; Malinova, P.; Atanasova, G.; Atanasov, N. Conductive Carbon Black/magnetite Hybrid Fillers in Microwave Absorbing Composites Based on Natural Rubber. *Compos. Part B: Eng.* **2016**, *96*, 231–241. © 2016 with permission from Elsevier.)

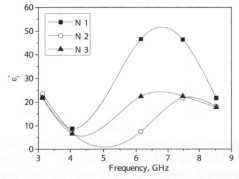

FIGURE 4.62   The frequency dependence of the imaginary part of the relative permittivity for the composites containing CCB/M hybrid fillers. (Reprinted from Al-Ghamdi, A. A.; Al-Hartomy, O. A.; Al-Solamy, F. R.; Dishovsky, N.; Malinova, P.; Atanasova, G.; Atanasov, N. Conductive Carbon Black/magnetite Hybrid Fillers in Microwave Absorbing Composites Based on Natural Rubber. *Compos. Part B: Eng.* **2016**, *96*, 231–241. © 2016 with permission from Elsevier.)

The resonance behavior of SE_T and SE_A for composites N 1, N 2 and N 3 in the 4–7 GHz range could be related to the resonance change in the dielectric loss tangent (tan δ) and magnetic loss tangent (tan δ_μ), having a peak at about 6 GHz (Figure 4.63) and 6.8 GHz (Figure 4.66), for tan δ_ε and tan δ_μ, respectively.

As known from the general EMI theory,[32,33] SE_R of a "good conductor" is a function of the ratio of conductivity (σ_V) and relative permeability (μ_r) of the shield material, that is quantity (σ_V/μ_r). The analysis of the results for the real and imaginary parts of permeability (Figures 4.65 and 4.66) and conductivity (Figure 4.64) of the studied composites shows that in the 4–9 GHz range the values for the relative complex permeability for the three composites change from 1.08 to 1.28 and from 0.13 to 1.07 for $\mu_r{}''$. The values for composites N 2 and N 3 are lower than that for composite N 1 in the entire frequency range. Hence, SE_R dependence on σ_V/μ_r ratio is valid for the conductive composites. So, the higher SE_R values for composite N 1 are chiefly due to the higher conductivity of the composite. The addition of magnetite to the matrix of the conductive composite narrows its SE_A dynamic range and widens the frequency range of its shielding properties. The convergence of SE_A curves demonstrates its being determined mainly by the composites matrix.

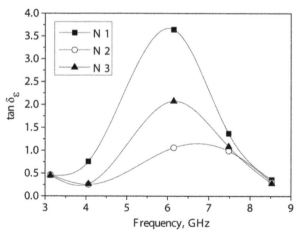

FIGURE 4.63   The frequency dependence of tan δ_ε of composites containing CCB/M hybrid fillers. (Reprinted from Al-Ghamdi, A. A.; Al-Hartomy, O. A.; Al-Solamy, F. R.; Dishovsky, N.; Malinova, P.; Atanasova, G.; Atanasov, N. Conductive Carbon Black/magnetite Hybrid Fillers in Microwave Absorbing Composites Based on Natural Rubber. *Compos. Part B: Eng.* **2016,** *96,* 231–241. © 2016 with permission from Elsevier.)

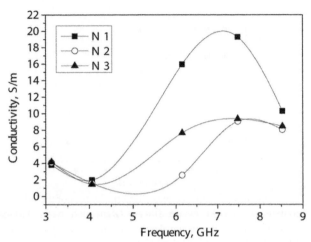

FIGURE 4.64   The frequency dependence of conductivity for composites containing CCB/M hybrid fillers. (Reprinted from Al-Ghamdi, A. A.; Al-Hartomy, O. A.; Al-Solamy, F. R.; Dishovsky, N.; Malinova, P.; Atanasova, G.; Atanasov, N. Conductive Carbon Black/magnetite Hybrid Fillers in Microwave Absorbing Composites Based on Natural Rubber. *Compos. Part B: Eng.* **2016,** *96,* 231–241. © 2016 with permission from Elsevier.)

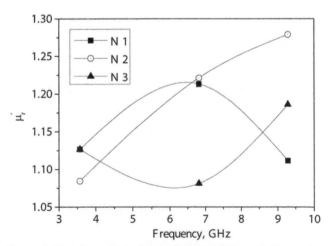

FIGURE 4.65   The frequency dependence of the real part of the relative permeability for composites containing CCB/M hybrid fillers. (Reprinted from Al-Ghamdi, A. A.; Al-Hartomy, O. A.; Al-Solamy, F. R.; Dishovsky, N.; Malinova, P.; Atanasova, G.; Atanasov, N. Conductive Carbon Black/magnetite Hybrid Fillers in Microwave Absorbing Composites Based on Natural Rubber. *Compos. Part B: Eng.* **2016,** *96,* 231–241. © 2016 with permission from Elsevier.)

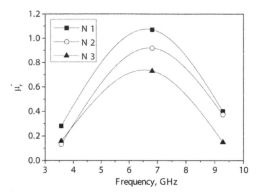

FIGURE 4.66   The frequency dependence of the imaginary part of the relative permeability for composites containing CCB/M hybrid fillers. (Reprinted from Al-Ghamdi, A. A.; Al-Hartomy, O. A.; Al-Solamy, F. R.; Dishovsky, N.; Malinova, P.; Atanasova, G.; Atanasov, N. Conductive Carbon Black/magnetite Hybrid Fillers in Microwave Absorbing Composites Based on Natural Rubber. *Compos. Part B: Eng.* **2016,** *96,* 231–241. © 2016 with permission from Elsevier.)

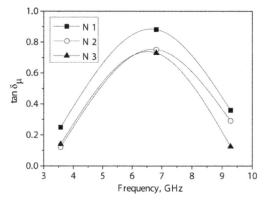

FIGURE 4.67   Frequency dependence of the loss tangent tan δ_μ for composites containing CCB/M hybrid fillers. (Reprinted from Al-Ghamdi, A. A.; Al-Hartomy, O. A.; Al-Solamy, F. R.; Dishovsky, N.; Malinova, P.; Atanasova, G.; Atanasov, N. Conductive Carbon Black/magnetite Hybrid Fillers in Microwave Absorbing Composites Based on Natural Rubber. *Compos. Part B: Eng.* **2016,** *96,* 231–241. © 2016 with permission from Elsevier.)

As shown in Figures 4.59 and 4.60, the data about the reflection and attenuation coefficients in the entire frequency range were analogous to those about SE_R and SE_A (Figures 4.57 and 4.58).

As Figure 4.59 shows, composite N 1 has the highest reflection coefficient values. Those of composite N 3 are lowest in the 1–6 GHz range, while in the 9–11 GHz range the lowest reflection coefficient values are those of composite N 0. It is of particular importance to note that, higher magnetite phase amounts in the hybrid filler cause a decrease of reflection coefficient values nearing about 11 dB at lower frequencies (for N 3) and remaining the lowest almost in the entire frequency range. The behavior of the control sample in 1–8 GHz range is close to that of the composite comprising a hybrid filler of 10% magnetite phase. Noteworthy is the fact that, the dependences have not a well pronounced resonance character, which is rather a monotonously decreasing one. The observed tendency of the reflection coefficient values to decrease with the increasing magnetite amount could be explained by inter- and intra-aggregate magnetite distribution in the carbon phase and by the conductive pathways it has built, producing an isolation effect and improving its discrete distribution. Other authors have also observed similar tendencies of decreasing the reflection coefficient modulus.[55]

Increasing the frequency from 4 to 6 GHz leads to an increase in the attenuation coefficient values of over 30 dB/cm while frequency changes from 10 to 12 GHz lead to an increase of more than 15 dB/cm. Maximum attenuation coefficient values of over 70 dB/cm have been observed at about 6 GHz for composite N 1, what ensures very good EM SE (Figure 4.60). It is also noteworthy that higher magnetite phase amounts decrease the attenuation coefficient values. The attenuation coefficient values over 30 dB/cm in the 1–12 GHz frequency range allow the statement that the NR based composite we have prepared and the hybrid CCB-magnetite fillers do possess microwave absorption properties.

Microwave absorbers are characterized by their permittivity and permeability. Both parameters express the interactions between materials and EM fields.[26] The permittivity is a measure of the materials effect on the electric field in the EM wave and the permeability is a measure of the materials effect on the magnetic component of the wave. The real part of the permittivity, ε_r', measures how much energy from an external electric field is stored in the material. The imaginary part, ε_r'', named loss factor, accounts for the loss energy dissipative mechanisms in the materials. The quantity ε_r'' is a measure of the attenuation of the electric field caused by the material. Both components contribute to wavelength compression inside the material. Additionally, due to coupled EM wave, loss in either the magnetic or electric field will attenuate the energy in the wave.

The complex permittivity, complex permeability and conductivity of the composites comprising hybrid CCB/M at various phase ratios have been investigated aiming at clarifying their shielding properties. Figures 4.61 and 4.62 plot the real part (ε_r') and imaginary part (ε_r'') of the relative permittivity for composites N 1, N 2 and N 3 in the frequency range from 3.1 to 8.5 GHz. The determination of the complex permittivity, complex permeability and conductivity is based on the theory of perturbation.[55]

As seen from Figures 4.61 and 4.62, ε_r' and ε_r'' of the three composites as a frequency function exhibit resonance behavior and the real part of the relative permittivity decreases at frequencies up to 6 GHz and afterwards increases. The measurements reveal a decrease of ε_r'' in the 4.0–7.5 GHz range with increasing Fe_3O_4 concentration in the conductive composites. In the 3–4 GHz range ε_r'' values decrease and afterwards start increasing. With the increasing magnetite amount the resonance character of the dependence lessens. The ε_r'' values, however, show resonance peaks at different frequencies—6.2 GHz for composites N 1 and N 3, and at 7.5 GHz for N 2. The dielectric properties of conductive composites are depended on many factors. The most important of these is the ratio between the particles of the filler and the host material.[56]

The dielectric loss tangent tan δ_ε varies in the 3–8.5 GHz region (Figure 4.63), reaching maximum value in the 6–7 GHz region. Composite N 1 has the highest tan δ_ε values. The curves pattern and their resonance character are close to those of the imaginary part of the relative permittivity.

Dependencies of the changes in the real part of the relative permittivity ε_r', similar to those we observed in the 3–6 GHz range, are described in.[57,58] According to,[57] in the 1 MHz to 1 GHz after reaching the percolation threshold the imaginary part of relative permittivity (ε_r'') almost does not change with the increasing concentration of the conductive phase. On the other hand, the real part (ε_r') values change drastically, thus $\varepsilon_r''/\varepsilon_r'$ ratio (tangent of the dielectric loss angle, tan δ_ε) increases and leads to improved microwave properties after reaching the percolation threshold. The properties of the dielectric matrix and the conductive filler, as well as the morphology and structure of the filler also affect the changes in the real and imaginary part of the relative permittivity. The authors of[58] have observed that, after reaching the percolation threshold ε_r' and ε_r'', as well as tan δ_ε increase unsystematically with the increasing concentration of the conductive filler and decrease at higher frequency. The reason for such unsystematic increase in $_r'$ and ε_r'' (around 160%) may be the variety of interfaces in each composite determined by variations in the size and shape of the nano-filler. Other authors explain the effect by the formation of a considerable number of nano-condensers,[59] and by interfacial polarization as well.

Hence, the dielectric losses in a multiple composite are the result of complex phenomena like natural resonance, dipole relaxation, electronic polarization as well as their relaxation and interfacial polarizations.

Interfacial polarizations occur in heterogeneous media due to accumulation of charges at the interfaces and the formation of large dipoles.[60] The chosen filler Printex XE2-B contains vanadium, nickel and iron,[61] while the addition of Fe_3O_4 at different concentrations presupposes the formation of various dipoles of different relaxation frequencies. The different relaxation frequencies of various dipoles formed in the composites, hopping of electrons (electron hopping between Fe^{+3} and Fe^{+2}) and the relaxation due to interfacial polarization are all responsible for the resonant behavior of ε_r' and ε_r''.

Figure 4.64 presents the changes in the effective conductivity of composites N 1, N 2 and N 3 in the 3.1–8.5 GHz frequency range. As seen, the composites exhibit high conductivity. Many factors determine the conductivity of the composites. The interaction between the filler particles and the matrix is amongst the most important ones. At a lower volume ratio, the composites are good dielectric materials, but the conductivity of the material can increase sharply at the percolation threshold.[56] That is due to the development of conductive paths at the percolation threshold of filler particles. In the present study the filler used has a volume ratio much above the percolation threshold what could explain the high conductivity values of the composites. The results allow the conclusion that, higher Fe_3O_4 concentrations in the composites change their conductivity, namely $\sigma_e(N\ 1) > \sigma_e(N\ 3) > \sigma_e(N\ 2)$ in the 4.0–7.5 GHz range. That reveals that, the linear increase in magnetite amount does not increase linearly the effective conductivity of the composite. As Figures 4.63 and 4.64 show, the microwave conductivity is a direct function of dielectric loss, exhibiting a variation with frequency similar to that of the dielectric loss factor.[62] Those results presuppose multiple dielectric relaxations in the conductive magnetic composites. They are consistent with the change in their complex permittivity. The resulting multiple relaxations can be on account of the interface polarization, which are caused by the heterogeneous interfaces between the hybrid filler and the rubber matrix, and on the other hand between the two phases of the hybrid filler. Also, the conductivity of the composite should be determined at a certain optimum,[63–65] so that to ensure the advantage and domination of the absorption over the reflection. The increase in the composite's conductivity to a certain degree leads to a reflection greater than its absorption.[66] A tradeoff between the filler concentration and its dispersion is always a necessary condition to keep the electrical conductivity at an optimum level for enhanced microwave absorption.[62]

Figures 4.65–4.67 show the real (μ_r') and imaginary (μ_r'') part of complex relative permeability, and the magnetic loss tangent (tan δ_μ) for all the conductive composites, respectively. The change in μ_r' of composites N 1 and N 3with frequency increasing in the 3.5–9.5 GHz range is resonance (Figure 4.65), while the change in μ_r' of composite N 3 is almost linear. In general, μ_r'' values decrease as Fe_3O_4 concentration in the conductive composites increases, such as μ_r'' (N 1)$>\mu_r''$(N 2)$>\mu_r''$ (N 3). The μ_r'' values, however, show resonance peaks for all the conductive composites but the resonance peak is not shifted with the increasing magnetite amount in the hybrid filler (Figure 4.66). The magnetic loss tangent varies in the 3.6–9.5 GHz frequency range reaching a maximum value at 6.8 GHz (Figure 4.67).

The authors of[67] have observed similar resonance behavior of μ_r'' and *tan* δ_μ values for Fe_3O_4/paraffin at about 6 GHz. A resonance behavior of μ_r' and μ_r'' depending on filler and its concentration has been also observed in.[68,69] The magnetic losses for magnetic materials originate mainly from domain wall resonance, hysteresis loss, eddy current loss, and natural resonance.[67] The domain wall resonance normally happens at a frequency lower than 100 MHz, hence it can be neglected in the microwave range. The hysteresis loss is produced in a very strong external magnetic field, and there is no hysteresis loss in the weak magnetic field derived from microwaves. In this study the relative permeability was measured at a low microwave power (≤ 5 mW) and over a frequency range of 3.6–9.3 GHz, so neither hysteresis loss nor domain wall resonance is the main contributor to magnetic loss. Therefore, the magnetic loss in the composites are results of complex phenomena like eddy current loss and natural resonance.

It is known that, dielectric and magnetic losses are primary responsible for microwave absorption. Those losses can be controlled by adjusting factors, namely nature, distribution and content of fillers, nature of the matrix, and matching thickness. As expected, when the material is produced using dielectric components only, the values of the real and imaginary magnetic permeability are approximately 1 and 0, respectively, and they do not have any contribution to the microwave properties. That explains all the effects we have observed and gives an answer to the question why the composites comprising hybrid fillers possess microwave properties better than those of the control sample N 0. Besides, it is obvious that some structure sensitive properties, first of all the volume resistivity, should have an optimum value ensuring the best (optimal) microwave properties. The pattern of EM energy attenuation by the absorbing materials suggests that the electrical conductivity of those materials is related to the quantity of absorbing centers (CCB-magnetite) and to the type of polymer matrix (NR), which modify the impedance of absorbing materials. Absorbing materials must attenuate a large portion of the incident, which is a consequence of the equilibrium between electric conductivity and electric losses. On the other hand, that explains why the increasing magnetite concentration in the hybrid filler leads to gradual worsening of the microwave properties of magnetite. In the case, that depends on the inter- and intra-aggregate distribution of the magnetite phase in the carbon one. One should keep in mind that the magnetite phase has permittivity lower than that of the carbon one, that is it acts as an isolator regarding the conductive carbon structures.

Some of the effects observed could be explained by the results from TEM investigations.

As seen from Figure 4.68(a), the filler structure is the most ordered in the absence a magnetite phase: the concentric microstructure is well developed and the spherulites are the largest (about 60 nm). The structure is densely packed and compact. The introduction of a magnetite phase, and with its increase, the spherulites shrink, the concentric microstructure is not so well pronounced.

In some cases, as shown in Figure 4.68(d), there is no order in the carbon phase. There are dark and bright zones, allowing the assumption that, at higher concentrations of the magnetite phase it penetrates into the carbon aggregates and insolates them. The analysis of the figures confirms that the magnetite phase (the darker one) is distributed amongst the carbon particles as well as in the carbon aggregates. The phenomenon is not observed at lower magnetite concentrations. Our hypothesis is that, the phenomenon mentioned is of great importance for determining the microwave properties of the composites comprising the particular hybrid fillers–on one hand it affects the formation of dielectric and magnetic losses and their contribution, on the other hand, that impacts the permittivity of the composites— factors crucial for the formation of the absorption and reflection of EM waves. These data have also been confirmed by electrical and microwave properties.

Sections 4.3.3.6 *Electric Properties of Natural Rubber-Based Composites Comprising the Studied Hybrid Fillers and 4.3.3.9 Microwave Properties of Natural Rubber-Based Composites Comprising the Studied Hybrid Fillers* have been reprinted from Al-Ghamdi, A.A., Al-Hartomy, O.A., Al-Solamy, F.R., Dishovsky, N., Malinova, P., Atanasova, G., and Atanasov, N. CCB/M hybrid fillers in microwave absorbing composites based on NR. Composites Part B: Engineering 96, 231-241 (2016). Used with permission from Elsevier, License Number 3972550803609.

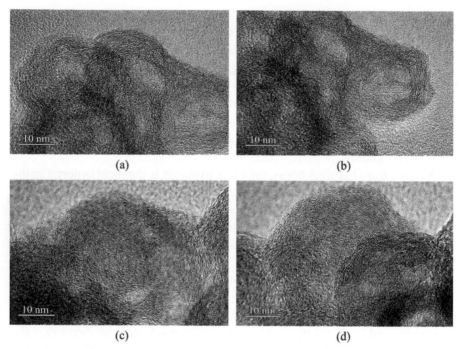

FIGURE 4.68   (a) High resolution TEM images of a filler comprising: 0 wt % magnetite; (b) 10 wt % magnetite; (c) 30 wt % magnetite; (d) 50 wt. % magnetite. (Reprinted from Al-Ghamdi, A. A.; Al-Hartomy, O. A.; Al-Solamy, F. R.; Dishovsky, N.; Malinova, P.; Atanasova, G.; Atanasov, N. Conductive Carbon Black/magnetite Hybrid Fillers in Microwave Absorbing Composites Based on Natural Rubber. *Compos. Part B: Eng.* **2016,** *96,* 231–241. © 2016 with permission from Elsevier.)

4.3.4   PYROLYSIS CARBON–SILICA HYBRID FILLER

4.3.4.1   CHARACTERIZATION OF THE PYROLYSIS CARBON–SILICA FILLER (PCSF)

Table 4.16 shows the ash content of the pyrolysis carbon–silica filler (PCSF) to be about 64%, that is PCSF comprises about 36% of organic matter (carbon black) and 64% of inorganic compounds. The availability of some amounts of carbon black in PCSF allowed determining its IA and oil absorption number. As seen, IA of the PCSF is much higher than that of the conventional Corax N220 carbon black. That means the specific surface area of PCSF is larger than the one of Corax N220 carbon black; what has been also confirmed by BET nitrogen specific surface area.

TABLE 4.16 Main Properties of the Fillers Studied.

	Corax N220	Ultrasil 7000 GR	PCSF
IA (mg/g)	121	–	215
OAN (ml/100 g)	114	–	140
S_{BET} (m²/g)	106	175	116
Ash content %)	<0.5	–	64
Mesopore volume, cm³ (STP)/g	–	–	91
Diameter of mesopores, nm	–	–	5

(Reprinted from Al-Hartomy O, Al-Ghamdi A, Al Said S, et al. Pyrolysed Carbon-Silica Filler Obtained by Pyrolysis-Cum-Water Vapour of Waste Green Tires vs. Conventional Fillers. Comparison of Their Effects Upon the Properties of Epoxidized Natural Rubber Based Vulcanizates. *International Review of Chemical Engineering*. 2014; 6: 160-8. © 2014 with permission from 2014 *Praise Worthy Prize*.)

The primary particles formed during the stage of initial carbon black formation fuse together building up three dimensional branched clusters called aggregates. High structure carbon black has a high number of primary particles per aggregate, while low structure carbon black exhibits only a weak aggregation. These aggregates again may form agglomerates linked by Van der Waals interactions. The empty space (void volume) between the aggregates and agglomerates, usually expressed as the volume of dibutylphthalate (DBP) absorbed by a given amount of carbon black, is described by the term "structure" (or "structurality") of the carbon black.[70] The data in Table 4.16 show the oil absorption number of PCSF to be significantly higher than that of the conventional carbon black. Hence, PCSF particles are able to form more pronounced secondary structures (aggregates and agglomerates).

The ash content of the PCSF and its silicate analysis data allow the conclusion that PCSF contains about 60% of silica, 36% of carbon black, 2.5% of zinc oxide and 1.5% of other metal oxides.[71].

As said above, the silica surface is hydrophilic—it has silanol groups. The silanol groups could be isolated, geminal (two hydroxyl groups on the same silicon atom), and vicinal (on adjacent atoms). Generally they are linked by strong hydrogen bonds[47] causing aggregation of the filler particles. That hampers the filler dispersion in the rubber matrix as well as the formation of stable bonds between the elastomer and the filler. Therefore the reinforcing effect of silica is less pronounced. A better dispersion of silica particles and stronger polymer-filler interactions are achieved using different types of silane coupling agents.[72]

Bifunctional organosilanes are organic compounds which improve the compatibility between the hydrophobic polymer matrix and the hydrophilic silica.[73] Bis(triethoxysilylpropyl) tetrasulfide (TESPT) is the most used of all and finds wide application in practice. During the mixing via hydrolysis and condensation the alkoxy groups of the bifunctional organosilane interact with the silanol groups of silica.[74–77] In the course of vulcanization the polysulfide groups of TESPT interact with the elastomer macromolecules.[78,79] The polymer-filler bonds thus formed have a significant effect upon the properties of rubber composites. Lowered surface energy of silica leads to a higher rate of rubber adsorption onto the filler surface and to a weaker filler-filler interaction.

The aforementioned reveals the crucial role that silanol groups over the surface of silica have in the reinforcing process. So, it is of particular importance those groups to be preserved when silica filled rubber items undergo pyrolysis. That is why both the PCSF and its ash contents were subjected to Fourier transform infrared (FTIR) spectroscopy (Figures 4.69(b) and 4.69(c)). The FTIR spectrum of conventional silica (Ultrasil 7000 GR) presented in Figure 4.69(a) is given for comparison.

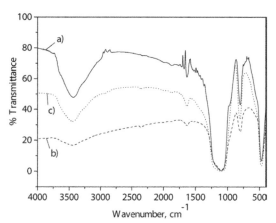

FIGURE 4.69 FTIR spectra of the studied fillers: (a) Conventional silica (Ultrasil 7000GR); (b) Pyrolysis carbon–silica filler (PCSF); (c) The ash content of PCSF. (Reprinted from Al-Hartomy O, Al-Ghamdi A, Al Said S, et al. Pyrolysed Carbon-Silica Filler Obtained by Pyrolysis-Cum-Water Vapour of Waste Green Tires vs. Conventional Fillers. Comparison of Their Effects Upon the Properties of Epoxidized Natural Rubber Based Vulcanizates. *International Review of Chemical Engineering*. 2014; 6: 160-8. © 2014 with permission from 2014 *Praise Worthy Prize*.)

As Figures 4.69(b) and (c) show, the absorption bands of the PCSF and the ash content are in full accordance with that of the conventional silica (Figure 4.69(a)). In all cases the following absorption bands have been observed: at 1200–1100 cm^{-1}—asymmetric valence vibrations of Si-O bonds; at 830–750 cm^{-1}—symmetric stretch vibrations of Si–O bonds; at 530–460 cm^{-1}— deformational vibrations of Si-O bonds; at 3600 - 3200 cm^{-1}—valence O-H vibrations assigned to hydrogen bonded OH groups; at 1660 - 1630 cm^{-1}— deformational O-H vibrations. All spectra have well pronounced absorption bands at about 3400 cm^{-1} corresponding to the hydrogen bonding interactions associated with silanols, which are available on the surfaces of PCSF, of PCSF ash and of the silica filler.[80] According to FTIR spectroscopy data, the particles of silica that had been used as filler in tires manufacturing have sustained their chemical activity in the tires pyrolysis product.

Representative HAADF images and compositional maps of the PCSF are shown in Figure 4.70. It is seen that carbon and silica are well dispersed in the obtained PCSF.

The maps give the impression that the carbon phase occupies the spaces between the silicon dioxide aggregates (mostly of spherical morphology). That is logical since at the time of the formation of this phase through elastomer's destruction, silica had already been existing as a phase. We consider that a positive effect, because thus the undesired formation of large silica aggregates as a result of the strong "filler–filler" interactions is hindered. It can be concluded that the carbon phase of the filler obtained as a pyrolysis result is located predominantly in the space among silica aggregates which have already been existing in the process of its formation by elastomer destruction in the time of pyrolysis.

The size of filler particles is of significant importance for its reinforcing effect. The smaller the particles, the more pronounced the reinforcing effect is. The morphology of the PCSF has been studied by transmission electron microscopy (TEM). Figure 4.71 present the micrographs of conventional carbon black Corax N220 (a), conventional silica Ultrasil 7000 GR (b), PCSF (c) and the ash content of the PCSF (d).

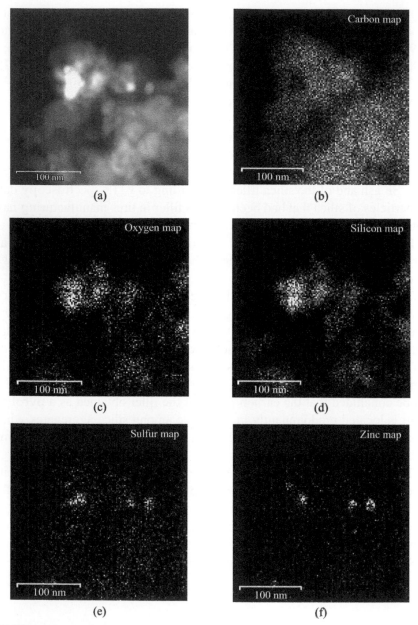

FIGURE 4.70 HAADF image and compositional maps of PCSF: (a) HAADF image; (b) carbon map; (c) oxygen map; (d) silicon map; (e) sulfur map; and (f) zinc map. (Reprinted from Al-Hartomy, O. A.; Al-Ghamdi, A. A.; Al Said, S. A. F.; Dishovsky, N.; Ward, M. B.; Mihaylov, M.; Ivanov, M. Characterization of Carbon Silica Hybrid Fillers Obtained by Pyrolysis of Waste Green Tires by the Stem–edx Method. *Mater. Charact.* **2015**, *101*, 90–96. © 2015 with permission from Elsevier.)

FIGURE 4.71 TEM micrographs of the investigated fillers: (a) Corax N220 (x100k); (b) Ultrasil 7000 GR (x100k); (c) Pyrolysis carbon-silica filler (x100k); (d) The ash content of pyrolysis carbon-silica filler (x150k). (Reprinted from Al-Hartomy O, Al-Ghamdi A, Al Said S, et al. Pyrolysed Carbon-Silica Filler Obtained by Pyrolysis-Cum-Water Vapour of Waste Green Tires vs. Conventional Fillers. Comparison of Their Effects Upon the Properties of Epoxidized Natural Rubber Based Vulcanizates. *International Review of Chemical Engineering.* 2014; 6: 160-8. © 2014 with permission from 2014 *Praise Worthy Prize.*)

It shows that the particles of the conventional carbon black (50 nm) are much larger than those of the conventional silica which are about 20–25 nm. TEM micrograph of the PCSF shows that its particles size varies from 25 to 50 nm. That means PCSF particles are of almost the same size as the particles of conventional carbon black and silica. However, Figure 4.71(d) shows that the ash of the PCSF has particles of uniform size which is commensurable with the size of the particles of conventional silica (about 20–25 nm). According to TEM analysis, the particles size of PCSF is commensurable with that of the conventional fillers used in rubber industry.

4.3.4.2   PREPARATION OF RUBBER COMPOSITES

The formulations of the compounds based on ENR—Epoxyprene 25 are presented in Table 4.17.

TABLE 4.17   Compositions of the Investigated Rubber Compounds (phr).

	E 1	E 2	E 3	E 4	E 5	E 6	E 7	E 8
ENR-Epoxyprene 25	100	100	100	100	100	100	100	100
Zinc oxide	3.0	3.0	3.0	3.0	3.0	3.0	3.0	3.0
Stearic acid	2.0	2.0	2.0	2.0	2.0	2.0	2.0	2.0
Carbon black Corax N 220	50	70	–	–	20	20	–	–
Silica	–	–	70	70	50	50	–	–
PCSF	–	–	–	–	–	–	70	70
TESPT	–	–	–	7.0	–	5.0	–	5.0
TBBS	1.5	1.5	1.5	1.5	1.5	1.5	1.5	1.5
Sulfur	2.0	2.0	2.0	2.0	2.0	2.0	2.0	2.0

(Reprinted from Al-Hartomy O, Al-Ghamdi A, Al Said S, et al. Pyrolysed Carbon-Silica Filler Obtained by Pyrolysis-Cum-Water Vapour of Waste Green Tires vs. Conventional Fillers. Comparison of Their Effects Upon the Properties of Epoxidized Natural Rubber Based Vulcanizates. *International Review of Chemical Engineering.* 2014; 6: 160-8. © 2014 with permission from 2014 *Praise Worthy Prize.*)

TABLE 4.18   Mixing Schedule of the Investigated Rubber Compounds.

Stage 1, Brabender Plasti-Corder PLE651, Rotor speed 40 rpm, Temperature 140°C			
Mixing order	Ingredients	Mixing time (min)	Cumulative time (min)
1	Epoxyprene 25	2	2
2	ZnO and stearic acid	2	4
3	Carbon black, silica or pyrolysis carbon–silica filler	5	9
Stage 2, Laboratory Two Roll Mills, Friction 1.27			
1	1st stage rubber batch	2	2
2	Sulfur and TBBS	5	7

(Reprinted from Al-Hartomy O, Al-Ghamdi A, Al Said S, et al. Pyrolysed Carbon-Silica Filler Obtained by Pyrolysis-Cum-Water Vapour of Waste Green Tires vs. Conventional Fillers. Comparison of Their Effects Upon the Properties of Epoxidized Natural Rubber Based Vulcanizates. *International Review of Chemical Engineering.* 2014; 6: 160-8. © 2014 with permission from 2014 *Praise Worthy Prize.*)

The rubber compounds were prepared at two stages according to the mixing schedule presented in Table 4.18. At the first stage the mixing was performed on a Brabender Plasti-Corder PLE651 fitted with a 300 cm³ cam type mixer. The silane coupling agent was mixed with the filler studied prior to their placing into the mixer camera. The amount of bis(triethoxysilylpropyl) tetrasulfide silane coupling agent (TESPT) used was 1/h per 10/h pure silica. At the second stage sulfur and the accelerator were added to the mixture compounded on an open two-roll laboratory mill L/D 320 × 160 and friction 1.27.

The vulcanization process of the NR-based compounds was carried out on an electrically heated hydraulic press using a special homemade mold at 150°C and 10 MPa.

4.3.4.3 CURING PROPERTIES OF THE INVESTIGATED RUBBER COMPOUNDS

Table 4.19 summarizes the main curing properties of the rubber compounds studied as determined from their cure curves presented in Figure 4.72.

TABLE 4.19   Curing Properties of the Investigated Rubber Compounds.

	M_L, dN.m	M_L, dN.m	ΔM, dN.m	T_{s2}, min:s	T_{90}, min:s	Cure rate, %/min
E1	1.69	38.02	36.33	2:11	6:12	25.0
E2	4.84	52.31	47.47	2:00	5:45	26.7
E3	15.88	46.04	30.16	9:00	27:30	5.4
E4	12.86	55.00	42.14	8:06	20:04	8.3
E5	11.18	42.50	31.32	7:30	20:00	8.0
E6	8.82	48.19	39.37	6:30	20:00	7.4
E7	2.59	32.66	30.07	4:18	12:23	12.1
E8	2.50	41.61	39.11	3:02	12:13	10.8

(Reprinted from Al-Hartomy O, Al-Ghamdi A, Al Said S, et al. Pyrolysed Carbon-Silica Filler Obtained by Pyrolysis-Cum-Water Vapour of Waste Green Tires vs. Conventional Fillers. Comparison of Their Effects Upon the Properties of Epoxidized Natural Rubber Based Vulcanizates. *International Review of Chemical Engineering*. 2014; 6: 160-8. © 2014 with permission from 2014 *Praise Worthy Prize*.)

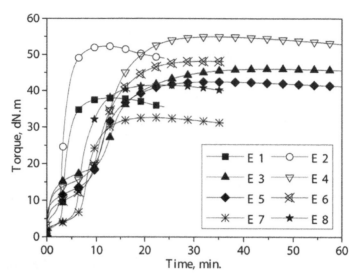

FIGURE 4.72 Cure curves of the investigated rubber compounds taken at 150°C. (Reprinted from Al-Hartomy O, Al-Ghamdi A, Al Said S, et al. Pyrolysed Carbon-Silica Filler Obtained by Pyrolysis-Cum-Water Vapour of Waste Green Tires vs. Conventional Fillers. Comparison of Their Effects Upon the Properties of Epoxidized Natural Rubber Based Vulcanizates. *International Review of Chemical Engineering.* 2014; 6: 160-8. © 2014 with permission from 2014 *Praise Worthy Prize.*)

As seen from Table 4.19, the increase in the amount of Corax N 220 carbon black from 50 phr to 70 phr leads to a regular increase in the minimum torque (M_L), in the viscosity of the studied rubber composites, respectively. M_L of the composite comprising Corax N220 at 50 phr (E 1) is 1.69 dN.m, while that of the composite with Corax N220 (E 2) at 70 phr is 4.84 dN.m. Obviously, the amount of carbon black does not affect considerably the scorch time (T_{s2}) and the cure time (T_{90}) of those two compounds. M_L of the composite filled with conventional silica Ultrasil 7000 GR at 70 phr without a silane coupling agent (E 3) is 15.88 dN.m and is quite higher than that of the composite comprising the same amount of carbon black (E 2). In the presence of TESPT at 7 phr (E 4), M_L values of the rubber composites filled with Ultrasil 7000 GR decrease about 20% and reach 12.86 dN.m. In the case neither the presence nor the lack of TESPT has a significant effect upon the scorch (T_{s2}) and cure time (T_{90}). Though, the scorch (T_{s2}) and cure time (T_{90}) are obviously much longer than those of the rubber compounds filled with carbon black (E 1, E 2). Naturally, that has been expected having in mind the adsorption of curing agents over silica surface which lowers the vulcanization rate. What has been said about M_L values, the scorch (T_{s2})

and the cure time (T_{90}) of the rubber compounds filled with conventional silica (E 3, E 4) is valid for the rubber compounds filled with a combination of conventional silica and carbon black (E 5, E 6). The presence or absence of a silane coupling agent does not lead to changes in M_L values for the rubber compounds comprising a PCSF (E 7, E 8). In the case the scorch (T_{s2}) and cure time (T_{90}) also do not change significantly because of the presence or lack of TESPT. As seen, M_L values of the rubber compounds comprising a PCSF both with and without TESPT are about 70–80% lower than those of the respective compounds filled with conventional silica (E 3, E 4) or with a combination of conventional silica and carbon black (E 5, E 6). Meanwhile, the scorch time (T_{s2}) and the cure time (T_{90}) are considerably shorter. Possibly the product yielded from the pyrolysis-cum-water vapor of waste green tires tread is similar to the dual-phase fillers CSDPF 2000 and CSDPF 4000 developed by Cabot Corporation. According to literature the filler-filler interactions in rubber compounds comprising similar fillers is less pronounced than in the case when the rubber compounds are filled with a physical mixture of carbon black and silica.[11] On its turn that leads to better dispersion of the filler particles along the rubber matrix and to lower viscosity of the rubber compounds investigated. As Table 4.19 shows, in all cases the presence of a silane coupling agent has not a very pronounced effect upon the viscosity of the rubber compounds filled with silica (E 3 to E 8). That is due to the polar character of the ENR. Epoxidation of NR increases the polarity of the polymer. That is why the interaction between ENR and silica is stronger and there is no need of coupling agents.[81]

4.3.4.4 MECHANICAL PROPERTIES OF THE INVESTIGATED RUBBER COMPOSITES

Table 4.20 presents the mechanical properties of the studied Epoxyprene 25 based rubber composites.

As Table 4.20, shows Modulus 100 (M_{100}) of the composites filled with Corax N220 carbon black (E 1) at 50 phr is about 3.5 MPa. M_{100} of the composites filled with Corax N220 carbon black (E 2) at 70/h is about 5.1 MPa or about 45% higher than that of the former composite, E 1. Modulus 300 (M_{300}) of composites E 1 and E 2 also increases about 45% with the increasing filler amount. The tensile strength (σ) values of the composites filled with carbon black (E 1, E 2) is about 25–26 MPa, i.e. the increase in the filler amount from 50 to 70/h does not affect much this parameter. However, the increase in the carbon black amount causes a decrease of the values of elongation at break (ε_{rel}) and residual elongation (ε_{res}), while Shore A hardness and abrasion values increase. The presence of a coupling agent has no significant effect upon M_{100} values of the composites filled with conventional silica Ultrasil 7000 GR (E 3, E 4), a combination of conventional silica and carbon black (E 5, E 6) or with PCSF filler (E 7, E 8). Though, in all three cases, the presence of TESPT leads to an increase in M_{300} and tensile strength values and to lowering of the abrasion. The mechanical properties of the composites comprising PCSF (E 7, E 8) are commensurable with those of the composites with conventional fillers.

TABLE 4.20 Mechanical Properties of the Investigated Rubber Composites.

	M_{100} (MPa)	M_{300} (MPa)	σ (MPa)	E_{rel} (%)	E_{res} (%)	Shore A hardness	Abrasion (mm³)
E1	3.5	16.8	25.8	420	30	75	130
E2	5.1	24.4	24.8	300	20	82	139
E3	3.7	14.1	16.5	300	10	80	150
E4	3.5	17.2	22.9	360	10	81	128
E5	4.1	20.4	21.9	320	10	77	137
E6	4.7	22.8	25.3	300	10	81	126
E7	3.8	18.8	23.8	420	25	76	137
E8	4.3	21.7	25.0	345	20	77	129

(Reprinted from Al-Hartomy O, Al-Ghamdi A, Al Said S, et al. Pyrolysed Carbon-Silica Filler Obtained by Pyrolysis-Cum-Water Vapour of Waste Green Tires vs. Conventional Fillers. Comparison of Their Effects Upon the Properties of Epoxidized Natural Rubber Based Vulcanizates. *International Review of Chemical Engineering.* 2014; 6: 160-8. © 2014 with permission from 2014 *Praise Worthy Prize.*)

The mechanical properties of the composites comprising PCSF and an organosilane are improved due to the occurrence of chemical polymer–filler interactions. Such interactions can proceed naturally, if the particles of silica preserve their primary size and surface activity under the chosen pyrolysis conditions. That has been confirmed by the data obtained via FTIR and TEM analyses.

4.3.4.5 DYNAMIC PROPERTIES OF THE INVESTIGATED RUBBER COMPOSITES

Filler dispersion along the rubber matrix has a significant impact upon the dynamic properties of the vulcanizates and mainly upon their heat build-up—a property related to tires rolling resistance. A flexometer Goodrich was used to determine the dependencies of the complex dynamic modulus and heat build-up on the dynamic deformation.

Figure 4.73 presents the complex dynamic modulus dependence on the dynamic deformation of the studied Epoxyprene 25 based composites. As the figure shows, at all studied deformations the dynamic complex modulus values of the composites comprising carbon black (E 1, E 2) increase with increasing the filler amount from 50 phr to 70 phr. In the case of the composites comprising carbon black at 70 phr (E 2), however, the drop of the complex dynamic modulus values at higher dynamic deformation (Payne effect) is more pronounced. This is due to the fact that at higher filler amounts the interactions between its particles are stronger. At low dynamic deformation values (up to 8%), the filler–filler interactions remain the same and the dynamic complex modulus values are relatively high. At higher dynamic deformations the filler–filler structure is destroyed and the dynamic complex modulus values decrease. That corresponds to a higher heat build-up.

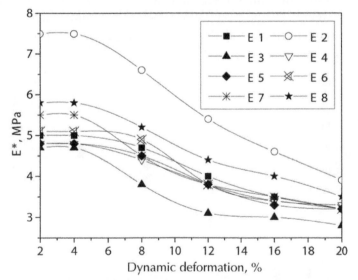

FIGURE 4.73 Dependence of the complex dynamic modulus of the composites studied on the dynamic deformation. (Reprinted from Al-Hartomy O, Al-Ghamdi A, Al Said S, et al. Pyrolysed Carbon-Silica Filler Obtained by Pyrolysis-Cum-Water Vapour of Waste Green Tires vs. Conventional Fillers. Comparison of Their Effects Upon the Properties of Epoxidized Natural Rubber Based Vulcanizates. *International Review of Chemical Engineering*. 2014; 6: 160-8. © 2014 with permission from 2014 *Praise Worthy Prize*.)

As seen from Figure 4.73, at all studied deformations the composites comprising conventional silica Ultrasil 7000 GR and no coupling agent (E 3) have the lowest dynamic complex modulus values. In the case of the composite comprising Ultrasil 7000 GR (E 4) the presence of TESPT leads to a significant increase in the dynamic complex modulus values, while E* values decrease less with the increasing dynamic deformation. The same is valid for the complex dynamic modulus (E*) values of the composites comprising a combination of conventional silica and carbon black (E 5, E 6), as well as for those of the composites comprising a PCSF (E 7, E 8). At all studied deformations the complex dynamic modulus (E*) values of the composites comprising a PCSF in the presence of TESPT (E 8) are the highest compared to the values of all studied silica filled composites.

Figure 4.74 plots the heat build-up as a function of the dynamic deformation of the ENR composites studied. The figure shows the heat build-up data to be in full accordance with those about the complex dynamic modulus (Figure 4.73). As seen, in the presence of TESPT the composites filled with conventional silica Ultrasil 7000 GR (E 4), a combination of silica and conventional carbon black Corax N220 (E 6), as well as those with a PCSF (E 8) have heat build-up values lower than those of the composites comprising no coupling agent. The heat build-up values of the composites comprising PCSF are commensurable with those of the composites comprising conventional fillers.

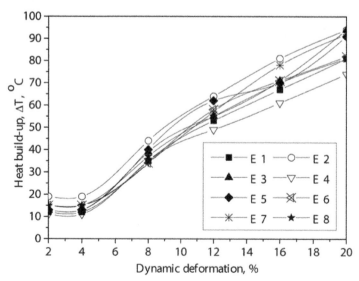

FIGURE 4.74 Dynamic deformation dependence of the heat-build up of the composites studied. (Reprinted from Al-Hartomy O, Al-Ghamdi A, Al Said S, et al. Pyrolysed Carbon-Silica Filler Obtained by Pyrolysis-Cum-Water Vapour of Waste Green Tires vs. Conventional Fillers. Comparison of Their Effects Upon the Properties of Epoxidized Natural Rubber Based Vulcanizates. *International Review of Chemical Engineering*. 2014; 6: 160-8. © 2014 with permission from 2014 *Praise Worthy Prize*.)

Dynamic mechanical testing is a powerful predictive tool that can provide valuable insight into the tire performance of tread compounds. Certain tire tread characteristics correlate with the mechanical dynamic properties measured on a dynamic mechanical thermal analyzer (DMTA) under defined frequency, strain (or stress) conditions, and temperature. The instrument can be used to differentiate instantly a series of tested compounds.[82]

Tire tread deformation during motion under load has been described as consisting of predominantly bending strains in which the curved tread is bent inwards and outwards as a function of the contact region. The total strain is composed primarily of the tangential strain in the plane of the tire and the shear strain between the radial and tangential directions. Strain and stress amplitudes depend on the magnitude of tire deflection. Medalia has put forth that the tread deformation can be resolved approximately into constant strain (bending) and constant stress (compression) conditions. Tension-compression, shear and torsional deformations are often utilized to test tire tread formulations under dynamic conditions. As a viscoelastic property *tan δ* deserves interest being in correlation with rolling resistance (due to hysteresis causing heat build-up) and with wet grip characteristics.[82]

Figure 4.75 presents the dependence of the dynamic storage modulus (E′) on temperature. According to the figure all studied composites are in the glass state in the range from −80 to −30°C. In the said temperature interval there are no considerable differences in the storage modulus values depending on the type of used filler or on the presence or absence of silane coupling agent. The considerable differences in storage modulus values are observed in the range from −10 to 80°C, wherein the studied composites are in the viscoelastic state. In the said temperature interval the increase in the carbon black amount from 50 phr (E 1) to 70 phr (E 2) leads to higher storage modulus values of the composites studied. All composites comprising conventional silica Ultrasil 7000 GR (E 4), a combination of silica and Corax N220 carbon black (E 6) or PCSF (E 8) in presence of TESPT have storage modulus values higher than those of the composites without a silane coupling agent.

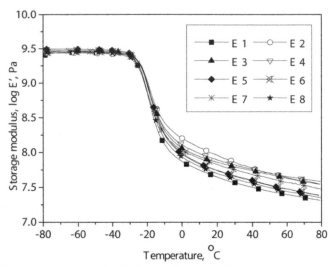

FIGURER 4.75 Storage modulus (E′) dependency on the temperature of the investigated composites. (Reprinted from Al-Hartomy O, Al-Ghamdi A, Al Said S, et al. Pyrolysed Carbon-Silica Filler Obtained by Pyrolysis-Cum-Water Vapour of Waste Green Tires vs. Conventional Fillers. Comparison of Their Effects Upon the Properties of Epoxidized Natural Rubber Based Vulcanizates. *International Review of Chemical Engineering*. 2014; 6: 160-8. © 2014 with permission from 2014 *Praise Worthy Prize*.)

Tan δ being the ratio between the dynamic loss modulus (E″) and dynamic storage modulus (E′) (tan δ = E″/E′) illustrates the macromolecules mobility as well as the phase transitions in the polymers. The temperature dependence of tan δ values deserves interest since it is considered that tan δ at 60°C corresponds to the tires rolling resistance while tan δ at 0°C—to the wet road grip of the tires.

Figure 4.76 presents the temperature dependence of the mechanical loss angle tangent (tan δ) of the composites studied. The figure does not show that the type of filler or organosilane used causes a significant difference in tan δ values of the composites at 0 and 60°C. The exception are the vulcanizates comprising carbon black at 70 phr (E 2) whose tan δ values at 0°C are the lowest while those at 60°C are the highest. In the case the wet road grip would be lower and the rolling resistance would be higher than that of the composites comprising conventional silica, a combination of silica and carbon black or PCSF.

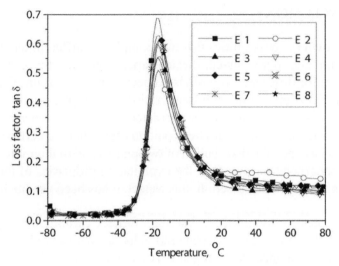

FIGURE 4.76 Mechanical loss angle tangent (tan δ) dependency on the temperature of the investigated composites. (Reprinted from Al-Hartomy O, Al-Ghamdi A, Al Said S, et al. Pyrolysed Carbon-Silica Filler Obtained by Pyrolysis-Cum-Water Vapour of Waste Green Tires vs. Conventional Fillers. Comparison of Their Effects Upon the Properties of Epoxidized Natural Rubber Based Vulcanizates. *International Review of Chemical Engineering.* 2014; 6: 160-8. © 2014 with permission from 2014 *Praise Worthy Prize.*)

The improved mechanical and dynamic properties of the composites comprising PCSF in the presence of organosilane are due to the formation of chemical bonds between the polymer and the filler. Such interactions are possible only provided that under the chosen pyrolysis conditions in the presence of organosilane the silica particles preserve their primary particle size and surface activity. That has been confirmed by the FTIR and TEM analyses. The filler–filler interaction in the rubber compounds comprising such fillers is markedly lower than that of the rubber compounds filled with a physical mixture of carbon black and silica. That leads to a better dispersion of the filler particles along the rubber matrix and to stronger polymer–filler bonds, what results into a more pronounced reinforcing filler effect. The analyses reported herein do not give a certain answer whether the pyrolysis product obtained is a dual-phase filler of the carbon–silica type or a mechanical mixture of those two components. The search of a definite answer to the question is going to be the scope of our future studies. The results obtained allow considering the PCSF tantamount to the conventional silica filler when used to develop elastomer composites based on ENR[83].

4.4   CONCLUSIONS

We have shown the possibilities that two completely different techniques—impregnation and pyrolysis-cum-water-vapor—provide for obtaining of hybrid fillers for different types of elastomers, namely: carbon black–silica and carbon black–magnetite at different ratios of their phases.

The fillers obtained have been characterized by a number of modern methods. Particular attention has been paid to effect that the certain preparation method has upon the distribution of two phases and to their interpenetration—factors crucial for output of the exploitation properties of the articles comprising the fillers discussed. In this aspect, it has been established that:

- In the case of hybrid fillers based on furnace carbon black and silica, the highest degree of interpenetration has been observed in the case of the filler obtained by impregnation of carbon black PM-75 comprising 3% silica (ICSF-75-3).
- In the case of hybrid fillers based on CCB and silica, the silica (dielectric) phase is distributed both over the surface and inside the carbon black aggregates (conductive phase). Thus, the interaction between the particles of the conductive filler (building of conductive pathways) is restricted. The effect is more pronounced with the increasing silica amount in the hybrid fillers.
- In the case of hybrid fillers based on CCB and magnetite, the number of parameters determined and the complex study on the structure of the filler have shown that the magnetite phase is distributed both over the carbon black particles (inter-aggregately) and intra-aggregately forming a true hybrid material and its amount influences all characteristics of the fillers such as OAN, IA number, specific surface area, texture, thermal, thermogravimetric properties, and so forth.
- The carbon phase of the filler obtained as a result of the pyrolysis is located predominantly in the space among silica aggregates which have already been existing in the process of its formation by elastomer destruction in the time of pyrolysis.

The effect of hybrid fillers obtained by different techniques on the properties of elastomer composites comprising them has been investigated. It has been found that:

- In the case of hybrid fillers based on furnace carbon black and silica, the higher quantity of the dielectric phase (silica) in the hybrid filler, the lower the real part of permittivity of the composites is. The fact evidences the lower conductivity of those composites confirmed by their volume resistivity and surface resistivity values. As a result, the shielding effectiveness of the composites changes, as well as their reflection and attenuation coefficients. The SE_A for the composites filled with non-modified conventional carbon black dominates in the frequency range from 1 to 5 GHz, while in the range from 5 to 12 GHz it is the SE_R that predominates. In the case of the composites comprising hybrid fillers of higher silica amount, the SE_A predominates in the entire frequency range studied. This is mainly due to the interpenetration of the two phases of the filler and the insulation of the conductive carbon phase by the dielectric silica one as well.

- In the case of hybrid fillers based on CCB and silica, the composites filled with the hybrid filler comprising the highest amount of silica possess the best SE_T and SE_A. In all cases absorption is the mechanism prevailing in effectiveness. Although the composites studied reflect the greater part of the power of the incident wave (about 66%), the wave penetrating the samples has power about 34% of the incident power. However, the transmitted power (P_T) is only 4% of the incident power (P_I) what in fact determines the absorption character of the achieved microwave shielding effectiveness.

- In the case of hybrid fillers based on CCB and magnetite, the ratio between the very active carbon phase and the much inert magnetite one (regarding the rubber macromolecules) in the hybrid fillers affects considerably the properties of the composites thus filled: vulcanization, physicochemical, mechanical, electric, dynamic, as well as the aging resistance and homogeneity of the composites. The microwave characteristics of composites comprising the hybrid fillers at loading over the percolation threshold reveal their promising properties and the possibility to use them for manufacturing elastomer-based microwave absorbers for several applications such as antennas, automobiles, sensors and military-related products. The good microwave characteristics of the composite are related to the dielectric and magnetic losses, as well as to the achieved optimal permittivity of the composite at the given ratio of the carbon and magnetite phase, the latter being chiefly inter-aggregately distributed. The percent content of the conductive filler has a significant impact upon EMI SE of the composites.

- It has been established that the vulcanization characteristics, mechanical, and dynamic properties of the studied rubber compounds and vulcanizates filled with PCSF are commensurable, even in some cases they are improved compared to those of the composites comprising the conventional fillers.

4.5   ACKNOWLEDGEMENTS

The work is the part of a project, funded by King Abdulaziz University, Jeddah and University of Tabuk, Tabuk, Saudi Arabia. The authors acknowledge the technical and financial support. The authors would also like to thank to Prof. Ahmed A. Al-Ghamdi, Prof. Omar A. Al-Hartomy, and Prof. Falleh R. Al-Solamy for useful and long-term cooperation.

KEYWORDS

- **furnace carbon black**
- **conductive carbon black**
- **magnetite**
- **hybrid dual-phase fillers**
- **impregnation technology**
- **pyrolysis-cum-water vapor**
- **microwave properties**
- **natural rubber**

REFERENCES

1. Donnet, J.-B.; Custodero, E. *Reinforcement of Elastomers by Particulate Fillers. The Science and Technology of Rubber,* 4th ed.; Academic Press: Boston, 2013; pp 383–416.
2. Wolff, S. Chemical Aspects of Rubber Reinforcement by Fillers. *Rubber Chem. Technol.* **1996,** *69,* 325–346.
3. Chameswary, J.; Sebastian, M. T. Effect of ba(zn1/3ta2/3)o3 and sio2 Ceramic Fillers on the Microwave Dielectric Properties of Butyl Rubber Composites. *J. Mater. Sci.: Mater. Electron.* **2013,** *24,* 4351–4360.
4. Kim, S.-T.; Kim, S.-S. Microwave Absorbance of Ni-fe Thin Films on Hollow Ceramic Microspheres Dispersed in a Rubber Matrix. *J. Alloys Compd.* **2016,** *687,* 22–27.

5. Sun, D.; Li, X.; Zhang, Y.; Li, Y. Effect of Modified Nano-silica on the Reinforcement of Styrene Butadiene Rubber Composites. *J. Macromol. Sci., Part B* **2011**, *50*, 1810–1821.

6. Zafarmehrabian, R.; Gangali, S. T.; Ghoreishy, M. H. R.; Davallu, M. The Effects of Silica/carbon Black Ratio on the Dynamic Properties of the Tread Compounds in Truck Tires. *E-J. Chem.* **2012**, *9*, 1102–1112.

7. Wolff, S.; Wang, M.-J. Filler—elastomer Interactions. Part IV. The Effect of the Surface Energies of Fillers on Elastomer Reinforcement. *Rubber Chem. Technol.* **1992**, *65*, 329–342.

8. Seyvet, O.; Navard, P. Collision-induced Dispersion of Agglomerate Suspensions in a Shear Flow. *J. Appl. Polym. Sci.* **2000**, *78*, 1130–1133.

9. Donnet, J.-B. Black and White Fillers and Tire Compound. *Rubber Chem. Technol.* **1998**, *71*, 323–341.

10. Sengloyluan, K.; Sahakaro, K.; Dierkes, W. K.; Noordermeer, J. W. M. Silica-reinforced Tire Tread Compounds Compatibilized by Using Epoxidized Natural Rubber. *Eur. Polym. J.* **2014**, *51*, 69–79.

11. Wang, M. J.; Kutsovsky, Y.; Zhang, P.; Mehos, G.; Murphy, L. J.; Mahmud, L. Using Carbonsilica Dual Phase Filler Improve Global Compromise Between Rolling Resistance, Wear Resistance and Wet Skid Resistance for Tires. *KGK-Kautsch. Gummi Kunstst.* **2002**, *55*, 33–40.

12. Al-Ghamdi, A. A.; Al-Hartomy, O. A.; Al-Solamy, F. R.; Dishovsky, N.; Malinova, P.; Lakov, L. Characterization of Hybrid Fillers Based on Carbon Black of Different Types Obtained by Impregnation. *Proceedings of the Institution of Mechanical Engineers, Part L: Journal of Materials Design and Applications,* 2017, v.231, №7, 584–599.

13. Al-Ghamdi, A. A.; Al-Hartomy, O. A.; Al-Solamy, F. R.; Dishovsky, N.; Malinova, P.; Nikolov, R. Comparative Study of the Phase Distribution in Carbon-silica Hybrid Fillers for Rubber Obtained by Different Impregnation Technologies. *KGK-Kautsch. Gummi Kunstst.* **2016**, *69*, 37–43.

14. Al-Ghamdi, A. A.; Al-Hartomy, O. A.; Al-Solamy, F. R.; Dishovsky, N.; Mihaylov, M.; Malinova, P.; Atanasov, N. Dielectric and Microwave Properties of Elastomer Composites Loaded with Carbon–silica Hybrid Fillers. *J. Appl. Polym. Sci.* **2016**, *133*, 42978(1)–42978(9).

15. Al-Ghamdi, A.; Al-Hartomy, O.; Al-Solamy, F.; Dishovsky, N.; Mihaylov, M.; Malinova, P.; Atanasov, N. Natural Rubber Based Composites Comprising Different Types of Carbon-silica Hybrid Fillers. Comparative Study on their Electric, Dielectric and Microwave Properties, and Possible Applications. *Mater. Sci. Appl.* **2016**, *7*, 295–306.

16. Al-Ghamdi, A. A.; Al-Hartomy, O. A.; Al-Solamy, F. R.; Dishovsky, N.; Mihaylov, M.; Atanasov, N.; Atanasova, G.; Nihtianova, D. Microwave Properties of Natural Rubber Based Composites Comprising Conductive Carbon Black/silica Hybrid Fillers. *J. Polym. Res.* **2016**, *23*, 180.

17. Al-Ghamdi, A. A.; Al-Hartomy, O. A.; Al-Solamy, F. R.; Dishovsky, N.; Zaimova, D.; Malinova, P.; Nihtianova, D. Preparation and Characterization of Natural Rubber Composites Comprising Conductive Carbon Black/magnetite Hybrid Fillers Obtained by Impregnation Technology. *Polym.-Plast. Technol. Eng.* **2016**, *55*, 1344–1356.

18. Al-Ghamdi, A. A.; Al-Hartomy, O. A.; Al-Solamy, F. R.; Dishovsky, N.; Malinova, P.; Atanasova, G.; Atanasov, N. Conductive Carbon Black/magnetite Hybrid Fillers in

Microwave Absorbing Composites Based on Natural Rubber. *Compos. Part B: Eng.* **2016**, *96*, 231–241.

19. Isayev, A. I. Recycling of Rubbers. In *The Science and Technology of Rubber*, 4th ed.; Erman, B., Mark, J. E., Roland, C. M., Eds.; Academic Press: Boston, 2013; pp 697–764.

20. Bulgarian Patent 65901 B1 Method and installation for pyrolysis of tires.

21. Kolev, D. N.; Ljutzkanova, R. B.; Abadjiev, S. T. European Patent EP1879978 A1. Method and installation for pyrolisis of tires.

22. Barrett, E. P.; Joyner, L. G.; Halenda, P. P. The Determination of Pore Volume and Area Distributions in Porous Substances. I. Computations from Nitrogen Isotherms. *J. Am. Chem. Soc.* **1951**, *73*, 373–380.

23. Flory, P. J.; Rehner, J. Statistical Mechanics of Cross Linked Polymer Networks i. Rubberlike Elasticity. *J. Chem. Phys.* **1943**, *11*, 512–520.

24. Norman, R. H. *Conductive Rubbers and Plastics; their Production, Application and Test Methods;* Elsevier Pub. Co.: Amsterdam, 1970.

25. Cohen, J. *Applied Multiple Regression/correlation Analysis for the Behavioral Sciences;* L. Erlbaum Associates: Mahwah, N.J., 2003.

26. Chen, L. F.; Ong, C. K.; Neo, C. P.; Varadan, V. V.; Varadan, V. K. *Microwave Electronics: Measurement and Materials Characterization,* 1st ed.; John Wiley & Sons, Ltd.: Chichester, 2004.

27. Sheen, J. Measurements of Microwave Dielectric Properties by an Amended Cavity Perturbation Technique. *Measurement* **2009**, *42*, 57–61.

28. Verma, A.; Dube, D. C. Measurement of Dielectric Parameters of Small Samples at x-band Frequencies by Cavity Perturbation Technique. *IEEE Trans. Instrum. Meas.* **2005**, *54*, 2120–2123.

29. Mi, L.; Wang, Y.; Afsar, M. N. Precision Measurement of Complex Permittivity and Permeability by Microwave Cavity Perturbation Technique. Infrared and Millimeter Waves and 13th International Conference on Terahertz Electronics, 2005 IRMMW-THz 2005 The Joint 30th International Conference on. 2005, vol. 61, pp 62–63.

30. Kumar, S. B.; Hohn, H.; Joseph, R.; Hajian, M.; Ligthart, L. P.; Mathew, K. T. Complex Permittivity and Conductivity of Poly Aniline at Microwave Frequencies. *J. Eur. Ceram. Soc.* **2001**, *21*, 2677–2680.

31. Jana, P. B.; Mallick, K.; De, S. K. Effects of Sample Thickness and Fiber Aspect Ratio on Emi Shielding Effectiveness of Carbon Fiber Filled Polychloroprene Composites in the x-band Frequency Range. *IEEE Transactions on Electromagnetic Compatibility* **1992**, *34*, 478–481.

32. Paul, C. R. *Introduction to Electromagnetic Compatibility,* 2nd ed.; John Wiley & Sons, Inc.: Hoboken, 2006.

33. Ott, H. W. *Electromagnetic Compatibility Engineering,* 1st ed.; John Wiley & Sons, Inc.: Hoboken, 2009.

34. Hernandez, B. *Effect of Graphitic Carbon Nanomodifiers on the Electromagnetic Shielding Effectiveness of Linear Low Density Polyethylene Nanocomposites;* The Graduate School of Clemson University: South Carolina, USA, 2013.

35. Srikanth, V. V. S. S.; Raju, K. C. J. Graphene/polymer Nanocomposites as Microwave Absorbers. In *Graphene-based Polymer Nanocomposites in Electronics,* 1st ed.; Sadasivuni, K. K., Ponnamma, D., Kim, J., Thomas, S., Eds.; Springer International Publishing: Switzerland, 2015; pp 307–343.

36. Sung-Hoon, P.; Theilmann, P. T.; Asbeck, P. M.; Bandaru, P. R. Enhanced Electromagnetic Interference Shielding Through the use of Functionalized Carbon-nanotube-reactive Polymer Composites. *IEEE Trans. Nanotechnol.* **2010**, *9*, 464–469.

37. Hong, Y. K.; Lee, C. Y.; Jeong, C. K.; Lee, D. E.; Kim, K.; Joo, J. Method and Apparatus to Measure Electromagnetic Interference Shielding Efficiency and its Shielding Characteristics in Broadband Frequency Ranges. *Rev. Sci. Instrum.* **2003**, *74*, 1098–1102.

38. Więckowski, T. W.; Janukiewicz, J. M. Methods for Evaluating the Shielding Effectiveness of Textiles. *Fibres Text. East. Eur.* **2006**, *14*, 18–22.

39. Dishovsky, N.; Tsenkov, G. *Handbook of Rubber,* 1st ed.; Es Print Ltd.: Sofia, Bulgaria, 2006.

40. Chen, C. H.; Koenig, J. L.; Shelton, J. R.; Collins, E. A. Characterization of the Reversion Process in Accelerated Sulfur Curing of Natural Rubber. *Rubber Chem. Technol.* **1981**, *54*, 734–750.

41. Shankar, U. Investigations of the Reversion of Vulcanized Rubber under Heat. *Rubber Chem. Technol.* **1952**, *25*, 241–250.

42. Coran, A. Y. Vulcanization. In *The Science and Technology of Rubber,* 4th ed.; Mark, J. E., Erman, B., Roland, C. M., Eds.; Academic Press: Boston, 2013; pp 337–381.

43. Al-Ghamdi, A. A.; Al-Hartomy, O. A.; Al-Solamy, F. R.; Dishovsky, N.; Malinova, P.; Lakov, L. Characterization of Hybrid Fillers Based on Carbon Black of Different Types Obtained by Impregnation. *Proceedings of the Institution of Mechanical Engineers, Part L: Journal of Materials: Design and Applications,* In press, 2015.

44. Puryanti, D.; Ahmad, S. H.; Abdullah, M. H. Effect of Nickel-cobalt-zinc Ferrite Filler on Electrical and Mechanical Properties of Thermoplastic Natural Rubber Composites. *Polym.-Plast. Technol. Eng.* **2006**, *45*, 561–567.

45. Al-Hartomy, O.; Al-Ghamdi, A.; Al-Solamy, F.; Dishovsky, N.; Mihaylov, M.; Ivanov, M.; El-Tantawy, F. Influence of Matrices Chemical Nature on the Dynamic Mechanical and Dielectric Properties of Rubber Composites Comprising Conductive Carbon Black. *J. Polym. Res.* **2012**, *19*, 1–8.

46. Al-Hartomy, O. A.; Al-Solamy, F.; Al-Ghamdi, A.; Dishovsky, N.; Ivanov, M.; Mihaylov, M.; El-Tantawy, F. Influence of Carbon Black Structure and Specific Surface Area on the Mechanical and Dielectric Properties of Filled Rubber Composites. *Int. J. Polym. Sci.* **2011**, *2011*, 8.

47. Rodgers, B.; Waddell, W. The Science of Rubber Compounding. In *The Science and Technology of Rubber,* 4th ed.; Mark, J. E., Erman, B., Roland, C. M., Eds.; Academic Press: Boston, 2013; pp 417–471.

48. Krishnan, Y.; Chandran, S.; Usman, N.; Smitha, T. R.; Parameswaran, P. S.; Prema, K. H. Processability, Mechanical and Magnetic Studies on Natural Rubber-Ferrite Composites. *Int. J. Chem. Stud.* **2015**, *3*, 15–22.

49. Joy, J.; Tresá Sunny, A.; Mathew, L. P.; Pothen, L. A.; Thomas, S. Micro and Nano Metal Particle Filled Natural Rubber Composites. In *Natural Rubber Materials: Volume 2: Composites and Nanocomposites.* 1st ed.; Thomas, S., Chan, C. H., Pothen, L. A., Joy, J., Maria, H., Eds.; The Royal Society of Chemistry: Cambridge, United Kingdom, 2014; pp 307–325.

50. Kang, G.-H.; Kim, S.-H. Electromagnetic Wave Shielding Effectiveness Based on Carbon Microcoil-polyurethane Composites. *J. Nanomater.* **2014**, *2014*, 6.

51. Kaynak, A.; Polat, A.; Yilmazer, U. Some Microwave and Mechanical Properties of Carbon Fiber-polypropylene and Carbon Black-polypropylene Composites. *Mater. Res. Bull.* **1996**, *31,* 1195–1206.
52. Kruželák, J.; Ušaková, M.; Dosoudil, R.; Hudec, I.; Sýkora, R. Microstructure and Performance of Natural Rubber Based Magnetic Composites. *Polym.-Plast. Technol. Eng.* **2014**, *53,* 1095–1104.
53. Cornell, R. M. The Iron Oxides: Structure, Properties, Reactions, Occurrences and Uses. In: Schwertmann, U., (ed.). VCH: Weinheim, 1996.
54. Ou, R.; Gerhardt, R. A.; Marrett, C.; Moulart, A.; Colton, J. S. Assessment of Percolation and Homogeneity in Abs/carbon Black Composites by Electrical Measurements. *Composites, Part B* **2003**, *34,* 607–614.
55. Hosseini, S. H.; Asandia, A. Polyaniline/Fe$_3$O$_4$ Coated on MnFe$_2$O$_4$ Nanocomposite: Preparation, Characterization, and Applications in Microwave Absorption. *Inter. J. Phys. Sci.* **2013**, *8,* 1209–1217.
56. Thomas, S.; Chan, C. H.; Pothen, L. A.; Joy, J.; Maria, H. J. *Natural Rubber Materials, Volume 2: Composites and Nanocomposites;* Royal Society of Chemistry: Cambridge, 2013.
57. Darius, T. R. Electromagnetic Properties of Microwire-Epoxy Composite. *Journal of Young Investigators* March 2008.
58. Deng, Y.; Zhang, Y.; Xiang, Y.; Wang, G.; Xu, H. Bi$_2$S$_3$-batio$_3$/pvdf Three-Phase Composites with High Dielectric Permittivity. *J. Mater. Chem.* **2009**, *19,* 2058–2061.
59. Micheli, D.; Apollo, C.; Pastore, R.; Marchetti, M. X-band Microwave Characterization of Carbon-based Nanocomposite Material, Absorption Capability Comparison and Ras Design Simulation. *Compos. Sci. Technol.* **2010**, *70,* 400–409.
60. Mirsha, M.; Singh, A. P.; Sambyal, P.; Teotia, S.; Dhawan, A. K. Facile Synthesis of Phenolic Resin Sheets Consisting Expanded Graphite/-Fe$_2$O$_3$/SiO$_2$ Composite and its Enhanced Electromagnetic interference Shielding Propertie. *Indian J. Pure Appl. Phys.* **2014**, *52,* 478–485.
61. Degussa. Technical information № 1261.
62. Lakshmi, K.; John, H.; Joseph, R.; George, K. E.; Mathew, K. T. Comparison of Microwave and Electrical Properties of Selected Conducting Polymers. *Microwave Opt. Technol. Lett.* **2008**, *50,* 504–508.
63. Du, J.; Zhao, L.; Zeng, Y.; Zhang, L.; Li, F.; Liu, P.; Liu, C. Comparison of Electrical Properties Between Multi-Walled Carbon Nanotube and Graphene Nanosheet/high Density Polyethylene Composites with a Segregated Network Structure. *Carbon* **2011**, *49,* 1094–1100.
64. Aguilar, J. O.; Bautista-Quijano, J. R.; Avilés, F. Influence of Carbon Nanotube Clustering on the Electrical Conductivity of Polymer Composite Films. *eXPRESS Polym. Lett.* **2010**, *4,* 292–299.
65. Qin, F.; Brosseau, C. A Review and Analysis of Microwave Absorption in Polymer Composites Filled with Carbonaceous Particles. *J. Appl. Phys.* **2012**, *111,* 061301.
66. Folgueras, L. d. C.; Alves, M. A.; Rezende, M. C. Dielectric Properties of Microwave Absorbing Sheets Produced with Silicone and Polyaniline. *Mater. Res.* **2010**, *13,* 197–201.
67. Song, N.-N.; Yang, H.-T.; Liu, H.-L.; Ren, X.; Ding, H.-F.; Zhang, X.-Q.; Cheng, Z.-H. Exceeding Natural Resonance Frequency Limit of Monodisperse Fe$_3$O$_4$ Nanoparticles via Superparamagnetic Relaxation. *Sci. Rep.* **2013**, *3,* 3161.

68. Jacquart, P.; Acher, O. Permeability Measurement on Composites Made of Oriented Metallic Wires from 0.1–18 ghz. *IEEE Trans. Microwave Theory Tech.* **1996**, *44*, 2116–2120.
69. Khurram, A. A.; Rakha, S. A.; Zhou, P.; Shafi, M.; Munir, A. Correlation of Electrical Conductivity, Dielectric Properties, Microwave Absorption, and Matrix Properties of Composites Filled with Graphene Nanoplatelets and Carbon Nanotubes. *J. Appl. Phys.* **2015**, *118*, 044105.
70. Niedermeier, W. J.; Fröhlich, J.; Luginsland, H. D. Reinforcement Mechanism in the Rubber Matrix by Active Fillers. *KGK-Kautsch. Gummi Kunstst.* **2002**, *55*, 356–366.
71. Al-Hartomy, O. A.; Al-Ghamdi, A. A.; Al Said, S. A. F.; Dishovsky, N.; Ward, M. B.; Mihaylov, M.; Ivanov, M. Characterization of Carbon Silica Hybrid Fillers Obtained by Pyrolysis of Waste Green Tires by the Stem–edx Method. *Mater. Charact.* **2015**, *101*, 90–96.
72. Dannenberg, E. M. Filler Choices in the Rubber Industry the Incumbents and Some New Candidates. *Elastomerics* **1981**, *113*, 30–50.
73. Evans, L. R.; Waddell, W. H. Ultra-high Reinforcing Precipitated Silica for Tire and Rubber Applications. *KGK-Kautsch. Gummi Kunstst.* **1995**, *48*, 718–723.
74. Wolff, S. Silanes in Tire Compounding After Ten Years—a Review. *Tire Sci. Technol.* **1987**, *15*, 276–294.
75. Hunsche, A.; Görl, U.; Müller, A.; Knaack, M.; Göbel, T. Investigations Concerning the Reaction Silica/Organosilane and Organosilane/Polymer—Part 1: Reaction Mechanism and Reaction Model for Silica/Organosilane. *KGK-Kautsch. Gummi Kunstst.* **1997**, *50*, 881–889.
76. Hunsche, A.; Görl, U.; Koban, H. G.; Lehmann, T. Investigations on the Reaction Silica/Organosilane and Organosilane/Polymer—Part 2. Kinetic Aspects of the Silica-Organosilane Reaction. *KGK-Kautsch. Gummi Kunstst.* **1998**, *51*, 525–533.
77. Goerl, U.; Hunsche, A.; Mueller, A.; Koban, H. G. Investigations into the Silica/Silane Reaction System. *Rubber Chem. Technol.* **1997**, *70*, 608–623.
78. Görl, U.; Parkhouse, A. Investigations on the Reaction Silica/Organosilane and Organosilane/Polymer Part 3: Investigations Using Rubber Compounds. *KGK-Kautsch. Gummi Kunstst.* **1999**, 493–500.
79. Görl, U.; Münzenberg, J.; Luginsland, D.; Müller, A. Investigations on the Reaction Silica/Organosilane and Organosilane/Polymer—Part 4: Studies on the Chemistry of the Silane Sulfur Chain. *KGK-Kautsch. Gummi Kunstst.* **1999**, *52*, 588–593.
80. Siriwardena, S.; Ismail, H.; Ishiaku, U. S. A Comparison of White Rice Husk Ash and Silica as Fillers in Ethylene—Propylene—Diene Terpolymer Vulcanizates. *Polym. Int.* **2001**, *50*, 707–713.
81. Chapman, A. V. Natural Rubber and Nr-based Polymers—Renewable Materials with Unique Properties. 24th International HF Mark-Symposium, Advances in the Field of Elastomers and Thermoplastic Elastomers. Vienna, Austria 2007.
82. Martens, J. E.; Terrill, E. R.; Lewis, J. T.; Pazur, R. J.; Hoffman, R. Effect of Deformation Mode in Prediction of Tire Performance by Dynamic Mechanical Analysis. *Rubber World* **2013**, *248*, 29–35.

CHAPTER 5

ELASTOMER-BASED COMPOSITE MATERIALS COMPRISING PYROLYSIS CARBON BLACK

CONTENTS

ABSTRACT

Waste tires and their deposition have been a serious environmental concern as they are resistant to moisture, oxygen, ozone, solar radiation, microbiological degradation, and so forth. Pyrolysis is an environment-friendly process for recycling used tires. The aim of this chapter is to present some studies on the effect that pyrolysis carbon black (PCB) has upon the curing and mechanical properties of styrene–butadiene rubber (SBR)-based composites, as well as to demonstrate a possibility for the surface modification of PCB, so that its ash content could be reduced. It has been established that modification of the PCB lowers its ash content two times and the zinc oxide concentration six times in its ash. The curing characteristics of the investigated SBR-based compounds are not affected significantly by that modification. The values of modulus100, tensile strength and Shore A hardness for the vulcanizates comprising modified PCB increase, while the elongation at break and residual elongation decrease slightly. The effects observed are related to the lowered ash content in PCB following its modification as well as to the dominating silica content in the remaining ash. Pyrolysis carbon black could substitute about 40–50% of conventional carbon black in the production of rubber goods requiring weaker mechanical properties. The results obtained have shown that the nature of the pyrolysis product studied is much closer to that of conventional semi-reinforcing carbon black.

5.1 INTRODUCTION

Non-decomposing organic waste has been one of the major environmental problems. Use of tires in large numbers has resulted into immense amounts of such waste. According to statistics, scrap tires in Europe only amounted 3.3 million in 2010 while in 2011 the figure was over 5.7 million. Being resistant to moisture, oxygen, ozone, microbiological factors, and so forth, waste tires are of great environmental concern. On the other hand, that waste is a valuable source of energy and materials.[1–5]

Usually waste tires are dumped in landfill sites or burned in cement or brick kilns. Disposal of waste tires is unprofitable since it requires large plots for building the depots. Usually such plots are valuable arable land.[6] Some of the ingredients that comprise waste tires are prone to migrating what is another environmental hazard. Therefore, the disposal of such waste was banned on the landfills in 1999 in the European Union by Council Directive 1999/31/EC of 26 April 1999.[7] The polycyclic aromatic hydrocarbons and heavy metals resulting from tire combustion pollute the environment and installation of expensive gas purification systems is needed as a result. Therefore, combustion *of tires* is also undesirable.[8]

There are several other ways to dispose such waste: tire retreading, mechanical grinding,[9] rubber reclaiming,[10] chemical devulcanization,[11] ultrasound devulcanization,[12,13] microwave devulcanization,[14] pyrolysis, and so forth.

Waste tire pyrolysis could be an environmentally friendly solution as its products have a significant potential for wide range of applications. Pyrolysis is an endothermic process during which materials are thermally decomposed due to being heated at high temperatures in an oxygen-free or oxygen-poor medium.[15] Many researchers also call it "thermal destruction" or "thermolysis." Under typical pyrolysis conditions, the organic matter of waste tires, in particular the elastomers, is heated to temperatures above 400–450°C in the presence of a water vapor, vacuum, inert gas (N_2, Ar) medium, wherein the various particularities comprised by the starting materials are decomposed and evaporated. Some of these processes are dehydration, cracking, isomerization, dehydrogenation, aromatization, and condensation.[15,16] During the pyrolysis process, the rubber hydrocarbon in the tires (about 60%) is decomposed to low-molecular-weight products (liquid and gaseous). Carbon black and inorganic substances (40%) remain in the so-called "solid residue" pyrolysis carbon black (PCB). The derived oils may be used directly as fuel or added to petroleum refinery feedstock. The gaseous products are also useful as fuel, and the carbon black, due to its high carbon content, may be used as reinforcing filler in rubber or as activated carbon.[8,17]

The ideal application of PCB is in the production of rubber items. However, the product could hardly be reused as neat carbon black due to the high concentration of inorganic particularities (ash) in it[18] as well as due to the structure of its particles (degree of aggregation).[19] In this regard, Faulkner and Weinecke[20] have shown that PCB obtained by a longer pyrolysis process in a rotary kiln reactor possesses properties similar to those of conventional semi-reinforcing carbon black. They drew this conclusion by comparing the oil number, iodine adsorption, the discoloration in toluene, the specific surface area by Brunauer–Emmett–Teller (BET) analysis, and average particle size of PCB and that of various types of conventional fillers. The conditions under which the pyrolysis process is run inevitably affect the properties and characteristics of the resulting PCB. Willams and co-authors[21] have demonstrated that the specific surface area of PCB increases simultaneously with the pyrolysis temperature and the heating rate. Kawakami[22] et al. reveal that the mechanical properties of rubber composites containing

PCB depend on the pyrolysis temperature. Kaminsky and co-workers[23] obtained carbon black of porosity lower than that of conventional carbon black, working at temperatures between 700 and 790°C. There is literature and studies on the possibilities of reducing the ash content of PCB. Zhou Jie and co-authors used nitric acid to reduce the ash content, then modified the washed PCB with titanate coupling agent to obtain recycled carbonaceous materials for use as inks in printing industry.[24]

There are a number of scientific publications that have examined the pyrolysis of waste tires in detail. The conditions of pyrolysis, process kinetics, type of experimental equipment used, and so forth are discussed thoroughly. Whereas, only a limited number of studies report on the effect that PCB has upon the properties of virgin rubber composites.

The aim of this chapter is to present some studies on the effect of PCB upon the curing and mechanical properties of styrene–butadiene rubber (SBR)-based composites, as well as to demonstrate an opportunity for the surface modification of PCB regarding reduction of its ash content.

5.2   EXPERIMENT

5.2.1   MATERIALS

The investigations were performed on SBR 1500 type (Kralex 1500) supplied from Synthos Kralupy a.s., Czech Republic. Two types of commercial carbon black were used for comparison—Corax N 550 and Corax N 330, both produced by Orion Engineered Carbons GmbH. Other ingredients such as zinc oxide (ZnO), stearic acid (SA), N-tert-butyl-2-benzothiazolesulfen-amide (TBBS), sulfur (S), and so forth, were commercial grades and used without further purification.

5.2.2   PREPARATION AND CHARACTERIZATION OF PYROLYSIS CARBON BLACK

The investigated PCB was obtained via pyrolysis-cum-water vapor[25,26] of waste tires under the following conditions: temperature—600°C, water vapor concentration—40%.

According to the method, the tires were subjected to pyrolysis in which the raw materials were gradually heated between 300 and 1000°C with constant purging of some of the following gases: air, smoke gases, carbon dioxide, nitrogen, and/or addition of vapor from 0–100%. It is also possible to run pyrolysis and activation simultaneously using ammonium as well as mixtures of the former gases. The final product was evacuated for a period of 1 min–48 h then cooled in an airless atmosphere.

The PCB studied was characterized according to following parameters:

- Ash content—ISO 1125:1999;
- Iodine adsorption—ISO 1304:2006;
- Oil absorption number—ISO 4656:2012;
- Specific surface area (BET)—ISO 4652:2012.

The ash from the fillers was studied by silicate analysis, weight analysis, atomic absorption spectroscopy (AAS; Perkin Elmer 5000), inductively coupled plasma-optical emission spectroscopy (ICP-OES; Prodigy High Dispersion ICP-OES, Teledyne Lemas Labs).

5.2.3   MODIFICATION OF PYROLYSIS CARBON BLACK

The solid pyrolysis product was placed into a round-bottom flask and 5% hydrochloric acid was added at 1:4 ratio. The mixture was boiled under a reflux condenser for 20 min. Then the product was washed with distilled water over a Buchner funnel till a clear filtrate was obtained and no white residue was obtained when silver nitrate was added to it. A strong odor of hydrogen sulfide evolved during the modification indicating that the sulfur and sulfur compounds were removed from the carbon black. The modification of the solid pyrolysis product was performed to reduce one of its drawbacks, that is, its ash content. The resultant solid pyrolysis product is called PCB.

5.2.4   CHARACTERIZATION OF THE STUDIED COMPOSITES

The vulcanization characteristics of the rubber composites were determined at 150°C on a moving die rheometer (MDR 2000 Alpha Technologies) according to ISO 3417:2008. The mechanical properties of the composites studied were determined according to ISO 37:2011. The heat aging resistance was determined according to ISO 188:2009 in an air thermo chamber at 100°C for 72 h, ventilated by forced air convection. Shore A hardness of the composites studied was determined according to ISO 7619-1:2010.

The swelling was carried out according to ISO 1817:2015 at room temperature in toluene. The average number of cross-links per macromolecule ($n_c = M_m/M_c$) was calculated by dividing the molecular weight of rubber (M_m) by the molecular weight of the segment between cross-links (M_c)which was estimated using Flory–Rehner[27] equation:

$$M_c = \frac{-\rho_r V_s \left(V_r\right)^{1/3}}{\ln\left(1 - V_r\right) + V_r + \chi V_r^2}$$ (5.1)

where:

M_c is average molecular weight of the rubber segments between two cross-links; ρ_r is rubber density; V_s is molar volume of the solvent; χ is the parameter characterizing rubber–solvent interaction; and V_r is volume fraction of the swollen network.

5.2.5 PREPARATION AND VULCANIZATION OF THE RUBBER COMPOUNDS

The rubber compounds were prepared on an open two-roll laboratory mill (L/D 320 × 160 and friction 1.27) according to a specific recipe and blending regime. The speed of the slow roll was 25 rpm. The specimens were vulcanized on an electrically heated hydraulic press at 10.0 MPa and 160°C according to the optimums of each rubber compound that were determined on an moving die rheometer (MDR 2000).

5.3 RESULTS AND DISCUSSION

5.3.1 EFFECT OF DIFFERENT AMOUNTS OF PYROLYSIS CARBON BLACK UPON THE PROPERTIES OF STYRENE–BUTADIENE RUBBER-BASED COMPOSITES

Table 5.1 shows the main properties of the studied PCB. The table shows, its ash content is about 16% which is indeed high as the ash content of the carbon black used in rubber industry is below 0.5%. The incandescence losses are about 1.5% demonstrating that moisture content in the filler studied is within an acceptable range.

The preliminary ground PCB has been studied as a filler of SBR-based composites prepared according to the formulations presented in Table 5.2 in order to establish its reinforcing effect.

TABLE 5.1   Properties of the Pyrolysis Carbon Black Investigated.

	PCB
pH	6.5
Losses on heating (%)	1.45
Ash content (%)	16.2
OAN (ml/100 g)	120
IA (mg/g)	182
SBET (m2/g)	32

The properties of SBR vulcanizates comprising various amounts of PCB were investigated. In order to compensate the ash content in carbon black, it was set at concentrations of about 20% higher than those chosen for the experiments. Conventional carbon black N330 and N550 was used as reference.

TABLE 5.2   Composition of the Studied Rubber Compounds Comprising Pyrolysis Carbon Black (phr).

	E 1	E 2	D 40	D 50	D 60	D 70
SBR Kralex 1500	100.0	100.0	100.0	100.0	100.0	100.0
Zinc oxide	4.0	4.0	4.0	4.0	4.0	4.0
Stearic acid	2.0	2.0	2.0	2.0	2.0	2.0
Carbon black Corax N 550	50.0	–	–	–	–	–
Carbon black Corax N 330	–	50.0	–	–	–	–
Pyrolysis carbon black	–	–	48.0	60.0	72.0	84.0
TBBS	2.0	2.0	2.0	2.0	2.0	2.0
Sulfur	2.0	2.0	2.0	2.0	2.0	2.0

The curing characteristics presented in Table 5.3 show that at equal amounts of PCB and conventional carbon black, no significant differences are observed in the minimum torque of the rubber composites studied. The parameter gives an idea of the viscosity of composites. The difference between maximum and minimum torque (ΔM), giving an idea of the density of the formed cross-link network for rubber compounds containing PCB at 50 parts per hundred of rubber (phr), is greater than that of the compounds containing the same amount of conventional carbon black. That means the curing network in the vulcanizates of rubber compounds containing the pyrolysis product is denser probably due to the greater amount of sulfur and zinc oxide present in PCB. The results are confirmed by data on the molecular weight of the segment (M_c) between two cross-links and the average number of cross-links per macromolecule (n_c), calculated by the Flory–Renner equation having determined the equilibrium swelling degree of the vulcanizates tested in an organic solvent. Obviously, the vulcanization process proceeds at a higher rate in PCB-filled rubber compounds. This result could be explained with the higher content of sulfur and zinc oxide therein, as well as with its acidic character (pH = 6.5).

TABLE 5.3   Curing Characteristics of the Studied Rubber Compounds.

	E 1	E 2	D 40	D 50	D 60	D 70
M_L (dN.m)	4.0	4.1	4.3	4.5	5.8	6.7
M_H (dN.m)	39.0	40.0	43.0	52.0	56.0	59.0
$\Delta M = M_H - M_L$ (dN.m)	35.0	35.9	38.7	47.5	50.2	52.3
T_{s2} (min/s)	3:20	3:30	3:10	2:40	2:30	2:30
T_{90} (min/s)	12:00	10:15	7:30	6:30	7:00	7:00
Cure rate, V_c (%/min)	11.5	14.7	23.3	26.3	22.2	22.2
M_c	3950	3450	2980	2100	2200	1870
n_c	50	58	67	95	91	107

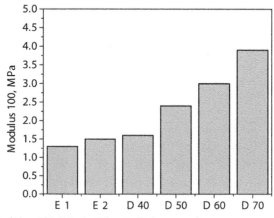

FIGURE 5.1   Modulus 100 (M_{100}) of the studied composites.

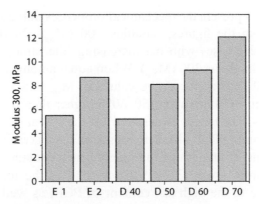

FIGURE 5.2 Modulus 300 (M_{300}) of the studied composites.

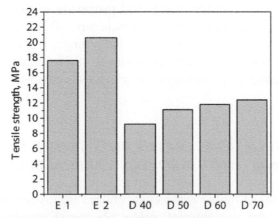

FIGURE 5.3 Tensile strength (σ) of the studied composites.

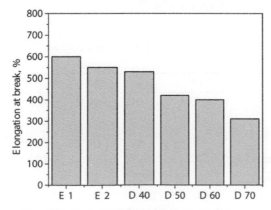

FIGURE 5.4 Elongation at break (ε_{rel}) of the studied composites.

Figures 5.1–5.4 present the mechanical properties of the studied composites. As seen from the figures, modulus 100 (M_{100}) for the vulcanizates containing PCB increases with the increasing filler amount. The same is observed for the modulus 300 (M_{300}). When equal amounts of conventional carbon black and PCB are used, the values of M_{100} and M_{300} of the vulcanizates containing PCB are about 50–80% higher than those of the reference vulcanizates. The tensile strength (σ) of the vulcanizates of the rubber compounds containing PCB also increases with the increasing filler amount. However, the tensile strength of the vulcanizates containing the pyrolysis filler is significantly lower (35–45%) than that of the reference vulcanizates. This is due to the high ash content of PCB, as well as to its lower surface area (32 m²/g vs. about 45 m²/g for Corax N 550 and 72 m²/g for Corax N 330). In all cases, the ash contained in PCB is completely inert and does not affect favorably the mechanical properties of the rubber vulcanizates. However, considering that the tensile strength of SBR vulcanizates comprising no active filler is in the 2–3 MPa range, it could be concluded that PCB improves the mechanical properties of the vulcanizates, although the properties achieved are not equivalent to those provided by conventional carbon black.

5.3.2 INVESTIGATIONS ON THE POSSIBILITIES TO SUBSTITUTE CONVENTIONAL WITH PYROLYSIS CARBON BLACK

The compositions of rubber compounds chosen for investigating the possibilities to substitute conventional carbon black with pyrolysis product are presented in Table 5.4. In the case 20, 40, 60, 80, and 100% of conventional carbon black were substituted by PCB.

TABLE 5.4 Composition of the Studied Rubber Compounds Comprising Conventional Carbon Black and PCB (phr).

	N 1	D 2	D 3	D 4	D 5	D 6
SBR-Kralex 1500	100.0	100.0	100.0	100.0	100.0	100.0
Zinc oxide	4.0	4.0	4.0	4.0	4.0	4.0
Stearic acid	2.0	2.0	2.0	2.0	2.0	2.0
Carbon black Corax N330	50.0	40.0	30.0	20.0	10.0	–
Pyrolysis carbon black	–	12.0	24.0	36.0	48.0	60.0
TBBS	2.0	2.0	2.0	2.0	2.0	2.0
Sulfur	2.0	2.0	2.0	2.0	2.0	2.0

Table 5.5 presents the vulcanization characteristics of the rubber compounds studied. The differences observed in the values of the main vulcanization properties of the compounds studied are not significant.

TABLE 5.5 Curing Characteristics of the Studied Rubber Compounds Comprising Conventional Carbon Black and PCB.

	N 1	D 2	D 3	D 4	D 5	D 6
M_L (dN.m)	4.9	5.1	5.9	6.1	5.9	5.9
M_H (dN.m)	61.0	62.0	65.0	65.0	65.0	65.0
$\Delta M = M_H - M_L$ (dN.m)	56.1	56.9	59.1	58.9	59.1	59.1
T_{s2} (min/s)	3:30	3:20	3:50	3:40	3:00	3:00
T_{90} (min/s)	6:40	7:10	7:50	7:10	6:30	7:00
Cure rate, V (%/min)	31.3	25.0	25.0	27.8	29.4	25.0
M_c	1760	1960	1730	1690	1780	1740
n_c	114	102	116	118	112	115

Figures 5.5–5.8 present the mechanical properties of rubber compounds comprising combinations of conventional carbon black Corax N330 and PCB. The figures show that the values of modulus100 increase, and those of the modulus300 and tensile strength of the vulcanizates from rubber compounds comprising a combination of conventional carbon black and PCB at different amounts decrease with the increasing PCB content in the total mass of the filler. Nevertheless, one can conclude that PCB could substitute about 40–50% of the conventional carbon black when rubber articles of weaker mechanical properties are manufactured. The results obtained reveal the resultant pyrolysis product produces an effect much close to the one of conventional semi-reinforcing carbon black.

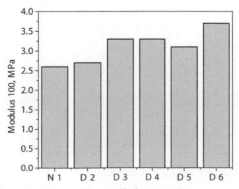

FIGURE 5.5 Modulus 100 (M_{100}) of the studied composites.

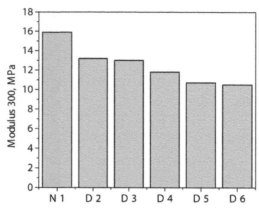

FIGURE 5.6 Modulus 300 (M_{300}) of the studied composites.

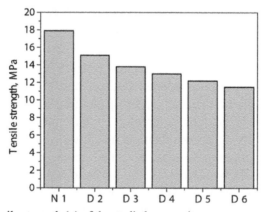

FIGURE 5.7 Tensile strength (σ) of the studied composites.

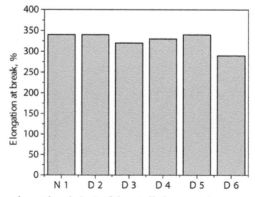

FIGURE 5.8 Elongation at break (ε_{rel}) of the studied composites.

5.3.3 CHARACTERIZATION OF VIRGIN AND MODIFIED PYROLYSIS CARBON BLACK

The obtained PCB has been subjected to surface modification in order to reduce its ash content. The results from the characterization of both—the virgin and modified PCB are summarized in Table 5.6.

TABLE 5.6 Properties of the Virgin and Modified Pyrolysis Carbon Black Investigated.

	Virgin pyrolysis carbon black	Modified pyrolysis carbon black
Losses on heating (%)	1.45	1.78
Ash content (%)	16.2	8.6
OAN (ml/100 g)	120	127
IA (mg/g)	182	171
S_{BET} (m²/g)	32	31

(Reprinted from N. Delchev, P. Malinova, M. Mihaylov, N. Dishovsky, Effect of the Modified Solid Product from Waste Tyres Pyrolysis on the Properties of Styrene-Butadiene Rubber Based Composites, Journal of Chemical Technology and Metallurgy, 49, 6, 2014, 525-534. Permission from University of Chemical Technology and Metallurgy.)

The compared properties of PCB before and after the modification reveal the following observations:

- The losses on heating of the PCB have changed slightly after the modification. These changes are due mainly to moisture and other volatile substances; therefore, the losses on heating are less than 1.8%.
- The ash content following the modification is almost two times lower (16.2% prior to the modification and 8.6% after that). In fact, this was the main purpose of treating the initial PCB. The result indicates that the experimental goal has been achieved.
- The modification does not affect the structure of PCB because the oil number values of virgin and modified PCB are close. The same is valid for the specific surface area and iodine number, that is the modification results predominantly into lowering the ash content of PCB. The ash in PCB has been subjected to additional analyses viewing its further lowering. The results from the weight, AAS, and ICP-OES analyses are summarized in Table 5.7.

As shown in Table 5.7, ZnO content in PCB is almost six times lower following its treatment with hydrochloric acid; hence, the main purpose of the modification has been achieved. The content of SiO_2 is the highest in the ash of modified PCB as it does not react with hydrochloric acid. SiO_2 is present in PCB since there had been green tires among those subjected to pyrolysis. Green tires are produced using silica as filler instead of carbon black. Another reagent, such as hydrofluoric acid, should be used to remove SiO_2 from the PCB. As a rule, silica removal is not a must, since it is an active filler. An appropriate solution might be introducing some amount of a silane coupling agent into the rubber mixture as in the case of filling the compounds with conventional silica. The bifunctional nature of silane improves the elastomer–filler interaction. On the other hand, the significantly lower zinc oxide amount in the ash might be considered a serious success. Lowering the concentration of zinc oxide in rubber compounds has been a challenge worldwide due to the ecotoxicity of zinc ions. As a result of modification, the most lowered concentrations are that of Na_2O, K_2O, Fe_2O_3, MgO, and CaO. There are no changes in the concentrations of TiO_2 and PbO.

TABLE 5.7 Ash Content Data as Obtained from Weight, AAS, and ICP-OES Analyses.

	Content (%)	
	Virgin pyrolysis carbon black	Modified pyrolysis carbon black
SiO_2	38.80	87.32
$Al2O_3$	0.77	0.51
CaO	0.31	0.20
MgO	0.50	0.27
Na_2O	2.41	0.49
K_2O	5.65	0.74
Fe_2O_3	0.53	0.26
TiO_2	0.03	0.03
ZnO	46.10	7.22
CuO	0.19	0.11
PbO	0.01	0.01
Losses on heating, 1000°C	4.69	2.83

(Reprinted from N. Delchev, P. Malinova, M. Mihaylov, N. Dishovsky, Effect of the Modified Solid Product from Waste Tyres Pyrolysis on the Properties of Styrene-Butadiene Rubber Based Composites, Journal of Chemical Technology and Metallurgy, 49, 6, 2014, 525-534. Permission from University of Chemical Technology and Metallurgy.)

Table 5.8 presents the compositions of the studied rubber compounds comprising modified and non-modified PCB. Four SBR-based compounds comprising virgin and modified PCB were prepared. In order to have an objective comparison, the amount of PCB to be added to the rubber during the compounding was estimated according to their carbon (active substance) content, that is the compensated ash content which was determined previously. Thus 60 and 83.3 g, respectively, of virgin PCB in the compounds denoted as SBR 1 and SBR 2, correspond to 50 and 70 g conventional carbon black. While 54.7 and 76.6 g, respectively, of modified PCB in the compounds denoted as SBR 3 and SBR 4, correspond to 50 and 70 g conventional carbon black.

TABLE 5.8 Composition of the Studied Rubber Compounds Comprising Modified and Non-modified PCB (phr).

	SBR 1	SBR 2	SBR 3	SBR 4
SBR 1500	100	100	100	100
Zinc oxide	3.0	3.0	3.0	3.0
Stearic acid	1.0	1.0	1.0	1.0
Non-modified pyrolysis carbon black	60.0	83.3	–	–
Modified pyrolysis carbon black	–	–	54.7	76.6
TBBS	1.0	1.0	1.0	1.0
Sulfur	1.75	1.75	1.75	1.75

Source: N. Delchev, P. Malinova, M. Mihaylov, N. Dishovsky, Effect of the Modified Solid Product from Waste Tyres Pyrolysis on the Properties of Styrene-Butadiene Rubber Based Composites, Journal of Chemical technology and Metallurgy, 49, 6, 2014, 525-534.
Permission from University of Chemical Technology and Metallurgy

The vulcanization isotherms of the SBR-based compounds comprising various amounts of virgin (SBR 1 and SBR 2) and modified (SBR 3 and SBR 4) PCB are shown in Figure 5.9.

The vulcanization isotherms of the studied SBR-based compounds have the same curve pattern. The curves possess a wide plateau of the vulcanization (equilibrium torque curves) revealing the suitability of the compounds for manufacturing thick-walled items.

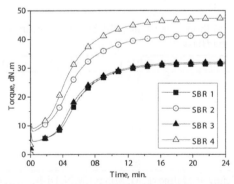

FIGURE 5.9 Cure curves of the investigated rubber compounds. (Reprinted from N. Delchev, P. Malinova, M. Mihaylov, N. Dishovsky, Effect of the Modified Solid Product from Waste Tyres Pyrolysis on the Properties of Styrene-Butadiene Rubber Based Composites, Journal of Chemical Technology and Metallurgy, 49, 6, 2014, 525-534. Permission from University of Chemical Technology and Metallurgy.)

The compounds comprising virgin and modified PCB at amounts equal to 50 phr carbon black have overlapping isotherms, while the isotherms of the compounds comprising virgin and modified PCB at amounts equal to 70 phr carbon black are distanced. The important fact in this case is that the usage of modified PCB does not worsen the curing characteristics of the compounds, that is there is no reason not to use the pyrolysis solid product as filler.

The curing characteristics (minimum torque—M_L, maximum torque—M_H, and the difference between them—ΔM; time to scorch—T_{s1}; optimum curing time—T_{90}; curing rate—V_c) of the studied SBR-based compounds obtained at 160°C are presented in Table 5.9.

Table 5.9 shows that the values of the vulcanization parameters increase at higher filler amounts, regardless whether it had been modified or not, that is treating the PCB has no effect upon the vulcanization process.

TABLE 5.9   Curing Characteristics of the Studied SBR-based Compounds Obtained at 160°C.

	M_L (dN.m)	M_H (dN.m)	ΔM (dN.m)	T_{s1} (min)	T_{90} (min)	Vc (%/min)
SBR 1	4.54	31.48	26.94	1:44	10:28	12.8
SBR 2	8.06	41.52	33.46	1:02	10:50	10.8
SBR 3	4.56	32.00	27.44	1:44	10:57	11.8
SBR 4	9.13	47.38	38.25	1:05	10:54	10.6

(Reprinted from N. Delchev, P. Malinova, M. Mihaylov, N. Dishovsky, Effect of the Modified Solid Product from Waste Tyres Pyrolysis on the Properties of Styrene-Butadiene Rubber Based Composites, Journal of Chemical Technology and Metallurgy, 49, 6, 2014, 525-534. Permission from University of Chemical Technology and Metallurgy.)

The mechanical properties of the SBR-based vulcanizates are shown in Table 5.10.

TABLE 5.10   Mechanical Properties of the SBR-based Composites Filled with Virgin and Modified Pyrolysis Carbon Black.

	SBR 1	SBR 2	SBR 3	SBR 4
M100 (MPa)	3.2	5.2	3.6	6.3
M300 (MPa)	12.0	–	–	–
σ (MPa)	12.2	11.8	12.8	12.3
εrel (%)	305	227	284	198
εres (%)	8	5	7	5
Shore A hardness	69	78	71	80

(Reprinted from N. Delchev, P. Malinova, M. Mihaylov, N. Dishovsky, Effect of the Modified Solid Product from Waste Tyres Pyrolysis on the Properties of Styrene-Butadiene Rubber Based Composites, Journal of Chemical Technology and Metallurgy, 49, 6, 2014, 525-534. Permission from University of Chemical Technology and Metallurgy.)

The table shows that the values for M_{100} of all studied SBR-based vulcanizates increase at higher filler amounts. In the case of vulcanizates comprising less filler due the modification, the increase in M_{100} values is lower (11%), while the increase in M_{100} values for the vulcanizates comprising higher filler amounts is more pronounced (17.5%).

The M_{300} is achieved only in the case of the sample comprising virgin PCB at an amount equal to 50 phr carbon black.

The slight changes in the tensile strength values demonstrate that the modification of PCB has a beneficial effect upon this parameter.

The elongation at break decreases with the increasing amount of the PCB having the same composition. However, in the case of vulcanizates comprising amounts of virgin and modified PCB equal to 50 phr active substance, the decrease is less than 10%, while the decrease for vulcanizates comprising amounts of virgin and modified PCB equal to 70 phr active substance is about 13%. Hence, the modification of the filler has no significant effect upon the values of this parameter, despite its slight tendency to decrease.

The residual elongation decreases with the increasing amount of PCB having the same composition, but does not change upon the filler modification.

Shore A hardness values increase slightly with the increasing filler amount for the vulcanizates comprising the same filler. The change is the same following the filler modification.

The results presented show that the values of M_{100}, tensile strength, and Shore A hardness for the vulcanizates comprising modified PCB increase slightly compared to those for vulcanizates comprising virgin PCB. Meanwhile, the values for the elongation at break and the residual elongation are a bit lower. The reason is the lower ash content in the modified PCB and the fact that the ash consists predominantly of silica. This leads to higher values of M_{100} and M_{300}, tensile strength, hardness and lower elongation values of the vulcanizates. That is characteristic of all compounds comprising silica.

The heat aging coefficients of the SBR-based vulcanizates are presented in Table 5.11.

TABLE 5.11   Changes in the Mechanical Properties of SBR-Based Vulcanizates Following Their Heat Aging at 100°C for 72 h.

	$K\sigma$ (%)	$K\varepsilon$ (%)	K_{Sh} (%)
SBR 1	− 31.1	− 54.4	10.1
SBR 2	− 23.9	− 55.5	2.6
SBR 3	− 25.8	− 50.0	7.6
SBR 4	− 21.0	− 47.5	1.3

(Reprinted from N. Delchev, P. Malinova, M. Mihaylov, N. Dishovsky, Effect of the Modified Solid Product from Waste Tyres Pyrolysis on the Properties of Styrene-Butadiene Rubber Based Composites, Journal of Chemical Technology and Metallurgy, 49, 6, 2014, 525-534. Permission from University of Chemical Technology and Metallurgy.)

Heat aging resistance of the vulcanizates comprising virgin and modified PCB (equal to 50 phr active substance) improves with regard to both the tensile strength and the elongation at break. The tendency remains with increasing the amount of PCB.

The heat aging resistance coefficient with regard to Shore A hardness has positive values for all vulcanizates studied. The modification of PCB affects the heat aging resistance of SBR-based vulcanizates. The change in Shore A hardness values is smaller in the case of modified PCB, hence those compounds are more aging resistant.

5.4 CONCLUSIONS

The effect of PCB on curing and mechanical properties of SBR-based composites has been investigated. The possibilities to substitute the conventional carbon black with PCB, as well as to modify the surface of PCB viewing reduction of its ash content have been presented and discussed. It has been established that about 40–50% of conventional carbon black could be substituted when rubber articles of weaker mechanical properties are produced. The results from the studies reveal the obtained pyrolysis product produces an effect much closer to the one of conventional semi-reinforcing carbon black.

PCB obtained via pyrolysis of waste tires has been modified successfully by a special method designed to lower its ash and zinc oxide content, in particular. It has been established that modification lowers two times the ash content and six times the zinc oxide concentration in the ash.

The studies on the curing characteristics, mechanical properties, and heat aging resistance of SBR-based vulcanizates comprising virgin and modified PCB have demonstrated that:

- The curing characteristics of the investigated SBR-based compounds are not affected by the modification of the PCB.
- The values of M_{100}, tensile strength, and Shore A hardness for the vulcanizates comprising modified PCB increase while the elongation at break and residual elongation decrease slightly. The effects observed are related to the lowered ash content in PCB following the modification.
- The modified PCB has a beneficial effect upon heat aging resistance of the vulcanizates.
- Lower the ash content in PCB could be achieved by a further modification with another reagent, such as hydrofluoric acid, which will lead to silica elimination. Another approach is using silane instead

of a secondary modification that is introducing silane into the rubber mixture thus enhancing the interaction between the rubber macromolecules and silica.

Although the mechanical properties of the elastomer composites comprising PCB are worse in comparison to those of the composites filled with conventional carbon black N 550 and N 330, the interest in such products has been intensive. This is due to the fact that pyrolysis is an environmentally friendly process for solving the ecological problem of waste tires.

KEYWORDS

- **waste tires recycling**
- **pyrolysis**
- **pyrolysis carbon black**
- **modification**
- **activation**

REFERENCES

1. Susa, D.; Haydary, J. Sulphur Distribution in the Products of Waste Tire Pyrolysis. *Chem. Pap.* **2013,** *67,* 1521–1526.
2. Quek, A.; Balasubramanian, R. Liquefaction of Waste Tires by Pyrolysis for Oil and Chemicals—A Review. *J. Anal. Appl. Pyrolysis.* **2013,** *101,* 1–16.
3. Williams, P. T. Pyrolysis of Waste Tyres: A Review. *Waste Manage.* **2013,** *33,* 1714–1728.
4. Galvagno, S.; Casu, S.; Casabianca, T.; Calabrese, A.; Cornacchia, G. Pyrolysis Process for the Treatment of Scrap Tyres: Preliminary Experimental Results. *Waste Manage.* **2002,** *22,* 917–923.
5. Mazloom, G.; Farhadi, F.; Khorasheh, F. Kinetic Modeling of Pyrolysis of Scrap Tires. *J. Anal. Appl. Pyrolysis.* **2009,** *84,* 157–164.
6. Koreňová, Z.; Haydary, J.; Annus, J.; Markoš, J.; Jelemenský, L. Pore Structure of pyrolyzed scrap tires. *Chem. Papers.* **2008,** *62,* 86–91.
7. Council directive 1999/31/EC of 26 april 1999 on the landfill of waste. *Official Journal of the European Communities L182/1* 42.
8. Shah, J.; Jan, M. R.; Mabood, F.; Shahid, M. Conversion of Waste Tyres into Carbon Black and Their Utilization as Adsorbent. *J. Chin. Chem. Soc.* **2006,** *53,* 1085–1089.
9. Carr, S. *Conventional Methods of Powder Production. Solid-State Shear Pulverization.* CRC Press: Lancaster, Pennsylvania, 2001; p. 155.

10. Myhre, M.; MacKillop, D. A. Rubber Recycling. *Rubber Chem. Technol.* **2002**, *75*, 429–474.

11. Hunt, L. K.; Kovalak, R. R. Devulcanization of Cured Rubber. U.S. Patent 5,891,926 A, 1999.

12. Isayev, A.; Chen, J. Continuous Ultrasonic Devulcanization of Vulcanized Elastomers. U.S. Patent 5,284,625 A, 1994.

13. Levin, V. Y.; Kim, S. H.; Isayev, A. I.; Massey, J.; Meerwall, E. v. Ultrasound Devulcanization of Sulfur Vulcanized SBR: Crosslink Density and Molecular Mobility. *Rubber Chem. Technol.* **1996**, *69*, 104–114.

14. Novotny, D. S.; Marsh, R. L.; Masters, F. C.; Tally, D. N. Microwave Devulcanization of Rubber. U.S. Patent 4,104,205 A, 1978.

15. Koreňová, Z.; Juma, M.; Annus, J.; Markoš, J.; Jelemenský, L. Kinetics of Pyrolysis and Properties of Carbon Black from a Scrap Tire. *Chem. Pap.* **2006**, *60*, 422–426.

16. Cheremisinoff, N. P.; Rezaiyan, J. Pyrolysis. Gasification Technologies. CRC Press: New York, 2005, pp. 145–164.

17. Juma, M.; Koreňová, Z.; Markoš, J.; Annus, J.; Jelemenský, L. Pyrolysis and Combustion of Scrap Tire. *Pet. Coal.* **2006**, *48*, 15–26.

18. Mastral, A. M.; Álvarez, R.; Callén, M. S.; Clemente, C.; Murillo, R. Characterization of Chars from Coal–Tire Copyrolysis. *Ind. Eng. Che. Res.* **1999**, *38*, 2856–2860.

19. González, J. F.; Encinar, J. M.; Canito, J. L.; Rodríguez, J. J. Pyrolysis of Automobile Tyre Waste. Influence of Operating Variables and Kinetics Study. *J. Anal. Appl. Pyrolysis.* **2001**, *58–59*, 667–683.

20. Faulkner, B. P.; Weinecke, M. Carbon Black Production from Waste Tires. *Miner. Metallurgical Processing.* **2001**, *18*, 215–220.

21. Williams, P. T.; Besler, S.; Taylor, D. T. The Pyrolysis of Scrap Automotive Tyres. *Fuel.* **1990**, *69*, 1474–1482.

22. Kawakami, S.; Inoue, K.; Tanaka, H.; Sakai, T. Pyrolysis Process for Scrap Tires. *Thermal Conversion of Solid Wastes and Biomass.* American Chemical Society, 1980; p. 557–572.

23. Kaminsky, W.; Mennerich, C.; Zhang, Z. Feedstock Recycling of Synthetic and Natural Rubber by Pyrolysis in a Fluidized Bed. *J. Anal. Appl. Pyrolysis.* **2009**, *85*, 334–337.

24. Zhou, J.; Wang, J.; Ren, X.; Yang, Y.; Jiang, B. Surface Modification of Pyrolytic Carbon Black from Waste Tires and its use as Pigment for Offset Printing Ink. *Chin. J. Chem. Eng.* **2006**, *14*, 654–659.

25. Bulgarian patent 65901 B1. Method and Installation for Pyrolysis of Tires.

26. Kolev, D. N.; Ljutzkanova, R. B.; Abadjiev, S. T. European Patent EP1,879,978 A1. Method and installation for pyrolisis of tires.

27. Flory, P. J.; Rehner, J. Statistical Mechanics of Cross-linked Polymer networks I. Rubberlike Elasticity. *J. Chem. Phys.* **1943**, *11*, 512–520.

CHAPTER 6

ELASTOMER-BASED COMPOSITE MATERIALS COMPRISING CARBON NANOSTRUCTURES

CONTENTS

ABSTRACT

Elastomeric nanocomposites based on natural rubber (NR) and fillers of different carbon nanostructures (fullerenes, fullerene soot, graphene, carbon nanotubes (CNTs), and their combinations with other carbon phases) at respective concentrations and/or ratios between them have been developed.

The morphology and microstructure of the materials have been characterized using different methods depending on the fillers used—scanning electron microscopy, transmission electron microscopy, selected area electron diffraction, X-ray spectroscopy, and Raman spectroscopy. The dielectric (permittivity, dielectric loss angle tangent ($tan\ \delta_\varepsilon$)) and microwave (reflection coefficient, attenuation coefficient, and shielding effectiveness (SE)) properties of the obtained elastomer composites have been determined within a wide frequency range— (1–12 GHz).

The influence of various amounts of the above carbon nanostructures (by themselves or in a combination with other carbon phases) on the dielectric and microwave properties of the elastomer composites containing them has been investigated both experimentally and theoretically.

The results achieved reveal how by using the selected combinations of CNTs and graphene nanoplatelets (GNPs) the microwave and dielectric properties of NR nanocomposites can be tailored depending on the requirements set by real practical applications. The developed nanocomposites could have commercial potential for lightweight shielding materials for electromagnetic radiation.

The summarized research results allow us to recommend CNTs and GNPs as second fillers in combination with carbon black for NR-based composites to afford specific absorbing properties.

6.1   INTRODUCTION

The term "nano" is familiar to the specialists who are associated with rubber industry. Since the early 1920s, carbon black and silica have been used as a reinforcing agent in rubber mixtures for various applications. The dimensions of the elementary particles of such fillers are in the range of nanometers. The main advantages of using nano-fillers are, however, not only in the reinforcing effect of the rubber matrix, but also in providing new barrier, electrical/electronic, microwave, and membrane properties.

Recently, carbon nanostructures with latest representatives fullerenes, graphene, and carbon nanotubes (CNTs), have been gaining importance in rubber technology. Due to their unique mechanical and electrical properties, production, and research—both experimental and theoretical, elastomer composites comprising them have been of great interest in terms of expanding opportunities for their application. Moreover, the prices of the nanostructures are constantly decreasing. Another aspect of their application, also offering great prospects, is the possibility to combine them with conventional or conductive carbon black to achieve an acceptable compromise between cost and desired performance. We had a particular interest in this type of nanostructures, especially in terms of using nanocomposites containing them for microwave applications. A key point in the research was to establish the condition in which carbon structures are after samples vulcanization, that is, whether they retain their nano-size (< 100 nm) or undergo aggregation. Establishing this fact was particularly important because a positive answer to the above question gives us the opportunity to speak about nanocomposites.

The thesis of this study is that nano-sized carbon structures (fullerenes, fullerene soot, graphenes, CNTs and their combinations with other carbon phases) are of considerable interest as a promising opportunity to obtain elastomer composites of good electric and microwave properties. Both experimental and theoretical investigations reveal the possibilities for expanding the application of elastomeric materials.

In connection with the aforementioned investigations, the aim of the study is to examine the effect of different amounts of the aforementioned carbon structures, by themselves or in combinations with other carbon phases, upon the dielectric and microwave properties of the composites comprising them.

6.2   EXPERIMENTAL

6.2.1   CHARACTERIZATION OF THE FULLERENE USED

The neat fullerene powder used in current study was manufactured by Alfa Aesar, Thermo Fisher Scientific, Heysham, Lancashire LA3 2XY, United Kingdom, comprising 99.5% of C_{60} fullerenes. Its density was 1.65 g/cm^3. As we have seen from the transmission electron microscopy (TEM) micrograph in Figure 6.1 and its marker 50 nm, fullerene particles are aggregates of several hundred nanometers, for example, about 300 nm wide and about 700 nm long, built of substrate particles about 40–50 nm large.

The electron diffraction image reveals both spot reflexes typical of the monocrystal structures and ring reflexes typical of polycrystal formations.

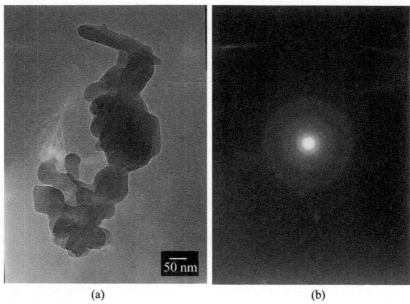

(a)   (b)

FIGURE 6.1 Micrographs of fullerene aggregates in a (a) transmission and in an (b) electron diffraction regime.

Figure 6.2 shows X-ray diffractogram of the fullerenes used, while Figure 6.3 shows their Raman spectrum.

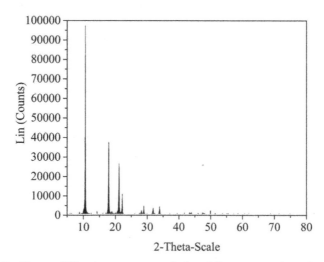

FIGURE 6.2 X-ray diffraction patterns of the fullerene powder. (Reprinted from Al-Hartomy, O.A.; Al-Ghamdi, A.A.; Al-Salamy, F.; Dishovsky, N.; Slavcheva, D.; Iliev, V.; El-Tantawy, F. Dielectric and Microwave Properties of Fullerenes Containing Natural Rubber-based Nanocomposites, *Fullerenes, Nanotubes and Carbon Nanostructures.* **2014,** *22,* 332–345. © 2014 with permission from Taylor & Francis.)

The main phase is C_{60} of cubic crystal structure: $a = 14.19470$ Å, with a Fm-3m spatial group, but some impurities have also been found: C_{70} (00-048-1206) with a hexagonal crystal structure and parameters: $a = 10.59340$ Å, $c = 17.26200$ Å, with a $P6_3/mmc$ spatial group and C_{70} (01-073-5048) c orthorhombic crystal structure and parameters $a = 17.30300$ Å, $b = 9.99000$ Å, $c = 17.92400$ Å, and with a Ccmm spatial group.

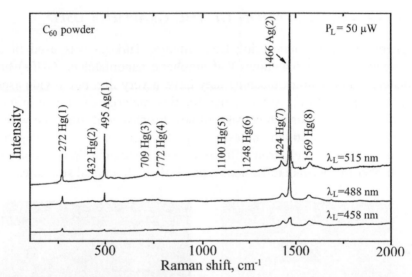

FIGURE 6.3   Raman spectrum of the fullerenes used. (Reprinted from Al-Hartomy, O.A.; Al-Ghamdi, A.A.; Al-Salamy, F.; Dishovsky, N.; Slavcheva, D.; Iliev, V.; El-Tantawy, F. Dielectric and Microwave Properties of Fullerenes Containing Natural Rubber-based Nanocomposites, *Fullerenes, Nanotubes and Carbon Nanostructures*. **2014**, *22*, 332–345. © 2014 with permission from Taylor & Francis.)

A large background only has been observed at 633 nm in the Raman spectrum recorded with He-Ne excitation probably due to luminescence. However, when the excitation is in the blue-green range, the Raman spectrum has a low background and a considerable number of lines (Figure 6.3). In accordance with literature[1], all the observed lines could be assigned to the active Raman bands of fullerenes C_{60}.

6.2.2 CHARACTERIZATION OF THE FULLERENE SOOT USED

The used fullerene soot were manufactured by Alfa Aesar and are characterized by a content of neat fullerenes C_{60} - 6%, fullerenes C_{70} - 1.5% (total amount of fullerenes - 7.5%), carbon black - 92.5%, and specific surface area of the particles 300 m^2/g.

6.2.3 CHARACTERIZATION OF THE GRAPHENE USED

Graphene (Hayzen Engineering Co., Ankara, Turkey) was used in our investigation. Figure 6.4 shows that graphene nanoplatelets (GNPs) have a "platelet" morphology, meaning they have a very thin but a wide aspect ("sheet"-like structure). Aspect ratios for this material can range in terms of thousands. Each particle consists of several sheets of graphene with an overall thickness of 50 nm and average plate diameter 40 μm. The pattern taken in selected area electron diffraction (SAED) regime shows a considerable number of spot-like reflections typical of single-crystal structures.

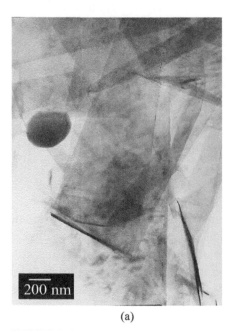

(a)   (b)

FIGURE 6.4 (a) TEM Micrographs of GNPs in transmission regime and (b) SAED regime. (Adapted from Al-Hartomy, O. A.; Al-Ghamdi, A; Al-Salamy, F; et al, Effect of carbon nanotubes and graphene nanoplatelets on the dielectric and microwave properties of natural rubber composites, Advanced Composite Materials , Volume 22, 2013 - Issue 5. © 2013 with permission from Taylor & Francis.)

6.2.4 CHARACTERIZATION OF THE CARBON NANOTUBES USED

Multi-walled carbon nanotubes (MWCNTs), as manufactured by Hayzen Engineering Co., Ankara, Turkey, were used in our investigations. The purity of the material was higher than 95%, density—150 kg/m³. CNTs were of average diameter measuring about 15 nm and length 1–10 μm (Figure 6.5(a)). The micrograph taken in an electron diffraction regime (Figure 6.5b) shows them to have a polycrystalline structure typical of CNT.

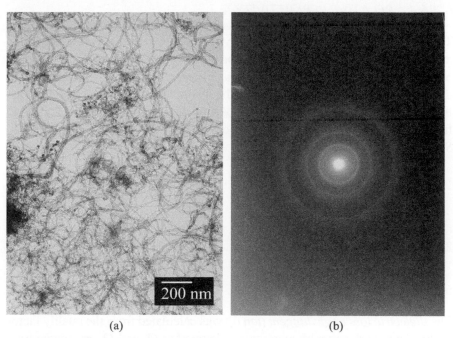

(a)   (b)

FIGURE 6.5   TEM micrographs of CNTs in a (a) transmission regime and a (b) SAED regime. (Adapted from Al-Hartomy, O. A.; Al-Ghamdi, A; Al-Salamy, F; et al, Effect of carbon nanotubes and graphene nanoplatelets on the dielectric and microwave properties of natural rubber composites, Advanced Composite Materials , Volume 22, 2013 - Issue 5. © 2013 with permission from Taylor & Francis.)

6.2.5 COMPOSITES PREPARATION AND CHARACTERIZATION

The rubber compounds were prepared on an open two-roll laboratory mill (L/D 320 × 160 and friction 1.27) by incorporating precharacterized carbon nanostructure into a natural rubber (NR) matrix at various loadings. The speed of the slow roll was 25 min^{-1}. The experiments were repeated to verify the statistical significance. The fabricated compounds in the form of sheets stayed 24 h prior to their vulcanization.

The optimal curing time was determined by the cure curves, which were taken on a moving die rheometer MDR 2000 (Alpha Technologies) at 150°C according to ISO 3417:2002.

6.2.6 MEASUREMENTS

6.2.6.1   RELATIVE PERMITTIVITY (ε_R)

The determination of relative permittivity was carried out by the resonance method, based on the cavity perturbation technique. The resonance frequency of an empty cavity resonator, f_r also was measured. After this measurement, the sample material was placed inside the resonator and the shift in resonance frequency f_ε was measured. The real part of permittivity E'_r was calculated from the shift in resonance frequency, cavity, and the sample cross-sections, S_r and S_ε, respectively.

$$\varepsilon'_r = 1 + \frac{S_r}{2S_\varepsilon} \cdot \frac{f_r - f_\varepsilon}{f_r}$$

(6.1)

The sample had the form of a disk with diameter of 10 mm and around 2 mm thickness.

6.2.6.2   DIELECTRIC LOSS ANGLE TANGENT (TAN δ_ε)

The *dielectric loss angle tangent (tan δ_ε)* was calculated from the quality factor of the cavity with and without a sample denoted as Q_S and Q_C, respectively:

$$\tan \delta_\text{a} = \frac{1}{4\varepsilon'_r} \frac{S_r}{S_\varepsilon} \left(\frac{1}{Q_S} - \frac{1}{Q_C} \right)$$

(6.2)

6.3.6.3   STUDIES ON THE MICROWAVE PROPERTIES IN THE 1–12 GHZ RANGE

The attenuation of electromagnetic waves by the sample was determined by the following equation:

$$L = 10\log\frac{P_a}{P_p}, dB \tag{6.3}$$

where P_a is the adopted power at the end of the coaxial line without a sample; P_p is the adopted power at the end of the coaxial line with a sample.

A portion of the incident power P_I of an electric microwave is reflected back. The power of the reflected microwave P_R depends on the reflection coefficient of the sample $|\Gamma|$. The rest of the microwave penetrates the material with absorbed power P_A. Upon attenuation L, the microwave leaves the sample with transmitted power P_T. The quantitative dependence of the described parameters is given by the following equation:

$$P_T = P_I\left(1 - |\Gamma|^2\right) \tag{6.4}$$

The power of the transmitted wave P_T is related to the absorbed power P_A via the attenuation L:

$$L = 10\log\frac{P_I}{P_T}, dB \tag{6.5}$$

The attenuation coefficient α is calculated from:

$$\alpha = L / d_{av}, \tag{6.6}$$

where $d_{av.}$ is the average thickness of the sample.

Reflection losses R_L, dB, are determined by the following formula:

$$R_L = -20\log|\Gamma|, \tag{6.7}$$

where $|\Gamma|$ is the reflection coefficient.

Electromagnetic SE is calculated by the formula:

$$SE = R_L + L, dB \tag{6.8}$$

6.2.7 SEM MICROGRAPHS

The micrographs were measured using a JEOL JSM-5510 scanning electron microscope (JEOL, Tokyo, Japan). The samples were frozen in liquid nitrogen prior to the observations. The observations were carried out at 10 kV.

6.2.8   TEM MICROGRAPHS

The particle size, filler dispersion, and the microstructure of the composite materials were determined on a TEM JEOL 2100 microscope at an acceleration voltage of 200 kV. The samples were ground in an agate mortar and were dispersed in ethanol by a successive ultrasound treatment for 6 min.

6.2.9   X-RAY SPECTROSCOPY

The X-ray diffraction patterns were taken on a Bruker D8 Advance Cu Kα radiation and Lynx Eye detector within the 5.3°–80° range 2θ with a constant step of 0.02° 2θ. Phase identification was performed with Diffrac *plus* EVA software, using ICDD-PDF2 database.

6.2.10   RAMAN SPECTROSCOPY

Raman spectra were recorded on a Lab RAM HR Visible (Jobin Yvon) spectrometer equipped with a microspore, filter and a Peltier-cooled charge-coupled device detector. *Laser* ((He–Ne) 458, 488, and 525 nm (Ar⁺)) lines with wavelength of *633 nm* were used for excitation. An objective × 100 magnification was used to *focus laser beam* to the spot of 1 μm in diameter. The power of the laser on the spot was kept very low (about 50 μW) to prevent eventual local heating of the sample.

6.3   RESULTS AND DISCUSSION

6.3.1   ELASTOMER-BASED COMPOSITE MATERIALS COMPRISING FULLERENES

In 1985, for their discovery of a new allotrope of carbon in which the atoms are arranged in closed shells, the Nobel Prize for Chemistry in 1996 had been awarded to Harold W. Kroto, Robert F. Curl, and Richard E. Smalley. The new form was found to have the structure of truncated icosahedrons and was named Buckminster fullerene, after the architect Buckminster Fuller who designed geodesic domes in the 1960s. Unlike the other two allotrope forms of carbon–graphite and diamond, which are volumetric materials that have the properties of three-dimensional crystals, the fullerenes are hollow molecules composed of carbon in sp^2-hybrid state of the valence electrons.

Fullerenes have a number of properties that make them distinct from the other forms of carbon: superconductivity,[2,3] low thermal conductivity,[4] greater hardness than the diamond in crystal state.[5] The most significant representative of the fullerenes is C_{60}. It consists of molecular balls made of 60 or more carbon atom clusters linked together.[2,3] Each of the carbon atoms has two single bonds and one double bond that attach to other carbon atoms. This causes C_{60} to act more like a "superalkene" than a "superaromatic".[4] The structure, the properties, and the possible applications of the fullerenes have been described in details in Ref. 6 Their possible applications as ingredients in rubber compounds and composites on the basis of these compounds have been described with expected positive influence on some characteristics—enhanced durability, lower heat-built-up, better fuel economy, improved wear and tear resistance.[6] The data of the influence of the fullerenes on the microwave properties of the elastomeric composites are very scarce. In Ref. 7, the influence of fullerenes in concentrations between 0.065 and 0.75 phr on the properties of composites on the basis of NR, as well as the influence on some microwave properties, has been investigated. An American patent[8] describes the usage of fullerenes in tire treads. Some other patents describe fullerenes C_{60} implemented in technical rubber goods.[9,10]

The formulations of the rubber compounds comprising fullerenes used in our study are presented in Table 6.1.

TABLE 6.1   Formulations of the Rubber Compounds Comprising Fullerenes (phr).

	F-1	F-2	F-3	F-4
NR	100	100	100	100
Zinc oxide	5	5	5	5
Stearic acid	2	2	2	2
Fullerenes	—	0.5	1.0	1.5
TBBS	0.8	0.8	0.8	0.8
Sulfur	2.25	2.25	2.25	2.25

NR: natural rubber; TBBS: N-tert-butyl-2-benzothiazolesulfenamide. (Reprinted from Al-Hartomy, O.A.; Al-Ghamdi, A.A.; Al-Salamy, F.; Dishovsky, N.; Slavcheva, D.; Iliev, V.; El-Tantawy, F. Dielectric and Microwave Properties of Fullerenes Containing Natural Rubber-based Nanocomposites, *Fullerenes, Nanotubes and Carbon Nanostructures*. **2014,** 22, 332–345. © 2014 with permission from Taylor & Francis.)

The frequency dependence of the real part of the relative permittivity ε_r' of the studied composites is presented in Figure 6.6.

As seen Figure 6.6, in the studied concentration interval the values of ε_r' are quite close, with a slight tendency to increasing at higher filler concentrations. A similar tendency has been reported by the author of Ref. 11 who has established an increase in ε_r' values from 4.8 to 5.2, with the increasing concentrations but the study refers to a frequency of 50 Hz. The values in our investigations are close to those described in Ref. 12 (2.40–2.70) obtained at a frequency of 1000 Hz.

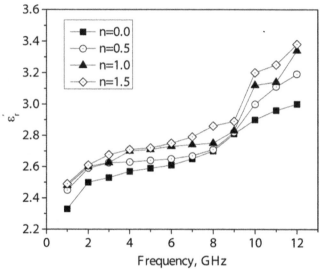

FIGURE 6.6 Frequency dependence of the real part of relative permittivity of the composites comprising fullerenes (n = phr of fullerene). (Reprinted from Al-Hartomy, O.A.; Al-Ghamdi, A.A.; Al-Salamy, F.; Dishovsky, N.; Slavcheva, D.; Iliev, V.; El-Tantawy, F. Dielectric and Microwave Properties of Fullerenes Containing Natural Rubber-based Nanocomposites, *Fullerenes, Nanotubes and Carbon Nanostructures.* **2014**, *22*, 332–345. © 2014 with permission from Taylor & Francis.)

As the figure shows, our studies also confirm that ε_r' increases by increasing the frequency. Three areas depending on the ε_r' of the frequency have been observed in the 1–3 GHz range; the permittivity increases slightly; relatively constant values are observed in the frequency range from 3 to 9 GHz and within the range 9–12 GHz, the increase in the permeability with frequency is relatively faster. In the latter case, the effect of various fullerenes concentrations is the most pronounced. Since ε_r' is related to the polarizability of the composite,[13] probably polarization in the NR matrix at frequencies above 9 GHz is harder to proceed that leads to a certain increase in its value.

The frequency dependence of the tan δ_ε of the composites at various loadings with fullerenes is shown in Figure 6.7.

The dielectric loss angle tangent (tan δ_ε) changes as a function of frequency in the 1–12 GHz range is much clearer (Figure 6.7) than those in the ε_r' discussed above (Figure 6.6). The values of tan δ_ε at various filler concentrations are well distinguishable at the lower frequencies. As expected, the dielectric losses increase at a higher frequency. The filler does not affect the course of the dependence, since the same one has been observed for the unfilled composite. Notably, the addition of fullerenes even at 0.5 phr significantly increases the dielectric losses. The higher the fullerene amount, the higher the dielectric losses. It is also seen that the dependence for the unfilled composite is almost linear, while when even 0.5 phr of fullerenes are added, the dependence of those composites loses its linearity. In correspondence to Ref. 7, the values for tan δ_ε for the unfilled NR-based composites are higher than those for composites filled with fullerenes at 0.065–0.75 phr and are within the 0.001–0.005 range, but are obtained at a relatively low frequency 1–10 MHz.

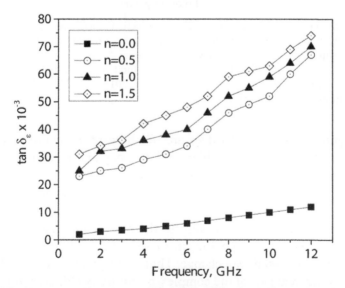

FIGURE 6.7 Frequency dependence of the tan δ_ε of the composites comprising fullerenes (n = phr of fullerenes). (Reprinted from Al-Hartomy, O.A.; Al-Ghamdi, A.A.; Al-Salamy, F.; Dishovsky, N.; Slavcheva, D.; Iliev, V.; El-Tantawy, F. Dielectric and Microwave Properties of Fullerenes Containing Natural Rubber-based Nanocomposites, *Fullerenes, Nanotubes and Carbon Nanostructures.* **2014,** *22,* 332–345. © 2014 with permission from Taylor & Francis.)

The frequency dependence of the coefficient of electromagnetic waves reflection of the composite at various loadings with fullerenes is presented in Figure 6.8.

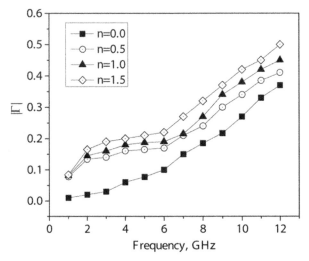

FIGURE 6.8 Frequency dependence of the reflection coefficient of the composites comprising fullerenes (n = phr of fullerene). (Reprinted from Al-Hartomy, O.A.; Al-Ghamdi, A.A.; Al-Salamy, F.; Dishovsky, N.; Slavcheva, D.; Iliev, V.; El-Tantawy, F. Dielectric and Microwave Properties of Fullerenes Containing Natural Rubber-based Nanocomposites, *Fullerenes, Nanotubes and Carbon Nanostructures.* **2014**, *22*, 332–345. © 2014 with permission from Taylor & Francis.)

As Figure 6.8 shows, $|\Gamma|$ values are not high in the entire range—they do not overestimate 0.50 at the highest frequency and loading. One should consider the values of tan δ_ε discussed above (Figure 6.7), which increase smoothly, but within a relatively narrow range. This explains to a great extent the small changes in the coefficient of reflection values. As a whole, the increase in frequency, especially in the 6–12 GHz range, and loading leads to an increase in the coefficient of reflection, while in the 1–6 GHz range $|\Gamma|$ values almost do not change. That corresponds to the similar effect observed for the real part of the complex permittivity (Figure 6.6.)

The frequency dependence of the attenuation coefficient of the composite at various loadings with fullerenes is presented in Figure 6.9.

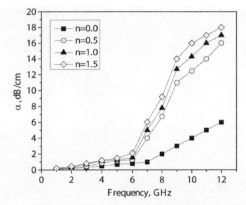

FIGURE 6.9 Frequency dependence of the attenuation of the composites comprising fullerenes (n = phr of fullerene). (Reprinted from Al-Hartomy, O.A.; Al-Ghamdi, A.A.; Al-Salamy, F.; Dishovsky, N.; Slavcheva, D.; Iliev, V.; El-Tantawy, F. Dielectric and Microwave Properties of Fullerenes Containing Natural Rubber-based Nanocomposites, *Fullerenes, Nanotubes and Carbon Nanostructures.* **2014,** *22,* 332–345. © 2014 with permission from Taylor & Francis.)

The changes in the attenuation coefficient (α) with frequency could be examined in three intervals: (1–6) GHz; (6–9) GHz, and (9–12) GHz (Figure 6.9). In the first interval, $\alpha \leq 2$; in the second, $2 \leq \alpha \leq 14$ and in the third, $\alpha > 14$, reaching 18 dB/cm at $n = 1.5$. The comparison of that dependence with the reflection coefficient and permittivity shows the similarity which is obviously due to the impact of the same factor, namely fullerene structure and its specifics.

The results from the studied dielectric and microwave properties of the composites comprising fullerenes are described in detail in Ref. 14.

It is known that the dispersion of the filler has a significant effect on all the properties of the vulcanizates obtained,[15] including the microwave properties. Good dispersion of the filler and achieving a high homogeneity of the composite is an important condition for attaining a good interaction of electromagnetic waves with it and its absorption in the entire volume, which is a fundamental requirement for creating microwave absorbers of high quality.[16] On the other hand, the increase in the degree of filler dispersion increases its surface area and surface energy. It has been found that the compounding begins with the formation of the so-called soft agglomerates, which are further shredded and dispersed in the rubber matrix.[17]

Figures 6.10 and 6.11 present the obtained scanning electron microscopy (SEM) and TEM images, respectively, of the composites filled with fullerenes.

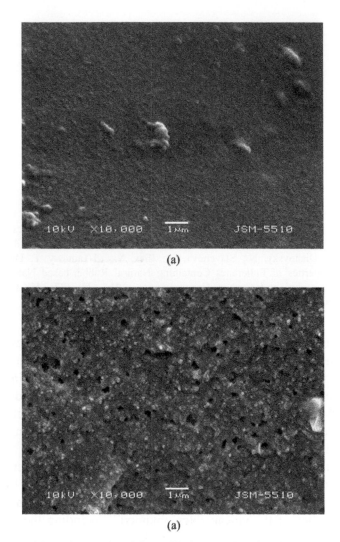

FIGURE 6.10   SEM micrographs of the unfilled composite (a) and a composite filled with fullerene at 1.5/h (b).

The difference between unfilled composites and those filled with fullerenes is obvious. Fullerene particles (black spots in Figure 6.10b) are well distinguishable from the particles of the other ingredients present in the composite. The resolution of SEM, however, is not sufficient to see the structure of the filler particles. Evidently, however, their dimensions are in the range of 0.1–0.2 μm. The particle size of the fullerenes is better visible in TEM images—Figure 6.11.

The micrographs presented in Figures 6.10 and 6.11 show that fullerene particles are relatively evenly dispersed in the elastomer matrix and it is characterized by a satisfactory homogeneity, ensuring good interaction with the electromagnetic waves. The particles do not undergo further agglomeration in the compounding process, what results into sustaining their dimensions within the 40–50 nm range, that is, we definitely can speak of polymer nanocomposites containing fullerenes.

The electronic diffraction in selected area (Figure 6.11(b)) confirms that these particles are really fullerenes, and no other ingredients.

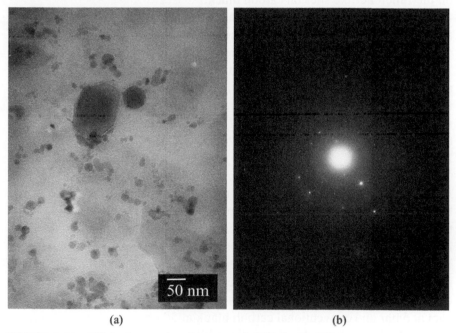

(a)   (b)

FIGURE 6.11   TEM micrographs of a composite filled with fullerenes at 1.5/h (a) and a SAED micrograph of the same sample (b).

The results of those studies are described in details in Ref.18.

6.3.2 ELASTOMER-BASED COMPOSITE MATERIALS COMPRISING FULLERENE SOOT

Fullerene soot as well as fullerene-containing products has been also studied, though in a few publications. The studies cover the structure, chemical properties, and applications of fullerene soot, considering that it is a carbon nanomaterial relatively cheaper than the neat fullerene. A typical characteristic feature of fullerene soot is its surface (about 300 m^2/g), which is more developed than that of conventional carbon.

The study[19] shows that the fullerene is a type of structure present on the surface of fullerene soot act as very active interaction sites between carbon black and rubber. The fullerene-like structures in this type of soot, in particular, pentagonal, or corannulene-type sites, allow us to explain the chemical interaction between filler and polymer (diene rubber) occurring during mixing in terms of free radical and Diels-Alder addition of polymer chains on the fullerene-like sites on carbon black.[20]

The significant increase in the amount of the bound rubber and in the amount of rubber grafted onto the soot compared to that in the case of conventional carbon black has also been established as a result.[21]

Fullerene soot can be subjected to modification, to hydroxylation and carboxylation, inclusive.[6] The investigations include hydroxylated fullerene soot at amounts up to 80 phr, while the results obtained show that some parameters, for example, M_H, M_L, and ΔM are twice higher compared to those of other fillers, while the vulcanization time T_{90} is three times longer.

It has been proven that the values of the vulcanizates prepared with fullerene soot at 80 phr are similar to those prepared with silica. It has been a considerable increase of the mechanical loss angle tangent at low temperatures and a considerable increase at high temperatures,[22] which is very suitable to be used in compositions for tire treads. The European Patent[23] describes the use of fullerenes, fullerene soot, and extracts thereof in combination with conventional carbon black in a SBR 1500-based compounds. The best reported results are for the combination of fullerene soot at 5 phr and conventional carbon black at 50 phr. A patent of *Goodyear Tire & Rubber company*[8] suggests the addition of 5 wt.% fullerene soot to 95 wt.% standard carbon black or silica in the production of tire treads. The authors state that the addition reduces the hysteresis loss, heat buildup, and rolling resistance.

The formulations of the rubber compounds used in our study are shown in Table 6.2.

TABLE 6.2   Formulations of the Studied Rubber Compounds Comprising Fullerene Soot (phr).

	FS-1	FS-2	FS-3	FS-4	FS-5	FS-6
NR	100	100	100	100	100	100
Zinc oxide	5	5	5	5	5	5
Stearic acid	2	2	2	2	2	2
Fullerene soot	—	1	2	3	5	7.5
TBBS	0.8	0.8	0.8	0.8	0.8	0.8
Sulfur	2.25	2.25	2.25	2.25	2.25	2.25

Figures 6.12 and 6.13 show the dependence of real part of permittivity and the tan δ_ε on frequency and on the concentration of the filler introduced. As seen from these figures, the values of the real part of relative permittivity and the tan δ_ε increase gradually with increasing filler concentration, as well as with increasing frequency at the same filler concentration. At a constant frequency, the increase in filler quantity leads to an almost linear increase in ε_r'. The same cannot be stated about the course of the curves representing the dependence of the tan δ_ε on the amount of introduced filler at a constant frequency.

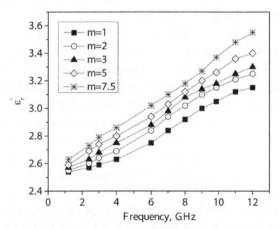

FIGURE 6.12   Frequency dependence of the real part of relative permittivity of the composites comprising fullerene soot (m = phr of fullerene soot).

An interesting fact is that the addition of fullerene soot even at 1 phr alters significantly the values of ε_r' and tan δ_ε (in particular), if compared to those of the unfilled composite.

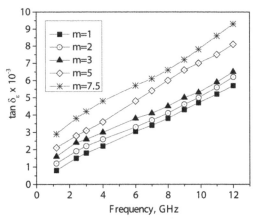

FIGURE 6.13   Frequency dependence of the tan δ_ε of the composites comprising fullerene soot (m = phr of fullerene soot).

Figures 6.14 and 6.15 summarize the microwave properties of the composites studied within the 1–12 GHz frequency range.

As seen from the results presented in Figure 6.14, the values of reflection coefficient increase with the increasing frequency and concentration of the filler introduced. The resulting curves have the same course. At a frequency of 1–8 GHz, the change in the values of the attenuation coefficient is minimal, while in the 8–12 GHz range a sharp increase in the values has been observed. The change observed within frequencies of 1–8 GHz is from 0.1 to 0.2, while at frequencies from 8 to 12 GHz, the average change in the values is from 0.2 to 0.7. Figure 6.15 shows the dependence of the coefficient of electromagnetic waves attenuation on the frequency and concentration of the filler introduced, the resulting curves are similar to those in Figure 6.14. Generally, the increase in frequency and concentration of the filler introduced leads to an increase in the coefficient of attenuation. The figure can be classified into two frequency ranges: from 1 to 8 GHz and from 8 to 12 GHz, within which the coefficient of electromagnetic waves attenuation increases drastically reaching values of 37–38 dB/cm. By increasing the frequency and quantity of filler introduced, one observes closer values of the coefficient of electromagnetic waves attenuation which is better pronounced than in the case of reflection coefficient of electromagnetic waves. Figures 6.14 and 6.15 show that the addition of a minimal filler amount (for example, at 1 phr) affects the values of the coefficient of electromagnetic waves attenuation (Figure 6.15) more than those of the coefficient of electromagnetic waves reflection (Figure 6.14).

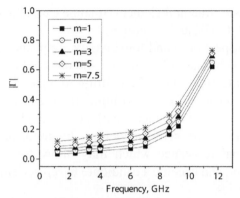

FIGURE 6.14   Frequency dependence of the coefficient of electromagnetic waves reflection of the composites comprising fullerene soot (m = phr of fullerene soot).

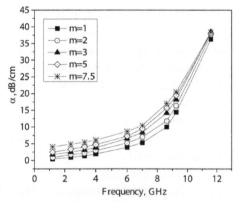

FIGURE 6.15   Frequency dependence of the attenuation of the composites comprising fullerene soot (m = phr of fullerene soot).

6.3.3   ELASTOMER-BASED COMPOSITE MATERIALS COMPRISING GRAPHENE

The discovery of graphene[24] and the subsequent development of graphene-based polymer nanocomposites is an important contribution to nanoscience and technology. Deservedly, the Nobel Prize in Physics for 2010 was awarded to Andre Geim and Konstantin Novoselov from the University of Manchester for their work on a single free-standing atomic layer of carbon (graphene). Graphene is an allotrope of carbon, whose structure is one-atom-thick planar sheets of sp^2-bonded carbon atoms packed in a honeycomb lattice. It is the basic structural element of some carbon allotropes including graphite, charcoal, CNTs, and fullerenes.

With its high aspect ratio and low density, graphene has attracted more attention than CNTs because of its unique and outstanding mechanical, electrical, and electronic properties[25]. In addition to good thermal conductivity, remarkable mechanical stiffness, and high fracture strength, graphene has been supposed to be a semiconductor with zero gap which is quite different from conventional silicon semiconductors. In graphene, electrons shoot along with minimal resistance which may allow for low-power, faster-switching transistor, and become a candidate to replace silicon in the area of microchip electronics.

All these unique properties in a single nanomaterial have made physicists, chemists, and material scientists excited about graphene potential. The history, chemistry, preparation methods, and possible applications of graphene are reviewed in Refs. 26–28. Another review focused on the trends and frontiers in graphene-based polymer nanocomposites that was published 5 years ago.[29] Not many years ago, the trustees of Princeton University received a patent for graphene–elastomer nanocomposites where functionalized graphene sheets had been dispersed in vulcanized NR, styrene butadiene rubber, Ps-isoprene-Ps, and PDMS (polydimethylsiloxane).[30] The patented work could find a wide range of industrial applications, including food packaging, gasketing, and automotive. The authors conclude that graphene–rubber nanocomposites possess qualities similar to those of CNT composites but are much cheaper to make. The data about graphene influence on the microwave properties of the elastomeric composites are very scarce. Chen et al. used functionalized graphene–epoxy composites as lightweight shielding materials for electromagnetic radiation.[31] De Rosa et al. have a wide expertise in the design of micro/nanocomposites based on carbon fibers and CNTs, for the realization of high-performing radar absorbing screens, with tailored properties.[32–34] In another paper,[35] the authors have accomplished a Salisbury screen that consists of three layers. The second layer (the spacer) is a low-loss-tangent nanocomposite based on a Bisphenol-A-based epoxy resin filled with GNPs at 0.5 and 1% wt.[35] The real and imaginary parts of the complex effective permittivity within the 8–18 GHz range of the nanocomposite filled with GNPs have been shown. It has been observed that the real part of the effective permittivity is nearly constant.

There are no literature data about the dielectric and microwave properties of nanocomposites comprising a higher amount of graphene nanoplatelets (GNPs), for example, over 1 phr, in a wider frequency range, first of all at frequencies lower than 8 GHz. Therefore, the aim of this work is to study the influence that graphene nanoparticles (in amounts of 2–10 phr) have on the dielectric (relative permittivity, tan δ_ε) as well as on the microwave properties (absorption and reflection of the electromagnetic waves, the effectiveness of the electromagnetic shielding) of NR-based composites in a significantly larger frequency range—1–12 GHz. Table 6.3 summarizes the formulation characteristics of the rubber compounds (in phr) used for the investigations.

TABLE 6.3   Compositions of the Rubber Compounds Comprising Graphene (phr).

	NR-1	NR-2	NR-3	NR-4
NR	100	100	100	100
Foaming agent	8	8	8	8
Stearic acid	1	1	1	1
Zinc oxide	4	4	4	4
Processing oil	10	10	10	10
GNP	0	2	6	10
MBTS	2	2	2	2
TMTD	1	1	1	1
IPPD 4020	1	1	1	1
Sulfur	2	2	2	2

GNP: graphene nanoplatelet; TMTD: tetramethyl thiuram sulfide; IPPD: isopropylphenidate.

Figures 6.16 and 6.17 present the dependence of the real part of relative permittivity (ε_r') and tan δ_ε on frequency and filler concentration for the studied composites filled with graphene nanoparticles.

As seen from the plots, ε_r' increases with an increasing frequency at constant filler concentration as well as with the increasing filler concentration at constant frequency.

The real part of permittivity is the ratio of the capacity of an electric capacitor filled with the substance investigated to that of the same capacitor in vacuum, at a definite external field of frequency[36]. The dependence of the real part of the relative permittivity on the frequency is shown in Figure 6.16. As seen, in the concentration interval studied, ε_r' values for the nonfilled and filled composites are relatively closer at lower frequencies (up to 7 GHz). At frequencies higher than 7 GHz, the dielectric constant increases with the increasing filler concentration. Moreover, there is a tendency of a more pronounced difference in the values for the nonfilled and filled composites.

The values we obtained for the non-filled composites are close to the ones for NR at 1000 Hz reported in literature, which are in the range 2.40–2.70.[13] As the figure shows, our investigations also confirm the fact that ε_r' increases at higher frequencies. Of particular interest is the 9–12 GHz range wherein there is a relatively fast increase in the ε_r' and its dependence on the GNP is the most prominent one. Due to the fact that the dielectric constant is related to the composites polarity,[36] probably at frequencies higher than 9 GHz, the polarization of the NR matrix is hampered. Hence, ε_r' increases. On the other hand, an increase in ε_r' proves the lower polarization of the composite at higher frequencies.

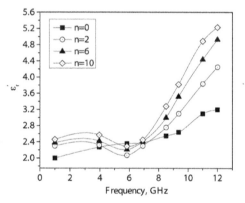

FIGURE 6.16 Frequency dependence of the real part of relative permittivity of the composites comprising graphene (n-phr of graphene).

The figure allows to calculate that the maximum difference within the range of 2 units is between the unfilled composite and the one comprising 10 phr GNP at 12 GHz. Besides these, at frequencies higher than 7 GHz the frequency dependence of ε_r' for the filled composites is much closer to the linear one.

Figure 6.17 shows the frequency dependence of the tan δ_ε in the 4–12 GHz range. As expected, the increase in frequency and filler amount leads to an increase in tan δ_ε values. The filler does not change the character of this dependence since it exists for the nonfilled composite as well. Noteworthy is the fact that the addition of 2.0 phr GNP has a relatively slight effect upon tan δ_ε values of the composite, which are close to the ones of the matrix. However, tan δ_ε increases considerably at 6.0 phr filler concentration. It is also seen that the frequency dependence of tan δ_ε both for the unfilled composite and the filled ones has an almost a linear character. In the case of composites comprising a higher filler amount, the frequency dependence of this parameter is more pronounced.

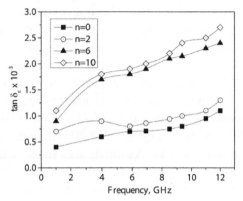

FIGURE 6.17   Frequency dependence of the tan δ_g of the composites comprising graphene (n = phr of graphene).

Figures 6.18–6.20 summarize the microwave properties of the composites studied within the 1–12 GHz frequency range.

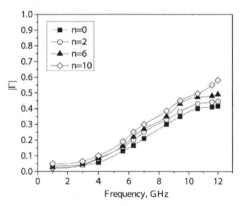

FIGURE 6.18   Frequency dependence of the reflection coefficient of the composites comprising graphene (n = phr of graphene).

Figure 6.18 shows the frequency dependence of $|\Gamma|$. As we have already discussed, the higher the frequency and filler amount, the higher $|\Gamma|$ is. The figure also reveals that $|\Gamma|$ values do not exceed 0.60 within the entire frequency range. A connection should be made between the dielectric losses mentioned above (Figure 6.17) which smoothly increase, though in a relatively small range. That to a great extent explains the small changes in the reflection coefficient. One should be aware of the fact that the composite comprises a foaming agent making its structure porous what lowers its reflection coefficient values. As a whole, the reflection coefficient slightly increases with the increasing frequency and filler amount and particularly in the 6–12 GHz range. In the 1–6 GHz range, the effect of frequency and filler amount on the reflection coefficient is less pronounced. The effect is in accordance with the one upon the real part of permittivity (Figure 6.16).

The frequency dependence of the attenuation coefficient (α) could be studied in three frequency ranges: (i) 1–6 GHz, (ii) 6–9 GHz, and (iii) 9–12 GHz (Figure 6.19). In the first range $\alpha \leq 2.2$ dB/cm; in the second $2.2 \leq \alpha \leq 6$ dB/cm and in the third $6 \leq \alpha \leq 17$. As the figures show, in the first frequency range the attenuation remains almost unchanged, especially for the nonfilled composites and for those with a small amount of filler. There is a statistically significant attenuation increase with the increasing frequency and filler amount. This increase is considerable in the third frequency range especially for the filled composites. The effect is less pronounced for the unfilled ones. The comparison of the aforesaid dependence with those of the reflection coefficient and permittivity reveals a similarity that is obviously due to the same factor (the structure of graphene and its specifics). It should be noted that the maximum attenuation of 17 dB/cm at 12 GHz is for a composite comprising 10 phr GNP.

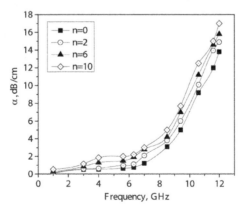

FIGURE 6.19 Frequency dependence of the attenuation of the composites comprising graphene (n = phr of graphene).

Figure 6.20 plots the frequency dependence of the electromagnetic SE. At lower frequencies, EMI SE values are higher but decrease with the increasing frequency. In the 1–7 GHz range, the dependence has a character close to the linear one. There is also a range (7–12 GHz) of a relatively small effect of frequency and filler amount on EMI SE values, which remain in the 7 and 10 dB range. The initial values are due mainly to the return loss which is of interference nature. At the highest frequencies in the said range, the attenuation in the sample is not high enough to compensate the increase in the reflection coefficient (Figure 6.18) and (Figure 6.19), hence the SE decreases.

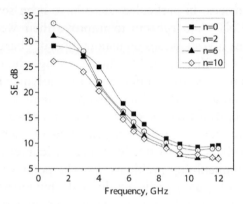

FIGURE 6.20 Frequency dependence of the shielding effectiveness of the composites comprising graphene (n = phr of graphene).

6.3.4 ELASTOMER-BASED COMPOSITE MATERIALS COMPRISING CNTS

In 1991, Iijima observed a graphitic tubular structure in an arc discharge apparatus that was used to produce C_{60} and other fullerenes. His realization[37] of the structural richness of these particles, which came to be known as nanotubes, has received considerable attraction; indeed, there are now in excess of 10,000 papers discussing the science of CNTs, including a large part on polymer composites.

There are two main types of CNTs—single-walled carbon nanotubes (SWCNTs) consisting of a single graphite sheet seamlessly wrapped into a cylindrical tube and MWCNTs comprising an array of such nanotubes that are concentrically nested-like rings of a tree trunk[38].

CNTs possess a whole range of remarkable properties. They have unique atomic structures, very high aspect ratios, and extraordinary electrical and mechanical properties, which make them ideal reinforcing fillers in nano-composites. CNTs reinforcing composites have been previously investigated to improve mechanical properties,[39-41] electrical conductivity and electro-static charging behaviors,[42-45] thermal performance,[46-48] magnetic proper-ties,[49] and crystallization and rheological properties.[50]

Even though elastomeric matrices have not been as widely studied as thermoplastic or thermoset matrices, there is a number of interesting reports in the literature. The addition of CNTs considerably increases the mechan-ical, electrical, thermal stability, inflammability, and tribological properties of elastomeric matrices. In particular, there have been several attempts to incorporate MWCNTs into rubbers to improve their wear resistance.[51,52] There are numerous investigations on nanocomposites based on elastomeric matrices and CNTs as filler in the literature, although the researchers atten-tion is focused mainly on the reinforcement of polymer matrices, but not on the influence that this unique filler has on the dielectric and microwave properties of the elastomeric composites. However, of late, there have been research works focusing on possible applications of such nanocomposites in microwave absorbers for solving electromagnetic interference (EMI) and electromagnetic compatibility problems.[53-62] The polymer matrices used in these cases are usually epoxy resin, acrylonitrile–butadiene rubber, styrene-butadiene rubber, silicone rubber, polyurethane rubber. Comparatively even in the past few years, it is rare to find investigations on nanocomposites based on NR and CNTs.[63-66]

The data about the influence of MWCNTs on the dielectric and micro-wave properties of the elastomeric composites are very scarce. There are no available reports on investigating those properties in a wider frequency range, first of all at frequencies lower than 8 GHz.

Therefore, the aim of this work is to study the influence that CNTs have on the dielectric (real part of permittivity, tan δ_{ε}) as well as on the microwave properties (absorption and reflection of the electromagnetic waves, the effectiveness of the electromagnetic shielding) of NR-based composites in a significantly higher frequency range—from 2 to 12 GHz. Some additional investigations on the morphology and microstructure of the CNT used and studied composites have been carried out by TEM and SAED.

The formulations of the rubber compounds used in the study and the vulcanizates based thereon are presented in Table 6.4.

TABLE 6.4   Compositions of the Rubber Compounds Comprising CNTs (phr).

	NR-1	NR-2	NR-3	NR-4
NR	100	100	100	100
Foaming agent	8	8	8	8
Stearic acid	1	1	1	1
Zinc oxide	4	4	4	4
Processing oil	10	10	10	10
CNTs	0	2	6	10
MBTS	2	2	2	2
TMTD	1	1	1	1
Vulkanox® 4020	1	1	1	1
Sulfur	2	2	2	2

CNTs: carbon nanotubes; MBTS = dibenzothiazyl disulfide; TMTD = Tetramethylthiuram disulfide. (Reprinted from l-Hartomy, O.A.; Al-Ghamdi, A.; Dishovsky, N.; Shtarkova, R.; Iliev, V.; Mutlay, I.; El-Tantawy, F. Effects of Multi-walled Carbon Nanotubes on the Dielectric and Microwave Properties of Natural Rubber-based Composites, *Fullerenes, Nanotubes and Carbon Nanostructures.* **2014,** *22,* 618–629. © 2014 with permission from Taylor & Francis.)

One of the primary purposes of the mixing process for rubber compounds is the distribution and disagglomeration, often called dispersion, of the reinforcing fillers. The filler dispersion is known to have a considerable effect upon the properties of the vulcanizates prepared.[15] On the other hand, the increasing filler dispersion increases its relative surface and surface energy. Occurrence of an agglomeration has been observed in the case of smaller particles which hinders the homogenization during the compounding. It has been established that the compounding begins with the formation of the so-called "soft" agglomerates, which further on breakup and disperse.[17] When the compounding cycle is a short one, the aforesaid initially formed agglomerates worsen the mechanical parameters since the size of the particles is too large. The agglomeration is probably the most important aspect of dispersion,[67] because of the detrimental effects of micro flaws on properties such as tensile strength, abrasion resistance, and fatigue life. We used TEM to investigate the CNTs distribution in the rubber matrix.

Figure 6.21 presents TEM micrographs of composites containing 6 phr and 10 phr of CNT.

The micrographs in Figure 6.21 show a relatively even distribution of the CNTs in the elastomeric matrix, which is marked by a satisfactory homogeneity. However, the CNTs have undergone an additional tearing during the compounding and thus their length has nearly decreased to 500 nm.

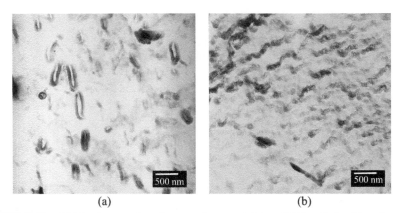

(a)   (b)

FIGURE 6.21   TEM micrographs of composites containing CNT at 6 phr (a) and at 10/h (b). (Reprinted from l-Hartomy, O.A.; Al-Ghamdi, A.; Dishovsky, N.; Shtarkova, R.; Iliev, V.; Mutlay, I.; El-Tantawy, F. Effects of Multi-walled Carbon Nanotubes on the Dielectric and Microwave Properties of Natural Rubber-based Composites, *Fullerenes, Nanotubes and Carbon Nanostructures*. **2014**, *22*, 618–629. © 2014 with permission from Taylor & Francis.)

Figures 6.22 and 6.23 present the dependence of relative permittivity (ε_r') and tan δ_ε (tan δ_ε) on the frequency and filler concentration for the CNTs filled composites under study.

As we have seen from the plots, the real part of relative permittivity increases with an increasing frequency at a constant filler concentration as well as with the increasing filler concentration at constant frequency. As seen, in the concentration interval studied $_r'$ values for the unfilled and filled composites are relatively close at lower frequencies (up to 7 GHz). At frequencies higher than 7 GHz the ε_r' increases with the increasing filler concentration. Moreover, there is a tendency of a more pronounced difference in the values for the nonfilled and filled composites. The values we obtained for the nonfilled composites are closer to the ones for NR at 1000 Hz reported in literature, which are in the range 2.40–2.70[12]. As the figure shows, our investigations also confirm the fact that ε_r' increases at higher frequencies. Of particular interest is the 7–12 GHz range wherein there is a relatively fast increase in ε_r' and its dependence on the CNTs is the most prominent. Considering the fact that the dielectric constant is related to the composites polarity,[36] probably at frequencies higher than 7 GHz, the polarization of the NR matrix is hampered. Hence, the dielectric constant increases. On the other hand, the increase in ε_r' proves the lower polarization of the composite at higher frequencies.

Figure 6.22 shows that at frequencies higher than 7 GHz, the frequency dependence of ε_r' for the filled composites is much closer to the linear one.

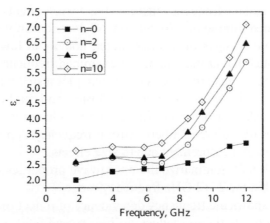

FIGURE 6.22 Frequency dependence of the real part of permittivity of the composites comprising CNTs (n = phr of CNTs). (Reprinted from l-Hartomy, O.A.; Al-Ghamdi, A.; Dishovsky, N.; Shtarkova, R.; Iliev, V.; Mutlay, I.; El-Tantawy, F. Effects of Multi-walled Carbon Nanotubes on the Dielectric and Microwave Properties of Natural Rubber-based Composites, *Fullerenes, Nanotubes and Carbon Nanostructures*. **2014**, *22*, 618–629. © 2014 with permission from Taylor & Francis.)

Figure 6.23 shows the frequency dependence of the tan δ_ε in the 2–12 GHz range. As expected, the increase in filler amount leads to an increase in tan δ_ε values.

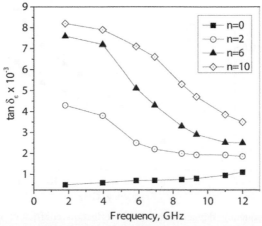

FIGURE 6.23 Frequency dependence of the tan δ_ε of the composites comprising CNTs (n = phr of CNTs). (Reprinted from l-Hartomy, O.A.; Al-Ghamdi, A.; Dishovsky, N.; Shtarkova, R.; Iliev, V.; Mutlay, I.; El-Tantawy, F. Effects of Multi-walled Carbon Nanotubes on the Dielectric and Microwave Properties of Natural Rubber-based Composites, *Fullerenes, Nanotubes and Carbon Nanostructures*. **2014**, *22*, 618–629. © 2014 with permission from Taylor & Francis.)

It is evident that the frequency increase has a weak effect on the dielectric losses of unloaded sample ($n = 0$), while CNT-loaded samples ($n = 2, 6, 10$) show appreciable changes of this parameter, notably at lower frequencies. It should be noted that at the end of the examined frequency range (11–12 GHz), the values of dielectric losses of the samples tested become closer.

CNT composite is a system comprising clusters and conducting networks, the microwave behavior of which depends on the cluster formation, instead of the properties of individual fibers. Strong frequency-dependent behavior may be observed in measurement due to size effects.

Figures 6.24–6.26 summarize the microwave properties of the composites studied within the 2–12 GHz frequency range.

Figure 6.24 shows the frequency dependence of reflection coefficient $|\Gamma|$.

The figure reveals that $|\Gamma|$ values do not exceed 0.90 within the entire frequency range. As a whole, the reflection coefficient increases with an increasing frequency and filler amount. The increase is particularly in the 7–12 GHz region. In the 2–4 GHz region, the effect of frequency and filler amount on the reflection coefficient is less pronounced. The effect is in accordance with the one upon the real part of the complex permittivity (Figure 6.22).

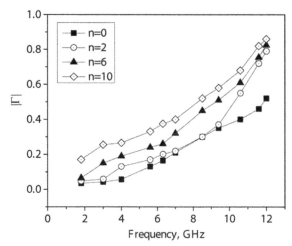

FIGURE 6.24 Frequency dependence of the reflection coefficient of the composites comprising CNTs (n = phr of CNTs). (Reprinted from l-Hartomy, O.A.; Al-Ghamdi, A.; Dishovsky, N.; Shtarkova, R.; Iliev, V.; Mutlay, I.; El-Tantawy, F. Effects of Multi-walled Carbon Nanotubes on the Dielectric and Microwave Properties of Natural Rubber-based Composites, *Fullerenes, Nanotubes and Carbon Nanostructures.* **2014,** *22,* 618–629. © 2014 with permission from Taylor & Francis.)

As Figure 6.25 shows when the frequency is lower, the attenuation remains almost unchanged for the nonfilled composite ($n = 0$) and for those with a small amount of filler ($n = 2$). Presumably the highest values of attenuation coefficient α in the whole frequency range are obtained for the samples with the largest loading ($n = 10$) and for 12 GHz, $\alpha = 22.3$ dB/cm. The comparison of the aforesaid dependence with those of the reflection coefficient and permittivity reveals a similarity that is obviously due to the specific structure of CNTs.

Figure 6.26 plots the frequency dependence of the electromagnetic shielding effectiveness (EMI SE). At lower frequencies, EMI SE values are higher but decrease with the increasing frequency. At lower frequencies, EMI SE values are higher but decrease with the increasing frequency. The character of the dependences is complicated because of EMI SE is defined as the sum of the reflection losses and attenuation in the material. But yet it makes an impression that with the increasing frequency, the values of this parameter for the samples with different loading almost draw level but they remain lower than the values for unloaded sample ($n = 0$). At the highest frequencies (7–12 GHz), the attenuation in the sample is not great enough to compensate the increase in the reflection coefficient (Figure 6.24) and (Figure 6.25), hence the SE decreases.

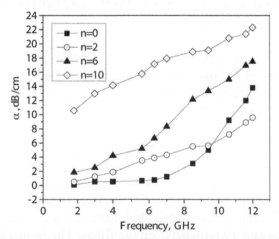

FIGURE 6.25 Frequency dependence of the attenuation of the composites comprising CNTs (n = phr of CNTs). (Reprinted from l-Hartomy, O.A.; Al-Ghamdi, A.; Dishovsky, N.; Shtarkova, R.; Iliev, V.; Mutlay, I.; El-Tantawy, F. Effects of Multi-walled Carbon Nanotubes on the Dielectric and Microwave Properties of Natural Rubber-based Composites, *Fullerenes, Nanotubes and Carbon Nanostructures.* **2014,** *22,* 618–629. © 2014 with permission from Taylor & Francis.)

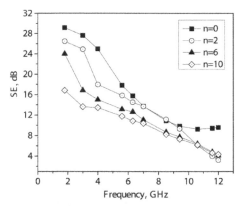

FIGURE 6.26 Frequency dependence of the SE of the composites comprising CNTs (n = phr of CNTs). (Reprinted from l-Hartomy, O.A.; Al-Ghamdi, A.; Dishovsky, N.; Shtarkova, R.; Iliev, V.; Mutlay, I.; El-Tantawy, F. Effects of Multi-walled Carbon Nanotubes on the Dielectric and Microwave Properties of Natural Rubber-based Composites, *Fullerenes, Nanotubes and Carbon Nanostructures.* **2014,** *22,* 618–629. © 2014 with permission from Taylor & Francis.)

The investigations carried out are described in detail in Ref. 68.

6.3.5   ELASTOMER-BASED COMPOSITE MATERIALS COMPRISING GRAPHENE AND CARBON BLACK[69]

NR-based nanocomposites containing a constant amount of standard furnace carbon black (50 phr) and GNP at concentrations 1–5 phr have been prepared. Their dielectric (dielectric constant, dielectric loss) and microwave properties (coefficients of absorption and reflection of the electromagnetic waves and EMI SE) have been investigated in the 1–12 GHz frequency range. The modification effect on the properties of the composites has been established. Our study showed that by changing the ratio of mass proportions of the fillers, the variation direction of the real and imaginary part of permittivity (depending on the particular application) can be controlled. The reflection coefficient of the investigated composites is strongly dependent on the frequency and the ratio between the fillers. The results for attenuation coefficient show that up to 7 GHz the composites with higher content of carbon black have higher values, and those with increased content of GNP— the lowest. The trend is just the opposite when the frequency is above 7 GHz. The frequency and filler amount ratio have an effect upon the reflection and attenuation coefficients as well as upon the electromagnetic SE.

The attenuation is not high enough to compensate for the increasing reflection. Therefore as a whole, the SE decreases gradually with an increasing frequency, especially at frequencies up to 7 GHz. The results achieved reveal how by using the selected combinations of carbon black and GNP the microwave and dielectric properties of NR nanocomposites can be tailored depending on the need for real practical application.

Adding small amounts of GNP to a constant amount of carbon black is one of the ways to control and primarily to improve the dielectric and microwave properties of composites based on NR in the high-frequency 1–12 GHz range. The results obtained from our research allow us to recommend graphene as second filler for NR-based composites to afford specific absorbing properties.

6.3.6   ELASTOMER-BASED COMPOSITE MATERIALS COMPRISING CNTS AND CARBON BLACK[70]

NR-based nanocomposites containing a constant amount of standard furnace carbon black (50 phr) and CNT at concentrations 1–5 phr have been prepared. Their dielectric (dielectric constant, dielectric loss) and microwave properties (coefficients of absorption and reflection of the electromagnetic waves and EMI SE) have been investigated in the 1–12 GHz frequency range. The modification effect on the properties of the composites has been established.

The values of relative permittivity increase with an increasing frequency and amount of the second filler (CNT) at a constant amount of carbon black (50 phr). The influence of the frequency and second filler amount is significantly pronounced in the 7–12 GHz range and the values of ε_r' are between 3.2 and 6.5.

The polarization mechanism operating in the gigahertz frequency range is purely electronic or orientational with relaxation times shorter than the time period of the applied signals. Interfacial polarization, which is the basic reason for the dispersion in permittivity at radio frequency regime, does not affect the microwave frequencies since it does not produce dispersion in ε_r' because of its much shorter relaxation time, though ε_r' has been found to increase at higher amounts CNTs in the composite. This phenomenon of increasing the permittivity at a higher second filler concentration can be attributed to the enhancement of electrical conductivity of the composites.

With the increase in frequency, the dielectric losses decrease smoothly throughout the entire frequency range studied (3×10^{-3}–9×10^{-3}). The influence of the second filler amount is not so well pronounced.

With the increase in frequency and CNT amount, the reflection coefficient and the coefficient of attenuation increase. When the amount of CNT is higher (for example, at 5 phr, the attenuation coefficient of the composites is in the range of 42–54 dB/cm).

The composite containing 5 phr CNT has the best values of EMI SE which remain within the 11–12 dB for the entire frequency range studied.

It has been found the percolation threshold for composites comprising the neat investigated fillers to be at 3 and 15 phr for CNTs and carbon black, respectively.[71] That might be due to the different geometry of CNT and carbon black. (The value of the aspect ratio for carbon black is 1 and for CNTs these values range from 120 to 200). According to Balberg,[72] the average interparticle distance can be different in the carbon nano-fillers with different aspect ratios. The more spherical-like structure, the larger the interparticle distance will be. The composites filled with elongated particles have a very narrow distribution of the distances between them due to the entangled particles structures and the volume resistivity of the composites decreases monotonically with increasing the aspect ratio. The interparticle distance, composites microstructure, and the volume resistivity have a strong influence on the microwave properties.[16] Presumably the increase in the ε_r' of the composites, containing CNTs and carbon black toward high frequencies, is induced by the dielectric relaxation which suggests that the established percolation network structure is not stable and easily affected by the external frequency disturbances.

Adding small amounts of CNTs to a constant amount of carbon black is a way to control and improve the dielectric and microwave properties of composites based on NR in the high-frequency range 1–12 GHz. The results achieved have allowed us to recommend CNTs also as the second filler for NR-based composites to afford specific absorbing properties.

6.4 CONCLUSION

1. It has been found that fullerenes and fullerene soot retard the vulcanization process, the effect being more pronounced for pure fullerenes. In the latter case, it affects the physical and mechanical properties, probably due to the incomplete vulcanization process. In the case of fullerene soot, physical and mechanical parameters are improved, probably due to the presence of other carbon structures in them, which are not radical acceptors.

    The influence of fullerenes in an amount of 0.5–1.5 phr on the dielectric properties in the frequency range 1–12 GHz of NR-based composites has been investigated. It has been found that the real part of permittivity increases slightly with the frequency and quantity of the filler. The effect is particularly pronounced in the 8–12 GHz range. A similar trend has the tan δ_ε. However, in its case the influence of the filler amount and the frequency is more pronounced, that is, the imaginary part of the complex relative permittivity is more sensitive to the changes in the frequency and filler amount than the real part.

The influence of fullerenes in an amount of 0.5–1.5 phr on microwave properties in the frequency range 1–12 GHz of NR-based composites has been studied. It has been found that regarding reflection in the 1–6 GHz range it almost does not depend on the frequency and filler quantity. In the 6–12 GHz range, it increases with their increase almost linearly. In terms of the coefficient of attenuation in the 1–6 GHz range, fullerenes have almost no influence, but in the 6–9 GHz and 9–12 GHz ranges, the attenuation increases with the increase in filler amount and frequency; and in the former case the increase is faster than in the latter. In the frequency and concentration ranges studied, the SE changes in the range of 10–20 dB, the increase in the attenuation is not large enough to compensate for the increase in the reflectance and therefore the overall effectiveness of shielding decreases gradually.

2.   The NR-based composites containing fullerenes at an amount of 0.5–1.5 phr have been characterized by SEM and TEM. Structurally, it was found that the tested composites can be classified as nanocomposites, since the size of fullerene particles dispersed in the rubber matrix after completion of the vulcanization process remains below 100 nm.

3.   The effect of fullerene soot containing fullerenes 6% of C_{60} and 1.5% of C_{70} fullerenes on the dielectric and microwave properties of composites based on NR containing fullerene soot at 1–7.5 phr has been investigated. It has been found that, in the 1–12 GHz range as the frequency fullerene soot amount increases, the values of permittivity, tan δ_ε, reflection, and absorption of electromagnetic wave get higher. The dielectric properties of the composites are affected more strongly by the changes in the concentrations of fullerene soot than the microwave ones.

4.   The effect that graphene nanoparticles have upon the dielectric and microwave properties of NR-based composites filled from 2.0 up to 10.0 phr has been studied in the 1–12 GHz range. It has been found that the real part of permittivity increases slightly with the increasing frequency and filler amount. The effect is pronounced especially in

the 9–12 GHz range. The tendency is the same in the case of tan δ_ε, although the impact of the filler amount is less marked, that is, the imaginary part of the complex relative permittivity is less sensitive to alternations of the filler amount and frequency than the real part.

5. The reflection coefficient increases with the increasing filler amount and frequency. The effects are not well pronounced. The availability of a foaming agent contributes to the low reflection coefficient.

6. The attenuation coefficient has been found to be almost independent of frequency and filler amount in the 1–6 GHz range, then in the 6–9 GHz range it increases slightly, while in the 9–12 GHz range its values increase drastically with the increasing frequency and filler amount. The dependence is almost a linear one.

7. The frequency and filler amount have an effect upon the reflection and attenuation coefficients as well as upon the electromagnetic SE. EMI SE values are in the 10–34 dB range. The attenuation is not high enough to compensate for the increasing reflectance. Therefore, as a whole, the SE decreases gradually with the increasing frequency, especially at frequencies up to 7 GHz.

8. The effect of graphene nanoparticles upon the dielectric and microwave properties of the composites studied occurs mainly in the 6–12 GHz range at a minimal filler amount of 6 phr.

9. The effect that CNTs have upon the dielectric and microwave properties of NR-based composites filled at 2 up to 10 phr has been studied in the 2–12 GHz range. It has been found that, the real part of permittivity increases slightly with the increasing frequency and filler amount. The effect is especially pronounced in the 7–12 GHz range. The tendency is not the same in the case of tan δ_ε, although the impact of the filler amount is marked.

10. The reflection coefficient increases with the increasing filler amount and frequency. The effects are not well pronounced.

11. The dependence of the attenuation coefficient on the frequency and filler amount is more apparent for the samples filled at 6 and 10 phr.

12. The frequency and filler amount have an effect upon the reflection and attenuation coefficients as well as upon the electromagnetic SE. The attenuation is not high enough to compensate the increasing reflectance. Therefore, as a whole, the SE decreases gradually with the increasing frequency, especially at frequencies up to 7 GHz.

13. The results achieved allow recommending CNTs as filler for NR-based composites to afford specific dielectric and microwave properties.

14. The results achieved allow us to recommend CNTs and graphene as second fillers in a combination with carbon black for NR-based composites to afford specific absorbing properties.

6.5 ACKNOWLEDGMENT

The authors acknowledge the support provided by Assoc. Prof. Vladimir Iliev, Ph.D. and Assoc. Prof. Nikolay Atanasov, Ph.D. for conducting the experiments and discussion of this work.

KEYWORDS

- **elastomer composites**
- **carbon nanostructures**
- **fullerenes**
- **graphenes**
- **dielectric properties**
- **microwave properties**

REFERENCES

1. Kuzmany, H.; Pfeiffer, R.; Hulman, M.; Kramberger, C. Raman spectroscopy of Fullerenes and Fullerene-nanotube Composites. *Philos. Trans. R. Soc., A.* **2004**, *362*, 2375–2406.
2. Murayama, H.; Tomonoh, S.; Alford, J. M.; Karpuk, M. E. Fullerene Production in Tons and More: From Science to Industry. *Fullerenes, Nanotubes, Carbon Nanostruct.* **2005**, *12*, 1–9.
3. Kroto, H. W.; Heath, J. R.; O'Brien, S. C., Curl, R. F., and Smalley, R. E. C60:Buckminsterfullerene. *Nature* **1985**, *318*, 162–163.
4. Birkett, P. R.; Hitchcock, P. B.; Kroto, H.W.; Taylor, R.; Walton, D. R. M. Preparation and Characterization of C60Br6 and C60Br8. *Nature.* **1992**, *357*, 479–481.
5. Thomas, S.; Stephen, R. *Rubber Nanocomposites Preparation, Properties and Applications*, Wiley: Singapore 2010.
6. Yadav, B. C.; Kumar, R. Structure, Properties and Applications of Fullerenes. *Int J Nanotechnol Appl.* **2008**, *2*, 15–24.
7. Jurkovska, B.; Kamrovski, P.; Pesetskii, S. S.; Koval, V. N.; Pinchuk, L. S.; Olkhov, Y. A. Properties of Fullerene-Containing Natural Rubber. *J Appl Polym Sci.* **2006**, *100*, 390–398.

8. Lukish, L. T.; Duncan, T. E.; Lansinger, C. M. Use of Fullerene Carbon in Curable Rubber Compounds. US Patent 5,750,615 A, May 12, 1998.

9. Ashiura, M.; Kawazura, T.; Yatsuyanagi, F. Rubber Composition Containing Modified Conjugated Diene-Based Polymer Bonded to Fullerene. US Patent 8,034,868 B2, Oct 11, 2011.

10. Thomann, H.; Brant, P.; Dismukes, J. P.; Lohse, D. J.; Hwang, J.-F.; Kresge, E. N. Fullerene-Polymer Compositions, US Patent 5,281,653, Jan 25, 1994.

11. Aoki, S. Rubber Composition and Tire Produced from the Same. US Patent 20,060,173,119 A1, March 08, 2006.

12. Kornev, A.; Bukanov, A.; Sheverdiaev, O. Technology of Elastomeric Materials, Istek: Moscow (in Russian). 2005.

13. Niedermeier, W.; Frohlich, J. Influence of Structure and Specific Surface Area of Soft Carbon Blacks on the Electrical Resistance of Filled Rubber Compounds. *Kaut Gummi Kunstst.* **2003**, *56*, 519–524.

14. Al-Hartomy, O.A.; Al-Ghamdi, A.A.; Al-Salamy, F.; Dishovsky, N.; Slavcheva, D.; Iliev, V.; El-Tantawy, F. Dielectric and Microwave Properties of Fullerenes Containing Natural Rubber-based Nanocomposites, *Fullerenes, Nanotubes and Carbon Nanostructures.* **2014**, *22*, 332–345.

15. Putman, M. Review of Dispersion Methods and the Relationships of Dispersion to Physical Properties. *Tire Technol Inter,* Annual Review, **2008**, 52–54.

16. Dishovsky, N. Rubber Based Composites with Active Behavior to Microwaves. *J Univ Chem Technol Metal.* **2009**, *44*, 115–122.

17. Payne, A. R. Effect of Dispersion on the Properties of Filler Loaded Rubbers. *J Appl Polym Sci.* **1965**, *9*, 2273–2284.

18. Al-Hartomy, O. A.; Al-Ghamdi; A. A.; Al-Solamy, F.; Dishovsky, N.; Slavcheva, D.; El-Tantawy, F. Properties of Natural Rubber-based Composites Containing Fullerene, *Int J Polym Sci.* **2012**, 8 pages, Article ID 967276.

19. Mathew, T.; Datta, R. N.; Dierkes, W. K.; van Ooij, W. J.; Noordermeer, J. W. M.; Gruenberger, T. M.; Probst, N. Importance of fullerenic Active Sites in Surface Modification of Carbon Black by Plasma Polymerization. *Carbon.* **2009**, *47*, 1231–1238.

20. Cataldo, F. Fullerene-Like Structures as Interaction Sites between Carbon Black and Rubber. *Macromol. Symp.* **2005**, *228*, 91–98.

21. Cataldo, F. The Role of Fullerene-Like Structures in Carbon Black and Their Interaction with Dienic Rubber. *Fullerene Sci Technol.* **2000**, *8*, 105–112.

22. Probst, N.; Fabry, F.; Grunberger, T.; Grivei, E.; Fulcheri, L.; Gonzalez-Aguilar, J. Method for Further Processing the Residue Obtained During the Production of Fullerene and Carbon Nanostructures. US Patent 20,080,279,749 A1, Nov 13, 2008.

23. Blank, V. D.; Mordkovich, V. Z.; Ovsjannikov, D. A.; Perfilov, S. A.; Pozdnjakov, A. A.; Popov, M. J.; Prokhorov, V. M. Method of Obtaining Carbon-based Composite Material and Composite Material. Patent RU2,556,673 C1, July 10, 2015.

24. Novoselov, K.S. Electric Field Effect in Atomically Thin Carbon Films. *Science.* **2004**, *306*, 666–669.

25. Salavagione, H. J.; Martínez, G.; Ellis, G. Graphene- Based Polymer Nanocomposites In: *Phys. Appl. Graphene: Exp.* Mikhailov, S. Ed.; In-Tech: Rijeka, 2011; pp. 169–192.

26. Allen, M. J.; Tung, V. C.; Kaner, R. B. Honeycomb carbon: A Review of Graphene. *Chem Rev.* **2010**, *110*, 132–145.

27. Zhu, Y.; Murali, S.; Cai, W.; Li, X.; Suk, J. W.; Potts, J.R.; Ruoff, R.S. Graphene and Graphene Oxide: Synthesis, Properties, and Applications. *Adv Mater.* **2010**, *22*, 1–19.

28. Compton, O. C.; Nguyen, S. T. Graphene Oxide, Highly Reduced Graphene Oxide, and Graphene: Versatile Building Blocks for Carbon-Based Materials. *Small.* **2010**, *6*, 711–723.

29. Mukhopadhyay, P.; Gupta, R. K. Trends and Frontiers in Graphene-Based Polymer Nanocomposites. *Plast Eng.* **2011**, *67*, 32–42.

30. Prud'Homme, R.; Ozbas, B.; Aksay, I.; Register, R.; Adamson, D. Functional Graphene-Rubber Nanocomposites. US Patent 7,745,528 B2, Jun 29, 2010.

31. Liang, J.; Huang, Y.; Ma, Y.; Liu, Z.; Cai, J.; Zhang, C.; Gao, H.; Y. Chen, Y. Electromagnetic Interference Shielding of Graphene/Epoxy Composites. *Carbon.* **2009**, *47*, 922–925.

32. De Rosa, I. M.; Sarasini, F.; Sarto, M. S.; Tamburrano, A. EMC Impact of Advanced Carbon Fiber/Carbon Nanotube Reinforced Composites for Next Generation Aerospace Applications. *IEEE T Electromagn.* **2008**, *C50*, 556–563.

33. De Rosa, I. M.; Mancinelli, R.; Sarasini, F.; Sarto, M. S.; Tamburrano, A. Electromagnetic Design and Realization of Innovative Fiber-Reinforced Broad-Band Absorbing Screens. *IEEE T Electromagn.* **2009**, *C51*, 700–707.

34. De Rosa, I. M.; Dinescu, A.; Sarasini, F.; Sarto, M. S.; Tamburrano, A. Effect of Short Carbon Fibers and MWCNTs on Microwave Absorbing Properties of Polyester Composites Containing Nickel-Coated Carbon Fibers. *Comp Sci Technol.* **2010**, *70*, 102–109.

35. De Bellis, G.; De Rosa, I. M.; Dinescu, A.; Sarto, M. S.; Tamburrano, A. Electromagnetic Absorbing Nanocomposites Including Carbon Fibers, Nanotubes and Graphene Nanoplatelets. *Proc IEEE Int Symp Electromagnetic Compatibility*, Fort Lauderdale, 25–30 July 2010, 202–207.

36. Banerjee, P.; Biswas, S. Dielectric Properties of EVA Rubber Composites at Microwave Frequencies. *J Microwave Power E E.* **2011**, *45*, 24–29.

37. Iijima, S. Helical Microtubules of Graphitic Carbon. *Nature* **1991**, *354*, 56–58.

38. Rupesh, K.; Suryasarathi, B. Carbon Nanotubes Based Composites-a Review. *J Miner Mater Charact Eng.* **2005**, *4*, 31–46.

39. Liu, T. X.; Phang, I.Y.; Shen, L.; S. Y. Chow; S.Y.; Zhang. W. D. Morphology and Mechanical Properties of Multiwalled Carbon Nanotubes Reinforced Nylon-6 Composites. *Macromol.* **2004**, *37*, 7214–7222.

40. Tang, W. Z.; Santare, M. H.; Advani, S.G. Melt Processing and Mechanical Property Characterization of Multi-walled Carbon Nanotube/High Density Polyethylene (MWNT/HDPE) Composite Films. *Carbon* **2003**, *41*, 2779–2785.

41. Kanangaraj, S.; Varanda, F. R.; Zhil'tsova, T. V.; Oliveira, M. S. A.; Simoes, J. A. O. Mechanical Properties of High Density Polyethylene/Carbon Nanotube Composites. *Compos Sci Technol.* **2007**, *67*, 3071–3077.

42. Grossiord, N. D.; Loos, J. C.; Regev, Q.; Koning, C. E. Toolbox for Dispersing Carbon Nanotubes into Polymers To Get Conductive Nanocomposites (Review).*Chem Mater.* **2006**, *18*, 1089–1099.

43. Calvert, P. Nanotube Composites: A Recipe for Strength. *Nature* **1999**, *399*, 210–211.

44. Curran, S. A.; Ajayan, P. M.; Blau, W. J.; Carroll, D. L.; Coleman, J. N.; Dalton, A. B.; Davey, A. P.; Drury, A.; McCarthy, B.; Maier, S.; Strevens, A. A Composite from Poly(m-phenylenevinyleneco-2,5-dioctoxy-p-phenylenevinylene and Carbon Nanotubes): A Novel Material for Molecular Optoelectronics. *Adv Mater.* **1998**, *10*, 1091–1093.

45. Meincke, O.; Kaempfer, D.; Weickmanm, H.; Friedrish, C.; Vathauer, M.; Warth, H. Mechanical Properties and Electrical Conductivity of Carbon-nanotube Filled Poly-amide-6 and its Blends with Acrylonitrile/Butadiene/Styrene.*Polymer.* **2004**, *45*, 739–748.

46. Yang, S. Y.; Castilleja, J. R.; Barrera, E. V.; Lozano, K. R. Thermal Analysis of an Acrylonitrile-Butadiene-Styrene/SWNT Composite. *Polym Degrad Stab.* **2004**, *83*, 383–388.

47. Yuen, S. M.; Ma, C. C. M.; Wu, H. H.; Kuan, H. C.; Chen, W. J.; Liao, S. H.; Hsu, C. W.; Wu, H. L. Preparation and Thermal, Electrical, and Morphological Properties of Multiwalled Carbon Nanotube and Epoxy Composites. *J App Polym Sci.* **2007**, *103*, 1272–1278.

48. Miyagawa, H.; Mohanty, A. K.; Drzal, L. T.; Misra, M. Nanocomposites from Bio-based Epoxy and Single-wall Carbon Nanotubes: Synthesis, and Mechanical and Ther-mophysical Properties Evaluation *Nanotechnol.* **2005**, *16*, 118–124.

49. Zilli, D.; Chiliotte, C.; Escobar, M. M.; Bekeris, V.; Rubiolo, G. R.; Cukierman, A. L.; Goyanes, S. Magnetic Properties of Multi-walled Carbon Nanotube–Epoxy Compos-ites. *Polymer.* **2005**, *46*, 6090–6095.

50. Wang, B.; Sun, G.P.; He, X.F.; Liu, J.J. *Polym. Eng. Sci.* **2007**, *47*, 1610.

51. Felhs, D.; Karger-Kocsis, J.; Xu, D. Tribological Testing of Peroxide Cured HNBR with Different MWCNT and Silica Contents Under Dry Sliding and Rolling Conditions Against Steel. *J Appl Polym Sci.* **2008**, *108*, 2840–2851.

52. Karger-Kocsis, J.; Felhs, D.; Thomann, R. Tribological Behavior of a Carbon-nano-fiber-modified Santoprene Thermoplastic Elastomer under Dry Sliding and Fretting Conditions Against Steel. *J Appl Polym Sci.* **2008**, *108*, 724–730.

53. Pacchini, S.; Idda, T.; Dubuc, D.; Flahaut, E.; Grenier, K. Carbon Nanotube-based Polymer Composites For Microwave Applications, Microwave Symposium Digest, 2008. *IEEE MTT-S International, Conference Publications*, 101–104 2008.

54. Micheli, D.; Pastore, R.; Apollo, C.; Marchetti, M.; Gradoni, G.; Moglie, F.; Primiani, V. M., Carbon Based Nanomaterial Composites in RAM and Microwave Shielding Applications. *IEEE-NANO 2009, 9th IEEE Conference on Nanotechnology, Confer-ence Publications*, 226–235 2009.

55. Saib, A.; Bednarz, L.; Daussin, R.; Bailly, C.; Xudong Lou; Thomassin, J.-M.; Pagnoulle,C.; Detrembleur, C.; Jerome, R.; Huynen, I. Carbon Nanotube Composites for Broadband Microwave Absorbing Materials, *IEEE T Microw Theory.* **2006**, *54*, 2745–2754.

56. Zhai, Y.; Zhang, Y.; Ren, W. Electromagnetic Characteristic and Microwave Absorbing Performance of Different Carbon-based Hydrogenated Acrylonitrile—Butadiene Rubber Composites, *Mater Chem Phys.* **2012**, *133*, 176–181.

57. Liu, Z.; Bai, G.; Huang, Y.; Li, F.; Ma, Y.; Guo, T.; He, X.; Lin, X.; Gao, H.; Chen, Y. Microwave Absorption of Single-Walled Carbon Nanotubes/Soluble Cross-Linked Polyurethane Composites, *J Phys Chem.* **2007**, *C 111,* 13696–13700.

58. Xu, Y.; Zhang, D.; Cai, J.; Yuan, L.; Zhang,W. Effects of Multi-walled Carbon Nano-tubes on the Electromagnetic Absorbing Characteristics of Composites Filled with Carbonyl Iron Particles, *J Mater Sci Technol.* **2012**, *28*, 34–40.

59. Li, H.; Wang, J.; Huang, Y.; Yan, X.; Qi, J.; Liu, J.; Zhang, Y. Microwave Absorption Properties of Carbon Nanotubes and Tetrapod-shaped ZnO Nanostructures Composites, *Mater Sci Eng.* **2010**, *B 175* 81–85.

60. Liu, L.; Kong, L. B.; Yin, W. Y.; Chen, Y.; Matitsine, S. *Microwave Dielectric Properties of Carbon Nanotube Composites*, www.intechopen.com/download/pdf/10007

61. H. Y. Zhang; G. X. Zeng; Y. Ge; T. L. Chen; L. C. Hu: Electromagnetic Characteristic and Microwave Absorption Properties of Carbon Nanotubes/Epoxy Composites in the Frequency Range from 2 to 6 GHz *J Appl Phys.* **2009**, *105*, 054314.

62. Liu, Z. F.; Bai, G.; Huang, Y.; Ma, Y. F.; Du, F.; Li, F. F.; Guo, T. Y.; Y. S. Chen; Y. S. Reflection and Absorption Contributions to the Electromagnetic Interference Shielding of Single-walled Carbon Nanotube/Polyurethane Composites, *Carbon.* **2007**, *45*, 821–827.

63. Gunasekaran, S.; Natarajan, R. K.; Kala, A.; Jagannathan, R. Dielectric Studies of Some Rubber Materials at Microwave Frequencies, *Indian J Pure App Phys.* **2008**, *46*, 733–737.

64. Shanmugharaj, A. M.; Bae, J. H.; Lee, K. Y.; Noh, W. H.; Lee, S. H.; Ryu, S.H. Physical and Chemical Characteristics of Multiwalled Carbon Nanotubes Functionalized with Aminosilane and its Influence on the Properties of Natural Rubber Composites. *Compos Sci Technol.* **2007**, *67*, 1813–1822.

65. Sui, G.; Zhong, W.; Yang, X.; Zhao, S. Processing and Material Characteristics of a Carbon-Nanotube-Reinforced Natural Rubber, *Macromol Mat Eng.* **2007**, *292*, 1020–1026.

66. Kolodziej, M.; Bokobza, L.; Bruneel, J.L. Investigations on Natural Rubber Filled with Multiwall Carbon Nanotubes. *Compos Interfaces.* **2007**, *14*, 215–228.

67. Dick, J.S., Ed. Rubber Technology: Compounding and Testing for Performance, 2nd ed. Hanser Publishers: Munich 2009.

68. Al-Hartomy, O.A.; Al-Ghamdi, A.; Dishovsky, N.; Shtarkova, R.; Iliev, V.; Mutlay, I.; El-Tantawy, F. Effects of Multi-walled Carbon Nanotubes on the Dielectric and Microwave Properties of Natural Rubber-based Composites, *Fullerenes, Nanotubes and Carbon Nanostructures.* **2014**, *22*, 618–629.

69. Al-Hartomy, O.A.; Al-Ghamdi, A.; Al-Solamy, F.; Dishovsky, N.; Shtarkova, R.; V Iliev, V.; El-Tantawy, F. Dielectric and Microwave Properties of Graphene Nanoplatelets/Carbon Black Filled Natural Rubber Composites, *Inter J Mater Chem.* **2012**, *2*, 116–122.

70. Al-Hartomy, O.A.; Al-Ghamdi, A.; Al-Solamy, F.; Dishovsky, N.; Shtarkova, R.; V Iliev, V.; El-Tantawy, F. Dielectric and Microwave Properties of Carbon Nanotubes/Carbon Black Filled Natural Rubber Composites. *Plast Rubber Compos.* **2012**, *41*, 408–412.

71. Li, Y.; Zhu, J.; Wei, S.; Ryu, J.; Wang, Q.; Sun, L.; Z. Guo, Z. Poly(propylene) Nanocomposites Containing Various Carbon Nanostructures, *Macromol Chem Phys.* **2011**, *212*, 2429–2438 2011.

72. Balberg, I. A Comprehensive Picture of the Electrical Phenomena in Carbon Black–Polymer Composites. *Carbon.* **2002**, *40*, 139–143.

CHAPTER 7

ELASTOMER-BASED COMPOSITE MATERIALS COMPRISING MODIFIED ACTIVATED CARBONS

CONTENTS

ABSTRACT

Use of active carbon as functional fillers in elastomer composites, designed to absorb electromagnetic waves, is evoked by its porous structure (its content of micro- and mesopores). This structure, particularly the size of pores, provides a significant inner surface and allows modification. The properties of activated charcoals can be controlled by the choice of initial material, selected method of activation, conditions of treatment and so on. On the other hand, the well-developed outer surface of the activated carbon, which contains a significant number of oxygen-containing groups, determines the presence of isoelectric points (IEPs). Those are a prerequisite for initiating an interaction between the elastomer macromolecules and functional filler. Last but not least, it is the specific porous structure of activated charcoal, which implies a significant uptake of insignificant reflection of electromagnetic waves

Six types of hybrid fillers have been used in our investigation. The adsorption-texture characteristics of the fillers have been determined by various methods. They have been used as functional fillers for natural rubber (NR)- and SBR-based composites. The dielectric (dielectric constant and dielectric loss angle tangent) and microwave (reflection coefficient attenuation coefficient and shielding effectiveness) properties of the vulcanizates thus obtained have been studied.

Hybrid dual-phase fillers based on activated carbon, and nano-sized magnetite, prepared in situ were obtained for the first time. The fillers have been characterized by X-ray diffraction and photoelectron spectroscopy. The influence of magnetite content in the hybrid filler on a number of properties of the resulting NR-based composites has been investigated. The vulcanization, mechanical and dynamic properties, electrical and thermal conductivity, dielectric and microwave properties of the new materials have been determined.

First, the observed effects could be explained by the impact that deposited magnetite has on the structure and properties of the very dual-phase filler (texture characteristics and adsorption activity) as well as by the change in the intensity of elastomer–filler interactions. It has been found that the total shielding effectiveness of the composite materials filled with activated carbon, modified with magnetite, exists mainly because of the mechanism of reflection of electromagnetic power (reflective shielding effectiveness). However, in the 5–6 GHz range, it is the mechanism of absorption (absorptive shielding effectiveness) that has a crucial role for all studied composites.

Another part of our study is focused on the modification of activated carbon to improve the absorption properties as a functional filler in elastomeric composites intended for microwave absorbers. Impregnation method has been used as the most popular among other procedures for depositing various phases over the support (in the production of catalysts) or to surface and texture modification of adsorbents. The method allows controlled deposition of a modifying agent in the macro- and the mesopores of the activated carbon. The chosen modifying agent is ZnO.

The fillers obtained have been characterized by Auger electron spectroscopy (AES) and have been used to prepare the SBR-based elastomer composites. It has been found that the modified fillers improve several important operational characteristics of the vulcanizates containing them.

7.1   INTRODUCTION

The published data about the utilization of active carbons (their content amounting to about 10–20%) as reinforcing fillers are quite a few,[1] whereas data about their usage as functional fillers in elastomeric composites, absorbing the electromagnetic waves, are practically missing.

The working hypothesis of the present investigation is as follows: Some types of active carbons, possessing suitable textural characteristics, could be applied for the above-mentioned purposes. The reasons for this assumption are the following:

Active carbons are defined as carbonaceous materials, possessing well-developed porous texture, containing about 87–97% carbon. Those are classified into the groups of the so-called mixed (transitional) forms of carbon, in which the atoms can participate in various combinations, corresponding to sp^3-, sp^2- and sp types of hybridization of the electron orbitals. Their turbostratic structure is shown schematically in Figure 7.1.

Activated carbon is defined as a microporous material, but more often as a micro- and mesoporous or as a meso- and microporous one. The determining factor for a specific application is the share of pores having a particular size (Figure 7.2).

The specific pore size of activated carbons is of particular importance. Their determining (characteristic) size is <2 nm, that is, micropores (super micropores, respectively), which indicates considerable (internal) surface area of the carbons. Moreover, the carbons also comprise mesopores, the effective radii of which vary between 2 and 50 nm and some macropores. The properties of the activated carbons can be controlled by the choice of a respective raw material; the activation method (physical or chemical), by varying the activation time interval and treatment conditions, keeping in mind that under some other conditions a number of factors could be affected. For instance, the size distribution of the pores (micro- and meso-) depends on the type of raw material, on the type of activation procedure and its conditions.

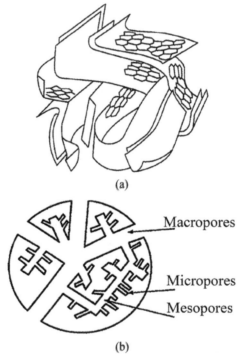

(a)

Macropores

Micropores

Mesopores

(b)

FIGURE 7.1   Schematic representation of (a) a three-dimensional and (b) a two-dimensional structure of activated carbon.[2] (Reprinted with permission from Francisco Rodríguez-Reinoso, The role of carbon materials in heterogeneous catalysis, Carbon, 1998, 36 (3) 159-175 © 1998 Elsevier.)

The availability of strongly developed (> 100 m^2/g) external surface, which usually contains a substantial number of surface oxygen-containing groups, determining the value of IEP, is a prerequisite for the interaction between elastomer macromolecules and the surface of active carbon. It is well known that the interaction "elastomer–filler", determined most of all by adsorption phenomena, is the core in the process of properties formation of elastomeric composites. On the other hand, it can be assumed that the specific structure of the active carbon will exert positive influence upon the dielectric losses. It will result in achieving a higher degree of absorbance and lower degree of reflection of the electromagnetic waves.

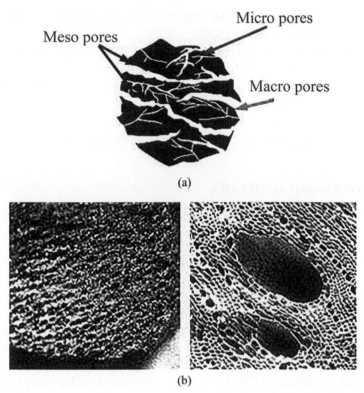

FIGURE 7.2 Theoretical model of the porous activated carbons texture (a) and TEM micrographs of the pores of a real activated carbon (b).[3] TEM transmission electron microscopy.

In this study, the complex dielectric constant (the real part of permittivity and dielectric loss tangent) and microwave properties (such as reflection coefficient, attenuation coefficient and shielding effectiveness) of rubber composites, containing activated carbons differing in their textural characteristics and properties, were investigated within the frequency range of 1–12 GHz with the aim to establish the following:

the values of the textural parameters and the IEPs of the active carbons, within such range of limits, in which they could be applied as fillers, without creating any technological problems during the vulcanization process;

- the influence of textural parameters of the activated carbon as well as their IEP values on the real part of permittivity and dielectric loss angle tangent and microwave properties (e.g. attenuation and reflection coefficients, the total, reflective and absorptive shielding effectiveness) of natural rubber (NR)-based composites in the frequency range of 1–12 GHz;

- the most suitable activated carbon for modification aimed at improving the microwave properties of composites containing it;
- the influence of different types of activated carbons on some other exploitation properties of the rubber composites;
- various methods for modification of active carbon viewing improvement in the properties of the composites thus filled.

7.2 EXPERIMENTAL DETAILS

7.2.1 ACTIVATED CARBONS

Six types of active carbon, differing in their adsorption-texture characteristics, parameters have been used. The types of active carbon were the following: Norit, AG-K, ART, AC-L, 207C and ACVM. AG-K and AC-L based on anthracite or on lignite brown coal, respectively. Norit and ART types were based on wooden material, whereas 207C and ACVM samples were prepared from different nutshells or fruit stones – cocoa nutshells and apricot stones, respectively.

The carbon samples ACVM and AC-L were prepared by using steam-gas activation, and the rest of the samples were used as commercial products.

7.2.2 ACTIVATED CARBONS CHARACTERIZATION

The active carbons were characterized by low-temperature nitrogen adsorption (at 77.4 K) using a Quantachrome Instruments NOVA 1200e (USA) apparatus.

On the basis of the adsorption–desorption equilibrium nitrogen isotherms, applying the specialized software, belonging to the apparatus, the following textural parameters were obtained:

- Specific surface area (S_{BET}) after the equation of Brunauer–Emmett–Teller, for the pressure range $P/P_O = 0.05–0.35$ (adsorbate N_2, at 77.4 K)
- Volume of the micropores (V_{MI}) using the density functional theory (DFT) (adsorbate N_2, at 77.4 K)
- Total pore volume (V_t) in accordance with the Rule of Gurvich for the pressure range $P/P_O = 0.95$ (adsorbate N_2, at 77.4 K)

- Volume of the mesopores (V_{MES}), estimated as the difference between the total pore volume and the volume of micropores (adsorbate N_2, at 77.4 K)
- Size distribution of the micropores based on the equation of Dubinin–Astakhov[4]
- Size distribution of the mesopores within the interval 1.7–15 nm on the basis of the adsorption branch of the adsorption–desorption equilibrium isotherm by using the density functional theory
- Average pore diameter (D_{AV}) as the ratio between V_t multiplied by 4 and the specific surface area S_{BET} (adsorbate N_2, at 77.4 K)
- The external surface area (S_{EXT}) based on the α_S method[5] from the adsorption branch of the isotherm (adsorbate N_2, at 77.4 K).

The IEPs of the carbons were determined by the method of Noh and Schwarz. [6] For this purpose, three different initial solutions having various pH factors were prepared for each type of carbon (respectively pH 3, 6 and 11) by using HNO_3 (0.1 M) and NaOH (0.1 M). Six flasks for each type of carbon were charged with 20 ml of the solutions and with various quantities of the studied carbon sample (0.05, 0.50, 0.75, 1.00, 5.00 and 10.00 g). The equilibrated pH was measured after 24 h. The dependency curves of pH on the mass of the carbon samples are passing over to a plateau, and the IEP was determined as the value at which the change in pH becomes zero.

7.2.3 RUBBER COMPOSITE PREPARATION

A rubber compound, taken as a reference, and six samples of different activated carbon powder of one and the same quantity were charged in an open two-roll laboratory mill (L/D 320 × 160 and friction 1.27). The speed of the slow roll was 25 min^{-1}. The formulations of the prepared samples are shown in Table 7.1. The inclusion of sulfur (vulcanization agent), zinc oxide (activator), stearic acid (dispersing agent and accelerating activator) and TBBS (*N-tert*-butyl-2-benzothiazolesulfonamide) was done with the aim to accomplish a normal vulcanization process.[7] The process of mixing was carried out as follows: After loading some raw rubber, the zinc oxide was added on the 5th minute as well as the stearic acid; after 3 min of homogenization, the activated carbon powder was added; and after 7 min of homogenization, the accelerator and the sulfur were added. Finally, the mixture was again homogenized for 4 min. The temperature of the rolls did not exceed 70°C. The experiments were repeated to check the reproducibility statistically. Samples in the form of sheets stayed for 24 h prior to their vulcanization. The optimal vulcanization time was determined by the vulcanization isotherms taken on a moving die rheometer MDR 2000 (Alpha Technologies) at 150°C according to ISO 3417:2002 standard.

The vulcanization was carried out in a hydraulic electric press with the size of the plate 400×400 mm at a pressure of 10 MPa. The obtained samples had dimensions 200×200×1.5 mm.

TABLE 7.1   Compositions of NR-Based Composites (phr).

	NR-1	NR-2	NR-3	NR-4	NR-5	NR-6	NR-7
Natural rubber SMR 10	100	100	100	100	100	100	100
Zinc oxide	5	5	5	5	5	5	5
Stearic acid	2	2	2	2	2	2	2
Activated carbon powder	0	Norit	AG-K	ART	ACL	207C	ACVM
TBBS	0.8	0.8	0.8	0.8	0.8	0.8	0.8
Sulfur	2.25	2.25	2.25	2.25	2.25	2.25	2.25

Note: The amounts of activated carbon in samples NR-2–NR-7 was 70 phr.

7.2.4   RUBBER COMPOSITE CHARACTERIZATION

7.2.4.1   MICROWAVE PROPERTIES MEASUREMENTS

Figure 7.3 indicates the mechanism of interaction of one composite with the electromagnetic wave, possessing incident power P_I. A portion of the power of the wave is reflected by the surface of the material P_R. Another portion passes P_A into the material and is absorbed by it, being transformed into a quantity of heat. The remaining part of the power is disseminated after the sample P_T.

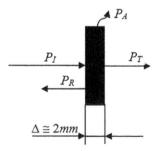

FIGURE 7.3   Mechanism of interaction of the composite with the incident power of the electromagnetic wave. (Reprinted from Al-Ghamdi, A. A.; Al-Hartomy, O. A.; Al-Solamy, F. R.; Dishovsky, N.; Mihaylov, M.; Malinova, P.; Atanasov, N. Dielectric and Microwave Properties of Elastomer Composites Loaded with Carbon–silica Hybrid Fillers. *J. Appl. Polym. Sci.* **2016,** *133*, 42978(1)–42978(9). © 2016 with permission from John Wiley.)

The shielding effectiveness was defined as the ratio between the power incident on the sample P_I and the power passing through the sample P_T in accordance with Equation 7.1[8-10] as follows:

$$SE_T = 10\log\frac{P_I}{P_T} \qquad (7.1)$$

The total shielding effectiveness (SE, in dB) and the reflective shielding effectiveness of the sample surface (SE_R, in dB) were determined by equations 7.2 and 7.3[11-13]:

$$SE_T = -10\lg T, \qquad (7.2)$$

where $T = |P_T / P_I| = |S_{21}|^2$

$$SE_R = -10\lg(1-R), \qquad (7.3)$$

where $R = |P_R / P_I| = |S_{11}|^2$.

The absorptive shielding effectiveness (SE_A) was calculated as the difference between (2) and (3), as is shown in equation (7.4) [14,15]:

$$SE_A = SE_T - SE_R. \qquad (7.4)$$

For the evaluation of the shielding effectiveness and measuring the coefficient of reflection from the surface of the studied composites, the apparatus represented in Figure 7.4 was used. It consists of coaxial reflectometric system (directed deviators Narda model 4222-16 and detectors Narda FSCM 998999 model 4503A, separating the incident power from the reflected power in the line); Ratio Meter HP Model 416A, calculating and depicting the amplitude of the coefficient of reflection from the sample; a series of radio wave frequency generators G4-37A, G4-79–G4-82, and HP 68A in the frequency interval from 1 to 12 GHz; the signal generator BM492 releasing a modulating signal of frequency 1 kHz directed towards the radio frequency generator, coaxial transmission line Orion type E2M for frequencies from 1 to 5 GHz; and coaxial measuring line APC-7 mm for frequencies from 6 to 12 GHz, power measuring device HP 432A. The critical frequencies for the coaxial measuring lines were determined by the formulas presented in Wieckowski and Janukiewicz and Chen et al.[16,17] The radio frequency generator and the reflection-measuring system (reflectometer) are interconnected by means of rugged phase stable cable N9910X-810 Agilent and through the connectors to avoid any interference.

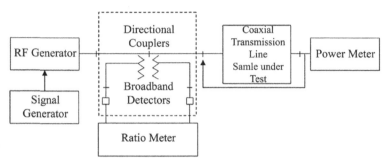

FIGURE 7.4 Apparatus for measuring the microwave properties. (Reprinted from Al-Ghamdi, A. A.; Al-Hartomy, O. A.; Al-Solamy, F. R.; Dishovsky, N.; Mihaylov, M.; Malinova, P.; Atanasov, N. Dielectric and Microwave Properties of Elastomer Composites Loaded with Carbon–silica Hybrid Fillers. *J. Appl. Polym. Sci.* **2016**, *133*, 42978(1)–42978(9). © 2016 with permission from John Wiley.)

The ratio-measuring device was calibrated prior to carrying out the actual measurements. Calibrating procedure of the type "Open-Short-Load" has been applied with calibrating kits of tools of the types Agilent N9330 and Agilent 1250.

The measurements were carried out observing the following procedure:

a) Calibration of the system for eliminating systematic errors using the abovementioned kits of calibrating tools and the standard procedures, described in[18,19]

b) Cutting out samples from the obtained vulcanized materials having dimensions as follows:

- External diameter of 20 mm and internal diameter of 7 mm using coaxial transmission line Oreon E2M
- External diameter of 7 mm and internal diameter of 3 mm using coaxial measuring line APC-7 mm

c) Carrying out measurements for determining the module of the coefficient of reflection using:

- Standard load of the type Agilent 1250, connected in the position of the coaxial line

- Blank coaxial line with a standard load at the end (calibrating kit of tools of the type Agilent N9330)
- Coaxial line with inserted standard material (polytetrafluoroethylene PTFE load 1 mm thick) for confirming the correctness of the measurements done

d) The studied sample is placed between the external and the internal conductor of the coaxial line.

Measurements were carried out at room temperature varying from 19 to 24°C, and incident power P_1 at the inlet of the coaxial measuring line varying from 800 to 1300 µW within the frequency range of 1–12 GHz.

7.2.4.2   DIELECTRIC PROPERTIES MEASUREMENTS

7.2.4.2.1   Complex Permittivity

The determination of complex permittivity was carried out by the resonance method on the basis of the cavity perturbation technique. The resonance frequency of an empty cavity resonator f_r was also measured and the sample material was placed into the resonator and the shift in resonance frequency f_ε was measured. The real part of permittivity ε_r' was calculated from the shift in resonance frequency, cavity, and the sample cross-sections S_r and S_ε, respectively:

$$\varepsilon_r' = 1 + \frac{S_r}{2S_\varepsilon} \cdot \frac{f_r - f_\varepsilon}{f_r} \tag{7.5}$$

The sample was in the form of a disc with a diameter of 11 mm and around 1.5 mm thickness and was placed where the maximum electric field of the cavity was highest.

7.2.4.2.2   Dielectric Loss Angle Tangent

The dielectric loss angle tangent was calculated from the quality factor of the cavity with Q_s and without a sample Q_c:

$$\tan \delta = \frac{1}{4\varepsilon_r'} \frac{S_r}{S_c} \left(\frac{1}{Q_c} - \frac{1}{Q_r} \right) \tag{7.6}$$

The measurement setup used several generators for the whole range: HP686A and G4-79 to 82, frequency meters: H 532A; FS-54, cavity resonator (Figure 7.5).

The dielectric properties were measured within the frequency range 1–10 GHz.

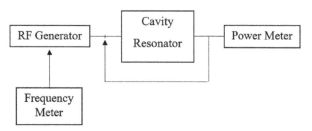

FIGURE 7.5   Scheme of the equipment for measuring the dielectric properties.

7.3   RESULTS AND DISCUSSION

7.3.1   CHOOSING THE ACTIVATED CARBON MOST APPROPRIATE FOR MODIFICATION

7.3.1.1   TEXTURAL PARAMETERS OF ACTIVATED CARBONS AND THEIR INFLUENCE ON THE RUBBER VULCANIZATION PROCESS

Figure 7.6 represents the adsorption-desorption equilibrium isotherms of the studied samples.

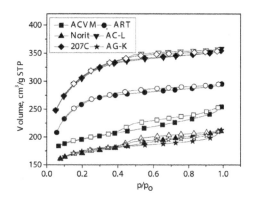

FIGURE 7.6   Adsorption–desorption equilibrium isotherms (N_2, at 77.4 K) of the studied samples of active carbon.

The textural parameters, calculated as per the above isotherms following the methods, described in the experimental section are listed in Table 7.2.

With regards to the usage of active carbon as reinforcing fillers, it should be noted that not all textural parameters are important. Most significant are the specific and the external surface areas and mainly mesopores volume. This is due to the fact that the pore size allowing large elastomer macromolecules to penetrate into the carbon particles are of significance, that is, the mesopores. As far as the size of the micropores is concerned, < 2.0 nm[20], these are not accessible to the molecules of the elastomer. The same is also valid for the specific surface area of the micropores. Therefore, the parameters, associated with the micropores, will be considered only from the viewpoint of comparison between the different samples of active carbon.

TABLE 7.2   The Main Textural Parameters of Activated Carbon Used and Its IEP Values.

	S_{BET} (m²/g)	S_{MI} (m²/g)	S_{EXT} (m²/g)	V_t (cm³/g)	V_{MI} (cm³/g)	V_{MES} (cm³/g)	D_{AV} (nm)	IEP	V_{MES}/A_{EXT}
Norit	518	400	118	0.33	0.21	0.12	2.5	7.2	10.2×10^{-4}
ART	957	840	117	0.46	0.36	0.10	1.9	7.1	8.6×10^{-4}
AG-K	652	569	83	0.33	0.23	0.10	2.0	6.4	12×10^{-4}
AC-L	1107	852	255	0.55	0.37	0.18	2.0	5.5	7.1×10^{-4}
207C	1111	890	222	0.55	0.39	0.16	2.0	5.3	7.2×10^{-4}
ACVM	688	555	133	0.39	0.25	0.14	2.3	9.7	10.5×10^{-4}

From the viewpoint of the type of texture, all the six studied samples of active carbons belong to the group of micro-mesoporous materials.

The specific surface areas of the commercial active carbon are varying within a comparatively wider range (Table 7.2), for example, that for Norit is more than twice smaller than for AC-L and 207C. This difference is demonstrated especially with regard to the value of the external surface area, this value in the case of AG-K compared to those of 207C and AC-L is 2.7 and 3.1 times lower, respectively.

There is a substantial difference in the performance of the studied active carbon during the vulcanization of NR-based samples.

The vulcanization process is occurs and ends normally with the samples containing active carbon Norit, ART, and AG-K as fillers, further on "the first group." It does not end in the case of the carbon fillers: AC-L, 207C, and ACVM (further on "second group").

The volume of the micropores and mesopores vary within comparatively narrower limits in the studied active carbon. Thus, the greatest difference observed in the volumes of the micropores—that between Norit and 207C is about 1.8 times. Analogously, with regard to the volume of the mesopores, the difference between ART and AG-K on the one hand and AC-L on the other (Table 7.2) is also 1.8 times.

The analysis of the first group of carbon samples shows that the specific surface area (except for ART sample) is comparatively low (Table 7.2). By analogy, the second group samples (except for ACVM samples having $S_{BET} = 688$ m^2/g) are characterized by specific surface area > 1100 m^2/g.

Analogously, the external surface of the active carbon of the first group is lower than that of the second group (Table 7.2).

The maximum in the curves of the size distribution of the micropores is an interesting fact (Figure 7.7). The comparison of the maxima of these curves, with respect to the maximum in the characteristic size of the micropores shows that for the first group they are characterized by larger values and vice versa, smaller values for the second group of active carbons.

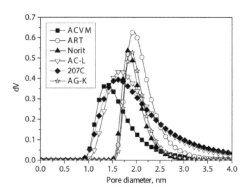

FIGURE 7.7   Size distribution of micropores for the studied active carbon.

The curves of mesopores size distribution (Figure 7.8) are similar in character (including close values of the maximum in the distribution curves) for both groups of carbons, but these features are more complicated in the second group of active carbon with larger number of ill-resolved peaks. However, no drastic difference is observed in the distribution curves that could explain the difference in the vulcanization properties of the carbons.

In this aspect, there is a possibility to explain the difference as based on the IEP of the active carbons (Table 7.2). It is obvious that the values of IEP for the carbons of the first group are positioned at about 7, while the values for the carbons of the second group are either under the value of the neutral point 7 (5.3 in the case of 207C and 5.5 for the AC-L) or it is 9.7 in the case of the ACVM. Therefore, the availability of superficial oxygen-containing functional groups, whatever their nature is (acidic or basic), is restricting the utilization of active carbons as fillers in view of the negative influence of some of them upon the vulcanization process. We can recommend only those having IEP ≈ 7 as appropriate fillers.

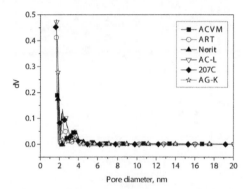

FIGURE 7.8   Size distribution of the mesopores for the studied active carbons.

Another interesting parameter, which, without a doubt, relates to the ability of carbons to be suitable reinforcing fillers, is the ratio between the volume of the mesopores and the external surface area. A logical prerequisite appears to be the fact that the external surface is easily accessible to the large elastomer molecules, while the internal surface of the mesopores (only that of the bigger ones) is not readily accessible to them.

The data in Table 7.2 suggest that the values of this ratio for the active carbons of the first group are higher than those for the second group, whereupon the only exception is ACVM, despite the appropriate value of the ratio V_{MES}/S_{EXT}, as a result of unfitting high value of IEP (Table 7.2), hindering the realization of the vulcanization process.

7.3.1.2   MICROWAVE PROPERTIES

Figure 7.9 presents the results for the reflection coefficient modulus of a standard material (PTFE) and the studied samples, measured within the frequency range 1–12 GHz. As seen, at the frequencies ranging from 1 to 5 GHz the module of the coefficient of reflection increases monotonously for all the materials, while in the 5–9 GHz range a resonance character is observed. The three studied materials have high values of the reflection coefficient modulus within the frequency range 5–11 GHz. At frequencies higher than 10 GHz a monotonous decrease in the values of the reflection coefficient modulus is observed, which is more pronounced in the case of AG-K sample.

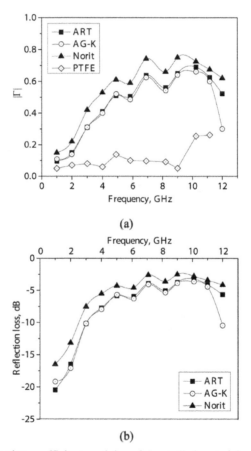

FIGURE 7.9 Reflection coefficient modulus of the studied materials: (a) on a linear scale, (b) as a logarithmic dependence.

Figure 7.10 presents the changes in the attenuation coefficient within the studied frequency range.

As seen from the results in Figure 7.10, the values of the attenuation coefficient are sensitive towards changes in the frequency. Within the 1–6 GHz range, the values of the coefficient of attenuation increase with the increasing frequency. Within the 5–12 GHz range, a strongly expressed resonance character is observed. The maximal values are observed at 6 GHz, 8 GHz, and 11 GHz, while the minimal ones are at 7 and 9–10 GHz. The highest values of the attenuation coefficient within the studied range belong to active carbon Norit, while the composites, containing the other two active carbons possess lower comparable values. The comparison with the sample of Teflon (PTFE), used as a standard, proves that indeed some of the studied active carbons can be applied as fillers in elastomeric composites designed for microwave application. It is also seen that the maximal absorbance of the electromagnetic waves is at about 22 dB/cm (Norit). The composite containing AG-K active carbon exhibits an interesting behavior. At a frequency of 11–12 GHz the attenuation coefficient raises sharply and reaches a value of 27 dB/cm. The Teflon sample (PTFE) does not exhibit any resonance behavior, which gives us the reason to suppose that the resonance behavior of the composites containing active carbons is due to their textural characteristics.

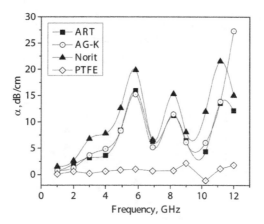

FIGURE 7.10   Attenuation coefficient of the studied materials.

7.3.1.3   TOTAL SHIELDING EFFECTIVENESS

Figure 7.11 presents the results on the total shielding effectiveness of the standard material (PTFE) and a function of frequency within the 1–12 GHz range. As seen, the Norit sample has values of the total shielding effectiveness (SE_T) around 2 dB higher in comparison to those of the other samples. Within the 1–11 GHz range, the total shielding effectiveness for all the three materials (ART, AG-K, and Norit) has identical behavior, while at frequencies of up to 6 GHz its values increase rapidly. Further, within the 7–10 GHz range the observed sensitivity is lower with respect to frequency changes. As expected, Teflon does not exhibit any shielding properties within the whole frequency band and its total *SE* tends to be zero dB.

Figure 7.12 illustrates the results of the reflective shielding effectiveness as a consequence of reflection from the sample surface (SE_R).

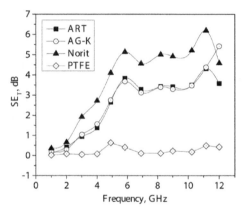

FIGURE 7.11   Total shielding effectiveness of the studied samples and PTFE.

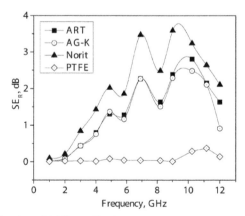

FIGURE 7.12   Reflective shielding effectiveness as a consequence of reflection from the tested samples and PTFE.

On the basis of the results in Figure 7.12, one can determine that the highest reflective shielding effectiveness comes is due to the reflection from the surface of the sample—the tested sample is Norit. For the other two tested samples the values of SE_R are very close to each other within the entire frequency range.

The results of the absorptive shielding effectiveness (SE_A) due to absorption of the electromagnetic radiation by the studied materials are presented in Figure 7.13. Within the 1–11 GHz range, Norit sample displays the greatest absorption. At frequencies higher than 11 GHz, AG-K material possesses higher values of the absorptive shielding effectiveness, exhibiting a behavior different from that of the other two materials. This determines the behavior of the total shielding effectiveness of Norit. For all the materials studied, the two components—reflective and absorptive shielding effectiveness—manifest resonance behavior at frequencies from 6 to 12 GHz. At a frequency of 6 GHz SE_A is prevailing, while within the 7–11 GHz range, the total shielding effectiveness is dominated by the reflective component (SE_R).

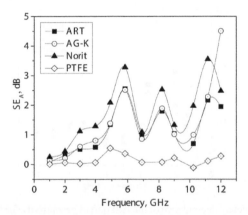

FIGURE 7.13 Shielding effectiveness due to absorption of the electromagnetic radiation for the tested samples and PTFE.

7.3.1.4   DIELECTRIC PROPERTIES

Figures 7.14 and 7.15 illustrate the dielectric parameters of the tested materials within a wide frequency range, 1–10 GHz. For all materials one can observe an increase in the real part of the relative permittivity (ε_r') upon increasing the frequency up to 3 GHz (Figure 7.14), while for the frequency range 5–10 GHz, the frequency influence is weak. Within the studied range, ART active carbon possesses ε_r' values higher than those of the other materials.

Figure 7.15 depicts through a logarithmic scale the results from the measurements of tan δ_ε. As seen, this parameter is sensitive to frequency changes. Within the 1–5 GHz range, all three materials have relatively high values of tan $_\varepsilon$, and probably for this reason, the values of the shielding effectiveness in this range are lower. Within the 5–10 GHz range, ART material possesses higher tan δ_ε values.

As a result of these investigations, it has been found that the active carbons exert serious influence upon the vulcanization process, which depends both on the chemistry of the carbon surface (respectively on the IEPs) as well as on their textural parameters (most of all on the ratio V_{MES}/S_{EXT}). To accomplish a normal vulcanization process without any technological problems it is necessary that the ratio V_{MES}/S_{EXT} should be within the limits $8.5–12.0 \times 10^{-4}$ and at the same time the IEP should be close to 7.

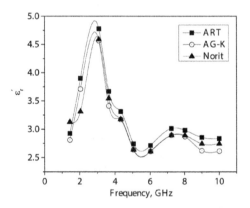

FIGURE 7.14   Frequency dependence of the real part of permittivity for the studied materials.

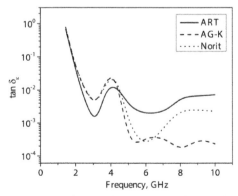

FIGURE 7.15   Frequency dependence of the dielectric loss angle tangent for the studied materials.

It has also been observed that the composite containing active carbon Norit possesses the highest value of the total shielding effectiveness in comparison with the composites containing carbons AG-K and ART. The same carbon displays the highest value of the reflective shielding effectiveness, but it is compensated by the exceeding value of the absorption shielding effectiveness.

The studied active carbons (Norit, ART, and AG-K), meeting the requirement for appropriate values of the IEPs and of the textural parameters (S_{EXT} and V_{MES}), can be used as fillers in elastomeric composites designed for microwave application.

7.3.2 ELASTOMER-BASED COMPOSITE MATERIALS COMPRISING ACTIVATED CARBONS/SYNTHESIZED IN SITU MAGNETITE AS FILLERS

7.3.2.1 PREPARATION OF THE FILLERS

Most of the present-day microwave-absorbing composites, which are finding wider applications, are produced from a dielectric rubber matrix and specific functional fillers. These fillers—carbon black, graphite, activated carbon, short carbon, or metal fibers, micro- and nano-sized metal powders—possess high values of the imaginary part of the complex permittivity and/or permeability, and absorb high frequency energy.[21-30] However, as the electromagnetic radiation has both dielectric and magnetic components, it is obvious that both dielectric and magnetic materials are effective for the absorption of microwave radiation. Therefore, it has been of interest to combine components of high dielectric and magnetic losses into a hybrid filler, which opens new opportunities of preparing modern microwave absorbers of specific properties.

In the Section 7.2, we have seen that under certain conditions, activated carbons possess properties of reinforcing and functional fillers for elastomeric materials with microwave application. For the activated carbon/rubber composites a reasonable assumption for $\mu^*=1\text{-}j0$ ($\mu_r'=1$ and $\mu_r''=0$) has been made. Only the real and imaginary part of the permittivity has been measured. This limits the absorptive properties of the composite regarding the electromagnetic waves.

On the other hand, magnetite (natural or synthetic) is used widely as a functional filler for elastomers for different microwave applications. Being very effective and often used in a broad frequency range for various applications, this filler has been intensively investigated.[31,32]

Synthetic magnetite is obtained predominantly from the following equation:

$$Fe^{2+} + 2Fe^{3+} + 8OH^{1-} \rightarrow Fe_3O_4 + 4H_2O$$

Our main hypothesis is that the porous texture of activated carbon contains a significant number of nano-sized pores. By steric reasons, if magnetite is synthesized in situ in such pores, it will be nano-sized too. Expectedly, after the modification of activated carbon with magnetite the microwave properties of composites containing hybrid carbon/nanosized magnetite fillers will be improved, besides that dielectric and magnetic losses will also contribute to the composites microwave absorption properties.

The purpose of this part of our investigation was to examine the impact of hybrid fillers based on activated carbon/nanoscale magnetite synthesized in situ on the properties of the filled NR-based composites, namely how their porous texture affects the vulcanization, mechanical and dynamic characteristics, electrical and thermal conductivities, dielectric characteristics, microwave properties, and homogeneity. We have found no data about a similar research. Special attention has been paid to the effect that the amount of magnetite phase has on composite's characteristics mentioned above.

Activated carbon Norit from wood was used in the research, since its properties described in Section 7.3.1 turned to be better than those of the studied six types of active carbon.

The hybrid fillers were synthesized using Karyakin and Angelov method.[33]

The detailed preparation steps of Karyakin and Angelov method were as follows:

- Different amounts of iron (III) sulfate and iron (II) sulfate at a mole ratio of Fe^{2+}/Fe^{3+} equal to 1/2 were dissolved in 200 or 300 ml of distilled water (according to the ferrous salts content).
- Solution at one of the adopted ferrous sulfate concentrations was poured over 100 g of activated charcoal (powder) and let stay for 2 h.
- Then the magnetite activated carbon precursor was treated with a 5% solution of KOH (80°C) for 15 min.
- Finally, after filtration the magnetite activated carbon was separated from the suspension and dried under vacuum at 70°C for 6 h.

The substrate virgin activated charcoal was marked as MAC-0.

The samples obtained from the above method comprising various amounts of magnetite phase were marked as MAC-1, MAC-2, and MAC-3, respectively.

According to the modified version of the method, the initial solution of ferrous sulfate was obtained by dissolving in advance the mixture of iron (III) sulfate and iron (II) sulfate in a little dH_2O. Then CH_3OH was added till the solution reached 300 ml.

The next steps were the same as those of the initial method.

The sample obtained by the modified version of the method was marked as MAC-4.

Reagents used were of analytical grade and were obtained from VWR Prolabo Chemicals.

7.3.2.2 CHARACTERIZATION OF THE FILLERS

The modified activated carbon was characterized according to the parameters described in Section 7.2. With regards to the modification carried out the following investigations were performed:

- X-ray diffraction (XRD) data were obtained using a Bruker D8 Advance diffractometer with Cu-Kα radiation and Sol X detector.
- The film composition and electronic structure were investigated by X-ray photoelectron spectroscopy (XPS). The measurements were carried out on an AXIS Supra electron- spectrometer (Kratos Analitycal Ltd.) using monochromatic AlK_α radiation with a photon energy of 1486.6 eV and a charge neutralization system. The binding energies (BE) were determined with an accuracy of ±0.1eV. The composition and chemical state of the films were determined by monitoring the areas and binding energies of C1s, O1s, and Fe2p photoelectron peaks using the commercial data-processing software of Kratos Analytical Ltd. The concentrations of the different chemical elements (in atomic %) were calculated by normalizing the areas of the photoelectron peaks to their relative sensitivity factors.

Table 7.3 summarizes the texture parameters of the fillers calculated on the basis of the adsorption-desorption isotherms.

TABLE 7.3 Magnetite Content and Main Texture Parameters of the Initial Activated Carbon and of Synthesized Samples.

	Content (mass.%)	S_{BET} (m²/g)	S_{EXT} (m²/g)	V_t (cm³/g)	V_{MI} (cm³/g)	V_{MES} (cm³/g)	D_{AV} (nm)	IEP	V_{MES}/A_{EXT}
MAC-0	-	541	154	0.37	0.21	0.16	2.7	7.2	10.4×10^{-4}
MAC-1	2.7	468	144	0.33	0.17	0.16	2.8	6.9	11.1×10^{-4}
MAC-2	3.5	451	128	0.30	0.16	0.14	2.7	7.0	10.9×10^{-4}
MAC-3	5.1	456	136	0.33	0.17	0.16	2.9	6.7	11.8×10^{-4}
MAC-4	5.4	431	138	0.39	0.15	0.24	3.6	7.1	17.4×10^{-4}

As Table 7.3 shows, with increasing the magnetite phase the specific surface area of the samples decreases, if compared to that of the substrate activated carbon. The decrease is most pronounced in the case of sample MAC-4 obtained by deposition of the precursors from organic medium.

Table 7.3 also shows that values of mesopore volume for all the samples, except MAC-4, are equal to the ones of the substrate activated carbon (MAC-0). This suggests that magnetite phase either does not penetrate the activated carbon or if it does, the amount of magnetite is negligible. On the other hand, the value of mesopore volume for sample MAC-4 is about 50% higher than that of MAC-0, which evidences that magnetite fills the fine carbon pores. This is also confirmed by the increase in the average diameter (D_{AV}) of MAC-4 (3.6 nm) in comparison to that of MAC-0 (2.7 nm) and rest of the samples.

The analysis of values of micropores volume (Table 7.3) proves indubitably that the decrease with the increasing magnetite amount being the lowest in the case of MAC-4.

All this leads to an assumption that in the case of MAC-4 the better wetting of carbon surface (in CH_3OH medium of the precursor) ensures greater penetration of the magnetite phase into the fine carbon pores (the micropores, inclusive) filling them partially or completely.

7.3.2.3 XRD MEASUREMENTS

According to the data of XRD patterns, the phase distributed into the nano-sized porous texture of the samples comprises predominantly magnetite. Probably maghemite and arcanite are also present. Magnetite and maghemite have similar XRD patterns and it is difficult to distinguish them by XRD method.

7.3.2.4   CHEMICAL ANALYSIS AND SURFACE ANALYSIS OF THE SAMPLES

The content of Fe in the synthesized samples was determined by atom absorption. The results are presented in Table 7.2.

The presence of a Fe phase in the samples as well as the dominating distribution (over the external surface and in their volume) was monitored by X-ray Photoelectron Spectroscopy (XPS).

The obtained photoelectronic spectra for Fe in the $2p$-excitation range for all samples have the binding energies for Fe in Fe_3O_4,[34] as well as there is a possibility of existing γ-FeOOH and Fe_2O_3 (Table 7.4).

TABLE 7.4   Binding Energies of Fe ($2p_{3/2}$) and O ($1s$) Peaks.

| | Fe($2p_{3/2}$) | | | O($1s$) | | |
	Fe_3O_4	γ-FeOOH	Fe_2O_3	Fe_3O_4	γ-FeOOH	Fe_2O_3
MAC-1	710.7	711.2	711	530.2	–	–
MAC-2	710.6	711.2	711	530.3	–	–
MAC-3	710.4	711.4	711	530.3	530.2	530.07
MAC-4	710.9	711.2	711	530.5	–	–

According to[35] the comparison of Fe/C ratios determined by XPS (Table 7.5) with those from the chemical analysis (cha) of the samples allows to locate the distribution of the impregnation Fe phase over the external surface or inside the sample's volume.

As Table 7.5 shows, the magnetite phase is distributed predominantly over the external surface and to a lesser extent in the volume (internal surface) of all samples. With the increasing magnetite amount the phase over the external surface also enlarges. It is worth comparing samples MAC-3 and MAC-4. According to the data from chemical analysis, they comprise close amounts of Fe. On the other hand, the methods of depositing the precursors (from aqueous and organic media, respectively) are different. Noteworthy is the fact that although the ratios of the values determined by XPS and by chemical analysis (cha) of the samples are similar (Table 7.5), the sample obtained by the modified method (MAC-4) has a lower ratio (2.438 against 2.574 in MAC-3). Accordingly, the precursor phase from the organic medium is distributed in the internal surface (volume) to an extent greater than that in the case of using an aqueous medium.

TABLE 7.5   Fe/C Ratios as Determined by XPS and Chemical Analysis (cha).

	Fe/C (XPS)	Fe/C (cha)	$\dfrac{(Fe / C)_{xps}}{(Fe / C)_{cha}}$
MAC-1	0.056	0.034	1.647
MAC-2	0.093	0.044	2.114
MAC-3	0.175	0.068	2.574
MAC-4	0.178	0.073	2.438

7.3.2.5 PREPARATION OF NR-BASED COMPOSITE MATERIALS

Five rubber compounds with formulations in correspondence to Table 7.6 were prepared in an open two-roll laboratory mill (L/D 320×160 and friction 1.27). The process of mixing was carried out as follows: the mill was loaded with some raw rubber, the zinc oxide as well as the stearic acid were added at the 5th minute. After 3 min of homogenization, activated carbon powder was added, and after 7 min of homogenization, the accelerator and sulfur were added. Finally, the mixture was again homogenized for 4 min. The temperature of the rolls did not exceed 70°C. Samples in the form of sheets stayed for 24 h prior to their vulcanization. The optimal vulcanization time was determined by the vulcanization isotherms taken on moving die rheometer MDR 2000 (Alpha Technologies) at 150°C according to ISO 3417:2002. The vulcanization was carried out on a hydraulic electric press, with the size of the plate 400×400 mm at a pressure of 10 MPa.

TABLE 7.6   Compositions of the NR-Based Rubber Composites (phr).

Composites	MAC-0	MAC-1	MAC-2	MAC-3	MAC-4
Natural rubber SMR 10	100	100	100	100	100
Zinc oxide	5	5	5	5	5
Stearic acid	2	2	2	2	2
Activated carbon powder	MAC-0	MAC-1	MAC-2	MAC-3	MAC-4
TBBS	0.8	0.8	0.8	0.8	0.8
LDA	1	1	1	1	1
Sulfur	3.25	3.25	3.25	3.25	3.25

Note: The amounts of activated carbons in all samples are 70 phr.

7.3.2.6   CHARACTERIZATION OF THE NR-BASED COMPOSITE MATERIALS

7.3.2.6.1   Vulcanization Characteristics

Table 7.7 summarizes the main vulcanization characteristics of the studied rubber compounds determined according to their vulcanization isotherms.

Usually, the minimum torque (M_L) is related to viscosity, while the maximum torque (M_H)—to the resistance of composites to stress-strain deformation. $\Delta M = (M_H - M_L)$ is related to the vulcanization network density. As seen from the results presented in Tables 7.3 and 7.7, with the increasing amount of magnetite synthesized and deposited in the activated carbon the minimum torque (viscosity, respectively) increases while the maximum torque (the resistance against deformation, respectively) decreases, and the density of the vulcanization network also decreases. Tables 7.3 and 7.7 also show that at higher amounts of deposited magnetite the scorch time T_{s1} and T_{s2} as well as the 50 and 90% vulcanization time is prolonged, however, the process is quick. Tand@ML decreases, while tand@MH increases with the increasing amount of deposited magnetite. Obviously, the presence of magnetite hinders the vulcanization process—the higher the magnetite amount the harder the vulcanization. Having in mind the mechanism of accelerated sulfur vulcanization, described in Mark et al.[36] we suppose that to a certain extent magnetite blocks the activity of zinc oxide used as a vulcanization activator. The deposition of magnetite in the activated carbon pores also blocks the active adsorption centers where macromolecules contact the sulfide complex, which in turn activate vulcanization. Comparing the vulcanization characteristics of samples MAC-3 and MAC-4, it is observed that the addition of CH_3OH to the aqueous solution of ferrous salts, demonstrated mainly as a certain elimination of the negative impact of magnetite, leads to shortening, although slight, of the vulcanization time. Evidently, CH_3OH addition is practically equivalent to a decrease of magnetite content in the activated carbon. Although being introduced as a medium of the precursor, magnetite improves the wetting of activated carbon surface and pores, which enables its penetration.

TABLE 7.7 Vulcanization Characteristics of Compounds MAC-0, MAC-1, MAC-2, MAC-3, and MAC-4, Determined by a Vulcameter at 150°C.

	MAC-0	MAC-1	MAC-2	MAC-3	MAC-4
M_L, dN.m	0.12	0.22	0.33	0.38	0.66
M_H, dN.m	30.46	18.99	14.40	10.93	13.53
$\Delta M = (MH–ML)$	30.34	18.77	14.07	10.55	12.87
T_{s1}, min:s	0:28	0:33	0:35	0:39	0:32
T_{s2}, min:s	0:32	0:37	0:39	0:46	0:36
T_{50}, min:s	0:48	0:53	0:57	1:10	0:58
T_{90}, min:s	1:11	1:34	2:28	5:57	5:45
tand@ML	2.417	2.364	1.794	1.538	1.373
tand@MH	0.015	0.034	0.054	0.056	0.049

7.3.2.6.2 Cross-link Density

The swelling was carried out as per the ISO 1817 at room temperature in toluene. The vulcanization network density ($v=1/2Mc$) was estimated using Flory-Rehner[37] equation for calculating the molecular mass of the segment between cross-links:

$$M_C = \frac{-\rho_r.V_s.(V_r)^{1/3}}{\ln(1-V_r)+V_r+\chi V_r^2} \qquad (7.7)$$

where:

M_c is average molecular mass of the segments of elastomer chains between cross-links of the vulcanization network .

ρ_r is rubber density; V_s is molar volume of the solvent; χ is the parameter characterizing rubber–solvent interaction

V_r is volume fraction of the swollen network

The density of the vulcanizates network (v) of the vulcanizates comprising the hybrid fillers are shown in Figure 7.16.

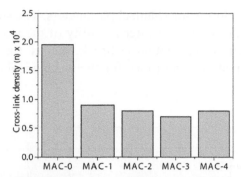

FIGURE 7.16 Cross-link density for equilibrium swelling degree of vulcanizates comprising filler activated carbon with different amounts of magnetite.

The results presented in Figure 7.16 confirm those obtained from the isotherms. As seen, the density of the vulcanization network decreases with the increasing magnetite amount deposited in the activated carbon. Similar results are obtained for hybrid fillers based on conductive carbon black and magnetite,[38] in which cases the increasing amount of magnetite phase leads to a loosened vulcanization network, while the molecular weight of the segment between cross-links increases, due to much inert character of magnetite.

7.3.2.6.3 Mechanical Properties

The mechanical properties of the vulcanized rubber composites were determined according to ISO 37:2002. Shore A hardness of the rubber composites was determined according to ISO 7619:2001.

The dependence of the vulcanizates mechanical parameters on the type of hybrid fillers studied is shown in Figures 7.17–7.20. The higher magnetite amounts in the activated carbons lead to lower modulus at 100% of elongation and Shore A hardness values (Figures 7.19 and 7.20), and to higher relative elongation at break and residual elongation (Figures 7.17 and 7.18). The results are in accordance with those about the swelling and vulcanization characteristics determination. The denser vulcanization network predetermines higher values of modulus at 100% of elongation and Shore A hardness, and lower values of the relative elongation at break and residual elongation, and vice versa. As seen, the magnetite deposited in activated carbon pores besides being inert filler, occupies the active centers which might be occupied by elastomer macromolecules. Thus, magnetite weakens the elastomer-filler interaction which is of considerable

importance for all physics-mechanical parameters. The weaker elastomer–
filler interaction also ensures greater mobility of the macromolecules, an
increase in the relative elongation at break, that is, the ratio of the two
phases is of the greatest significance for the mechanical characteristics of
the vulcanizates.

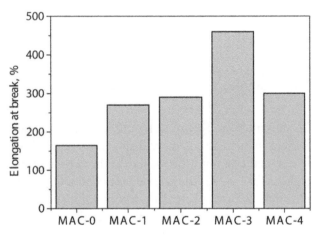

FIGURE 7.17   Elongation at break of vulcanizates comprising filler activated carbon with
different amounts of magnetite.

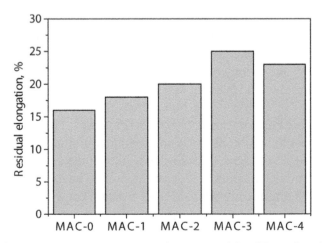

FIGURE 7.18   Residual elongation of vulcanizates comprising filler activated carbon with
different amounts of magnetite.

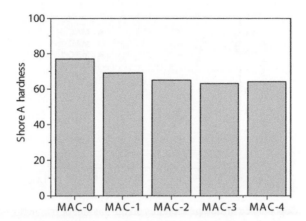

FIGURE 7.19   Shore A hardness of vulcanizates comprising filler activated carbon with different amounts of magnetite.

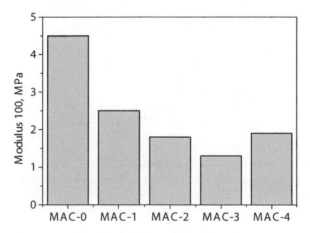

FIGURE 7.20   Stress at 100% of elongation of vulcanizates comprising filler activated carbon with different amounts of magnetite.

7.3.2.6.4   Dynamic Properties

Figures 7.21 and 7.22 present the storage modulus and tangent of mechanical loss angle of the vulcanizates as a function of the amount of magnetite deposited in the activated carbon.

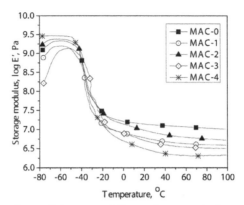

FIGURE 7.21   Dependence of storage modulus on the amount of magnetite deposited in the pores of activated carbon in the range of −80 to +100°C.

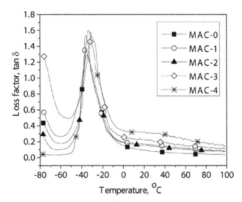

FIGURE 7.22   Dependence of mechanical loss angle tangent on the amount of magnetite deposited in the pores of activated carbon in the range of −80 to +100°C.

As shown in Figure 7.21, when the samples studied are in the viscoelastic state (over −20°C) the higher amount of magnetite in the activated carbon leads to lower storage modulus values. This is due to magnetite being and inert filler, on the other hand, to its distribution into the activated carbon pores, which blocks the adsorption of rubber molecules over the carbon surface. The final effect is a weakened elastomer-filler interaction. The presumption is confirmed completely by the changes in tangent of mechanical loss angle with the increasing amount of deposited magnetite in the pores of activated carbon (Figure 7.22). It is known that the intensity of the peak assigned to the glass transition temperature, reveals macromolecules mobility, the intensity of the activated carbon-elastomer interaction.[36] The peak is small when the filler is virgin activated carbon (MAC-0), but it increases with the increasing magnetite amount in the activated carbon. Obviously, in the case of virgin activated carbon the macromolecules mobility is the lowest, because of the strong adsorption interactions. Being introduced into activated carbon magnetite blocks the active centers of the former, which leads to enhanced mobility of elastomer macromolecules and a higher peak of glass transition temperature. It should be also noted that because of the reasons mentioned above, the increase in magnetite amount leads to a slight increase (of about 2–3°C) in the glass transition temperature as well.

7.3.2.6.5 Thermal Conductivity

The coefficients of thermal conductivity and their change with the changes in magnetite amount deposited in the activated carbon (Table 7.8) were evaluated by the stationary method at 20°C using poly(methyl methacrylate) with a known coefficient of thermal conductivity $\lambda=0.186$ [W/m.K] as an etalon. The dimensions of the measured specimens were width—140 mm; length—170 mm; and thickness—4.15 mm.

TABLE 7.8   Thermal Conductivity Coefficient of Natural Rubber-Based Composites.

	λ, W/m.K
MAC-0	0.16
MAC-1	0.21
MAC-2	0.23
MAC-3	0.25
MAC-4	0.20

As seen in Table 7.8, the increasing magnetite amount leads to an increase in the thermal conductivity coefficient of the composites, if compared to that of the composite without magnetite (MAC-0). This could be considered a regularity, having in mind what the available data show; the thermal conductivity coefficient of NR is 0.15 W/m.K,[39] of virgin activated carbon—0.17÷0.28 W/m.K,[40] and of virgin magnetite—around 4.0 W/m.K.[41] Comparing the thermal conductivity of composites MAC-3 and MAC-4, we can observe the effect that the mode of magnetite deposition and the degree of its penetration into carbon pores have upon the thermal conductivity coefficient. Obviously, introducing CH_3OH as a medium of the precursor used improves the wetting of activated carbon surface and favors the penetration of magnetite phase into the finest pores (micropores, inclusive) filling them entirely or partially. However, thus the magnetite effect upon the thermal conductivity coefficient lessens since it has been entirely insulated by activated carbon, which has a lower coefficient.

7.3.2.6.6 Volume Resistivity and Homogeneity

Volume resistivity (ρ_v, Ω.m) of the obtained flat rubber-based samples was measured using two electrodes (2-terminal method) and calculated by the equation:

$$ñ_v = R_v S / h \tag{7.8}$$

where:

R_v is Ohmic resistance between the electrodes; A *Wheatstone bridge* was used to measure the resistances;

h is sample thickness between the electrodes, m is the measurements were performed on a Mitutoyo micrometer (0.001 mm);

S is cross-sectional area of the measuring electrode, m^2.

The surface resistivity (ρ_s, Ω) of the samples was calculated using the formula:

$$\rho_s = 100 R_s \tag{7.9}$$

where R_s is the Ohmic resistance between the electrodes, measured in Ω.

The average squared deviation σ_{asd} was calculated by the formula:

$$\sigma_{asd} = \sqrt{\frac{1}{n} \sum_{i=1}^{n} \left(x_i - \bar{x} \right)^2} \tag{7.10}$$

where n is the number of volume resistivity measurements (10 in our case).

Table 7.9 presents the results obtained from the volume resistivity measurements of vulcanizates comprising a filler of activated carbon with nano-sized magnetite deposited in it.

TABLE 7.9   Volume Resistivity of Vulcanizates Comprising a Filler of Activated Carbon with Nano-Sized Magnetite Deposited in it.

	Magnetite content (%)	ρ_v $(\Omega.m)$	Log ρ_v $(\Omega.m)$	Average square deviation $(\sigma_{asd}, \%)$
MAC-0	–	$2.6.10^4$	4.42	2.91
MAC-1	2.7	$2.9.10^4$	4.46	2.52
MAC-2	3.5	$3.5.10^4$	4.54	2.34
MAC-3	5.1	$9.0.10^4$	4.95	2.04
MAC-4	5.4	$3.2.10^6$	6.50	1.95

The results show that with increasing amount of magnetite the volume resistivity increases gradually. According to the data, activated carbon and magnetite have quite different volume resistivity values (activated carbon— $2.1–2.6 \times 10^{-5}$ $\Omega.m$,[42] magnetite—$10^0–10^{-1}$ $\Omega.m$[43]), that is, those of magnetite are much higher. Therefore, the increase in volume resistivity of the composite with increasing amount of magnetite could be considered regular. Obviously, the sites of magnetite deposition are also important and similarly to the described case of thermal conductivity, one notices the effect of deeper magnetite penetration and isolation of the conductive activated carbon structures by the externally located magnetite. This finally results into a significant increase in volume resistivity when using methanol, regardless of the small difference in magnetite concentration in the activated carbon comprised by samples—MAC-3 and MAC-4 (Table 7.3).

We used the volume resistivity measurements and the average squared deviation in the measurements, to estimate the homogeneity of the compounds comprising activated carbon with different amounts of magnetite (Table 7.9). There exists data about the use of results from different investigation methods, for example, intrinsic viscosity measurements, DSC, SEM,[44] dielectric constant (permittivity) and dielectric loss,[45] DC and AC electrical conductivity/resistivity,[46,47] for homogeneity evaluation, and the fillers dispersion in rubber compounds, respectively. Since, volume resistivity is an exceptionally structure-sensitive parameter, being an intrinsic material property, which therefore is independent of the sample geometry, its measurements may assess fundamental information about the filler distribution/dispersion[47] and about the filler network morphology/density. Our hypothesis is that the average squared deviation in volume resistivity measurements at a great number of points could give the real homogeneity, the filler dispersion in the rubber compound. The lesser the average squared deviation, the more homogeneous the compound and the better the filler dispersion. As seen from the results in Table 7.9, at a higher amount of magnetite deposited in the activated carbon the average squared deviation gets smaller, that is, the filler dispersion in the composites and their homogeneity improve. The homogeneity improvement at higher amounts of magnetite could be explained by its weakening the adsorption rubber-activated carbon interaction, distortion of activated carbon structures, and their insulation by magnetite phase, which leads to a weak particles interaction and to their better dispersion. This assumption is confirmed by the data in Table 7.5, which reveals that magnetite is distributed predominantly over the external surface and not in the volume (internal surface) of all fillers. With increase in the amount of magnetite the phase over the external surface also enlarges.

The dependence of average squared deviation in volume resistivity measurements on magnetite amount for composites filled with activated carbon black comprising a different magnetite content is shown in Figure 7.23. As seen, the dependence is linear, which reveals that it is following the additivity law. This can be considered as a proof of the improving compatibility between the elastomer matrix and filler with the increasing magnetite amount. On the other hand, the latter fact evidences the beneficial effect of magnetite amount on composites homogeneity and dispersion of its filler.

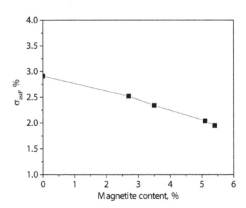

FIGURE 7.23 Dependence of average squared deviation on the amount of magnetite deposited in activated carbon.

7.3.2.6.7 Dielectric Properties

Figure 7.24 shows the change in dielectric constant as a function of the amount of magnetite in the activated carbon pores at 30°C and Figure 7.25— the dielectric loss angle tangent.

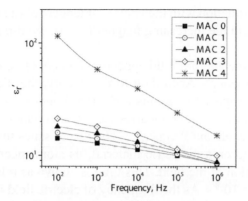

FIGURE 7.24   Dependence of real part of permittivity on the frequency at different amount of magnetite deposited in activated carbon at 30°C.

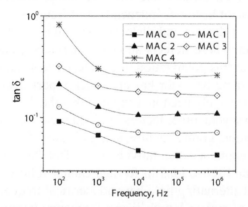

FIGURE 7.25   Dependence of tan δ_ε on the frequency at different amount of magnetite deposited in activated carbon at 30°C.

The dielectric constant of NR is 2.11; that of activated carbon—12, and of magnetite—between 33 and 80. Obviously, the higher magnetite concentration in activated carbon leads to higher dielectric constant values of the composites thus filled. Noteworthy is the well pronounced decrease of these values at higher frequency. At the end of the frequency range studied, these values become close and equal. It is also notable that at a higher magnetite concentration the dependence is more affected by frequency, that is, the changes in the dielectric constant with increasing frequency are more considerable. This is seen from the slope of the curves, which is steeper for the composite comprising a higher amount of magnetite.

Similar is the situation with the values for tangent of dielectric loss angle, which increase with the increasing magnetite concentration and decrease at higher frequency.

According to the authors,[48] this phenomenon could be explained on the basis of Koop's theory.[49] According to this theory, the dielectric structure was said to be composed of grains and grain boundaries in which grains were conductor while grain boundaries were non-conductive. This was also suggested by Maxwell and Wagner.[50,51] The theory states that when the electric field is applied on a dielectric material, its atoms need a finite time to align in the direction of the field. This time is known as relaxation time and its precise value is 10^{-9} s. As the frequency of electric field increases, a point is reached when charge carriers of dielectric do not align in accordance with the field, hence polarization cannot achieve its saturation value. The value of dielectric constant decreases and when we further increase the frequency, the dielectric constant becomes independent of frequency. To explain these trends, the effect of defects and dislocations in the samples might also be taken into account. These defects activate interfacial polarizations at low frequencies. Due to this polarization, the dielectric constant is higher at low frequencies. The same behavior has also been reported in literature.[52,53] In addition, it has intensively been investigated that the dielectric properties of ferrites depend on several other factors, including the method of preparation, chemical composition, grain structure, and grain size of the ferrite nanoparticles. This may be explained qualitatively by the fact that electronic exchange between ferrous and ferric ions in ferrites cannot follow the frequency of externally applied alternating field beyond a critical frequency value.[54]

Decrease in[55] ε_r' and $tg\ \delta$ with rising frequency is explained by decelerating of dipoles and decreasing the number of particles, participating in polarization, that is, with deterioration of polarizing process. At low concentration of Fe_3O_4 and, consequently, at a smaller size of ferro-filler particles the volume density of defects is considerably lower than in the samples with large Fe_3O_4 nanoparticles, which affects not only the dielectric, but also other electrophysical properties. Obviously, the changes in permittivity and dielectric loss tangent in dependence of frequency of the same magnetic fillers can be connected with weakened interaction between neighboring molecules of polymeric matrix and interphase action. It is also possible that the magnetic field of a polymer molecule can be partially oriented, and as a result dropping the potential barrier of shear acts and decreasing the value of inductive capacity of the nanocomposite. Permittivity decrease by frequency is connected with deterioration of the polarization process in nanocomposite.

7.3.2.6.8 Microwave Properties

Figure 7.26 presents the results for the total shielding effectiveness of the control material polytetrafluoroethylene (PTFE) and the studied fillers as a frequency function in the 1–12 GHz range. As seen in the entire frequency range, all samples exhibit the same behavior dependent on frequency changes. In the range of 5–12 GHz the tendency of changing the total shielding effectiveness of the sample studied remains the same in a diapason of 2.5 dB and reaching a maximum value of 7.16 dB for sample MAC-3 at 10 GHz. Total shielding effectiveness values start at about 0.5 dB at 1 GHz and increase monotonously with the increasing frequency. In the frequency range, studied sample MAC-0 is the least sensitive, which is probably due to the internal (in the carbon volume) and not external distribution of the magnetite phase. There is an obvious drop of the values for all samples at 9 GHz. Teflon does not exhibit shielding properties in the entire frequency range and its total shielding effectiveness tends to zero.

Figure 7.27 presents the results for the reflective shielding effectiveness (SE_R). Samples MAC-3, MAC-2, and MAC-1 possess the highest reflective shielding effectiveness values. With the increasing magnetite amount, the reflective shielding effectiveness improves. As seen, SE_R increases monotonously in the studied frequency range for all samples, in most cases SE_R values drop at 11 and 12 GHz. Sample MAC-0 is the least sensitive to the frequency changes and sample MAC-4 exhibits similar behavior.

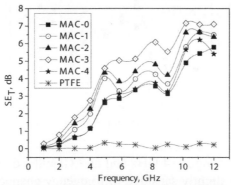

FIGURE 7.26   Frequency dependence of the total shielding effectiveness of the studied composites.

Figure 7.28 presents the results for the absorptive shielding effectiveness (SE_A) of the studied composites. It concludes that the absorption by the samples is affected strongly by the frequency changes. The achieved dynamic diapason of the values is less than 2 dB at frequencies 5.8, 11, and 12 GHz, and the maximum values are 2.9, 3.8, and 3.7 dB, respectively. Sample MAC-3 has the highest SE_A values, while sample MAC-1 has the lowest values at the highest frequencies.

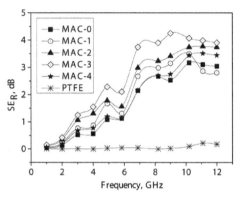

FIGURE 7.27 Frequency dependence of the reflective shielding effectiveness of the investigated composites.

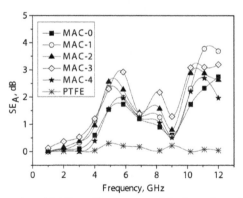

FIGURE 7.28 Absorptive shielding effectiveness of the investigated composites.

In the frequency range 1–4 GHz and at 7 GHz the dynamic diapason of the values remains slightly sensitive to frequency changes (less than 1 dB for all the samples).

As seen from Figures 7.26–7.28, the total shielding effectiveness of the composites studied is due mainly to the mechanism of electromagnetic power reflection. The absorption mechanism predominates only at frequencies in the 5–6 GHz range. As far as the location of the oxide phase is concerned, it is chiefly in the volume of the finest pores and does not have a well-pronounced positive effect.

7.3.2.6.9   Complex Permittivity and Permeability

Absorbers are characterized by their permittivity and permeability. The permittivity is a measure of the materials effect on the electric field in the electromagnetic wave and the permeability is a measure of the materials effect on the magnetic component of the wave. The real part of the permittivity, ε_r', named dielectric constant, measures how much energy from an external electric field is stored in the material. The imaginary part, ε_r'', named loss factor, accounts for the loss energy dissipative mechanisms in the materials, quantifying it as a measure of the attenuation of the electric field caused by the material. Both components contribute to wavelength compression inside the material. In addition, due to coupled EM wave, loss in either the magnetic or electric field will attenuate the energy in the wave.

Figures 7.29 and 7.30 present the frequency dependence of the real and imaginary parts of the permittivity for the composites comprising hybrid fillers with an in situ synthesized magnetite phase at various concentrations. Figure 7.31 plots the frequency dependence of the tangent of dielectric loss angle.

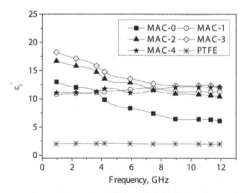

FIGURE 7.29   Frequency dependence of the real part of the permittivity for the composites investigated.

The real part of the relative permittivity for samples MAC-0, MAC-2, and MAC-3 decreases with the increasing frequency in the studied range, while for samples MAC-1 and MAC-4 practically there is no such dependence.

Sample MAC-3 possesses the highest values of the parameter in the 1–7.5 GHz range, while in the range over 7.5 GHz the values become practically equal for all materials except for sample MAC-0. Regarding the imaginary part of permittivity as a frequency function, there are two ranges: 1–7.5 GHz, wherein MAC-3 and MAC-2 possess the highest values decreasing rapidly with the increasing frequency; and 7.5–12 GHz wherein the values for all samples are very close and the frequency dependence of the parameter is slightly pronounced. As far as the dependence of the tangent of dielectric loss angle is concerned, sample MAC-3 has distinguishable values in 1–5 GHz range, while the values for MAC-0 are the highest in 5–12 GHz range.

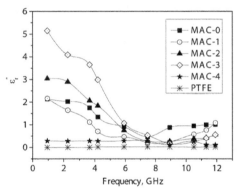

FIGURE 7.30   Frequency dependence of the imaginary part of the permittivity for the composites investigated.

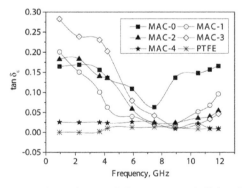

FIGURE 7.31   Frequency dependence of the tangent of dielectric loss angle for the composites investigated.

From the results displayed in Figures 7.29–7.31 one can conclude that the dielectric properties of most materials vary considerably with the frequency of the applied electromagnetic field. An important phenomenon contributing to the frequency dependence of the dielectric properties is polarization, which arises from the orientation of molecules that have permanent dipole moments with the applied electric field. The mathematical formula developed by Debye to describe the permittivity for polar materials[56] can be expressed as:

$$\varepsilon_r = \varepsilon_\infty + \frac{\varepsilon_s - \varepsilon_\infty}{1 + j\omega\tau} \qquad (7.11)$$

where $\omega = 2\pi f$ is the angular frequency, ε_∞ represents the dielectric constant at frequencies so high that molecular orientation does not have time to contribute to the polarization, ε_s represents the static dielectric constant, that is, the value at zero frequency (dc value), and τ is the relaxation time in seconds, the period associated with the time for the dipoles to revert to random orientation when the electric field is removed. Separation of last Equation into its real and imaginary parts yields:

$$\varepsilon_r' = \varepsilon_\infty + \frac{\varepsilon_s - \varepsilon_\infty}{1 + (\omega\tau)^2} \qquad (7.12)$$

$$\varepsilon_r'' = \frac{(\varepsilon_s - \varepsilon_\infty)\omega\tau}{1 + (\omega\tau)^2} \qquad (7.13)$$

Obviously, ε_∞, ε_s and, in particular, τ are specific parameters for each of the fillers used, which depends on the chemical nature and crystallographic structure of the two different filler phases (the active carbon one and that of Fe_3O_4). The dependence of ε_∞, ε_s and τ on activated carbon/magnetite ratio in the synthesized fillers explains why in some frequency ranges (usually at lower frequencies) the dielectric properties change monotonously, while in some higher frequency range, when relaxation is hindered, the increase is drastic. According to the Debye theory of dielectric properties,[56] ε_r'' is generally determined by relaxation and electrical conductivity losses. It is clear that both polarization relaxation and electrical conductance can affect ε_r''. As all other conditions are identical, obviously, that peculiarity is due to the specific chemical nature and crystallographic structure of the activated carbon and magnetite phases.

On the other hand, since the elastomer matrix is non-polar, apparently, the relaxation and polarization processes are greatly dependent on the fillers used and on their specific features. With regard to these specifics, the polarization may proceed according to three different mechanisms: electronic,

ionic, and orientational. All non-conducting materials are capable of electronic polarization. Therefore, we consider the polarization of the elastomer
matrix used for the studied composites to proceed according to this mechanism. The more polarization mechanisms of a composite are, the higher its
dielectric constant will be. The more easily the various polarization mechanisms can proceed, the higher the dielectric constant will be. For example,
among elastomers, the more mobile the chains are (i.e., the lower the degree
of crystallinity), the higher the dielectric constant will be. It is important for
the composites investigated because the NR crystallizes and the chemical
nature of the fillers used and their amounts can change the degree of crystallinity and the dielectric constant.

Figures 7.32 and 7.33 present the frequency dependencies of the real and
imaginary part of permeability. Figure 7.34 plots the frequency dependency
of tangent of magnetic loss angle.

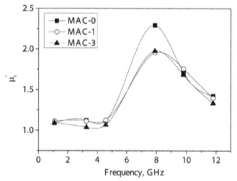

FIGURE 7.32 Frequency dependence of the real part of permeability for the composites
investigated.

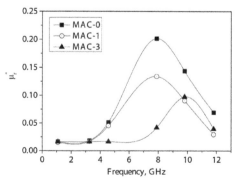

FIGURE 7.33 Frequency dependence of the imaginary part of permeability for the
composites investigated.

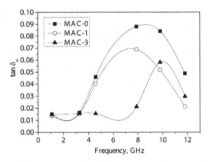

FIGURE 7.34 Frequency dependence of the tangent of magnetic loss angle of the studied composites.

In the 1–4 GHz range μ_r' is close to 1, and μ_r'' is almost zero, this means that there are no magnetic losses. In the 4–12 GHz range an increase in μ_r' and μ_r'' with the increasing frequency up to about 8 GHz has been observed. The values of μ_r' and μ_r'' decrease with the increasing frequency. The frequency fluctuation of the permeability curves reveals natural resonances in the composite, which can be ascribed to small size of Fe_3O_4 nanoparticles.[57] At a higher amount of the magnetite phase the real part of magnetite permeability increases negligibly, while the imaginary part and tangent of magnetic loss angle increase considerably.

As seen from the comparison of the results in the 4–12 GHz range SE_A values exhibit resonance properties, while in the ranges 4–6 GHz, 7–8 GHz, and 9–10 GHz the absorption is the predominant shielding mechanism. It could be related to the increase in the values of tangent of magnetic loss angle plotted in Figure 7.32. Fe_3O_4 nanoparticles act as tiny dipoles that get polarized by the activation of electromagnetic field and result in better microwave absorption. The magnetic loss μ_r'' is a combined result of eddy current effect, natural resonances, and anisotropy energy of the sheet. In the microwave ranges, the eddy currents are attributed to ferrite nanoparticles presence in the sheet.[58]

7.3.3 ELASTOMER-BASED COMPOSITE MATERIALS COMPRISING ACTIVATED CARBONS MODIFIED WITH ZNO AS FILLERS

7.3.3.1 PREPARATION OF THE FILLERS

Impregnation is the most used and popular among methods for applying various phases on supports (in the production of catalysts) or for surface and texture modification of adsorbents.

Mostly porous materials or those forming a pseudo-porous texture, for example, those made up of aggregates or agglomerates, undergo impregnation. In general, during impregnation, the phase is deposited from the solution of the precursor (often salts or any soluble complexes). After a certain technological procedure, for example, evacuation for a certain time in a controlled enclosure under vacuum or a in gas atmosphere, the impregnated materials are subjected to chemical or physical treatment—most often to thermal treatment, having the character of thermal activation.

In practice, the impregnation is carried out by "soaking" of the material in a suitable impregnating solution (at a certain concentration and temperature, stirring—mass exchange, and impregnation time) or by "spraying". The second method is more economical, but depends on the wettability of the material; on the interphase tension at the boundary "solid—liquid" and stirring in the reactor during the spraying, on the way the solution is sprayed. Usually, optimal uniformity of deposition of the impregnation phase is achieved when the process is repeated twice or more times. The so-called aqueous absorption (in aqueous impregnation solutions) or in the more general case "absorption capacity" of the respective solvent should be determined before carrying out the impregnation.

Complete removal of the solvent after the activation procedure is considered a technological requirement for impregnation. The possibility of easy destruction of the soluble impregnation complex after its application in the process of thermoactivation is also an important concern. To a large extent, the distribution of the impregnation phase is determined by the concentration of the impregnation solution and the treatment of the material after application of the impregnation precursor (e.g., duration of the evacuation in a controlled enclosure, purge of air with a certain humidity or of a gas). In all cases, the incorporation of a suitable surfactant in the impregnation solution has a positive effect. Similarly, for materials with low wettability, applying the precursor phase from an organic medium has a positive effect on the impregnation process. The latter, however, is associated with more complicated equipment and additional safety requirements.

For practical and operational reasons, it was absolutely necessary to seek an opportunity to enhance the activity of activated carbon as a filler while maintaining or improving its absorption activity regarding ultra-high frequency electromagnetic waves.

Impregnation, allowing a controlled penetration of the phase into the macro- and the mesopores of the activated carbon, was chosen for deposition of the modifying agent. Activated carbon obtained by steam-gas activation of lignite coal, meeting the requirements set out in Section 7.2 has been selected as an object of modification.

The preferred modifying agent was ZnO—on the one hand because of its key role[36] in processing of elastomeric materials (mainly in vulcanization), and because of the ability of ZnO to be converted easily into a soluble metal-ammine complex, which after application and appropriate thermal treatment can be converted easily back into ZnO. Moreover, it contains additional hydrosilane groups, facilitating the occurrence of "elastomer–filler" interactions. We have found no evidence of this kind of modification of activated carbon with ZnO in the literature.

The procedure of modifying impregnation involved the following stages:

- Drying of the active carbon at 393 K for 3 h
- Obtaining an impregnation Zn-ammine solution by mixing ZnO, (NH4) 2CO3, ammonium hydroxide (25%), and dest. H_2O at a ratio of 1:1.26:2.26:3.39. The solution was kept at pH ~ 10.5 not lower than pH 10
- Depositing the solution of Zn-ammine complex by a TLC sprayer, at continuous stirring of the carbon (in an amount of about 25–27g). The volume of solution required for a single application was determined on the basis of the total pore volume calculated from the nitrogen isotherm (77 K) of activated carbon
- Having being impregnated with the solution, the samples were kept in an enclosure for about 6 h. Then they were treated thermally for 2 h at a temperature not higher than 443 K.
- Following the second thermal treatment, the samples were ground and then transferred to the reactor for decomposition of the applied Zn-ammine complex
- Decomposition of Zn-complex was carried out at a residual pressure (vacuum) of 5 Torr, a heating rate of up to 673 K (3°/min) and duration of the thermal treatment at 673 K for 6 h The cooling of the samples was carried out in the reactor at room temperature and the above residual pressure.

The modified sample was labeled AC.1, the substrate one—with AC.0.

7.3.3.2   CHARACTERIZATION OF THE FILLERS

The parameters for fillers AC.0 and AC.1 calculated by the experimental adsorption-desorption isotherms are presented in Table 7.10.

TABLE 7.10   Content of the Modification Agent (Determined with Regard to Zn) and Main Texture Characteristics of the Fillers–Substrate (AC.0) and Zn-Modified (AC.1).

	Zn* content (mass.%)	S_{BET} (m²/g)	S_{MES} (m²/g)	W_0 (cm³/g)	V_{MI} (cm³/g)	V_{SMI} (cm³/g)	V_{MES} (cm³/g)	V_t (cm³/g)	r_P (nm)
AC.0	–	257	66	0.124	0.093	0.031	0.173	0.297	5.2
AC.1	6.13	282	29	0.156	0.125	0.031	0.101	0.257	7.1

*Determined by atom force spectroscopy.

As seen from the data in the table, filler AC.0 (substrate active carbon) is characterized by relatively underdeveloped specific surface area (257 m²/g) and a total volume of micropores 0.124 cm³/g. In contrast, the surface and the volume of mesopores are developed, which makes it suitable material for modifying a metal oxide phase.

Carbon filler AC1 obtained after modification by the described procedure has a 10% higher specific surface in comparison with that of AC.0. The probable explanation of this fact is the reactivation of basic carbon texture in the process of thermal decomposition of Zn-ammine complex. There is also an increase in the total volume of micropores, and the change in the ratio of volumes of micro- and supermicropores of substrate AC.0. In addition to reactivation, this can be explained by changes in mesopores texture resulting from the applied Zn-phase. Therefore, the reduction in the volume of the mesopores and the surface is 1.7 and more than 2 times, respectively in comparison with the same parameters of the substrate carbon. Thus, the increase in the total volume of the micropores results from the partial filling of the volume of the finer mesopores with a zinc phase, while the zinc phase deposited in larger mesopores during the process of impregnation and subsequent thermal treatment, is responsible for increasing their average radius up to 7.1 nm.

7.3.3.3   PREPARATION OF THE COMPOSITES

The modified and unmodified fillers were introduced into elastomeric SBR-based compounds with the following composition (in phr): SBR—100; stearic acid- 1,5; zinc oxide 5; N-isoprpyl-N-phenyl-p-phenylenediamine-1,5;

filler—50;   glycerol—3;   N-cyclohexyl-2-benzothiazolesulfenamide—1,3; sulfur—2.

The elastomer blends were prepared on an open laboratory rubber mixing mill (L/D 320×160 and friction 1.27). The speed of the slower roll was 25 min⁻¹. *They were vulcanized in the optimal time determined in correspondence to ISO 3417:2002 by the vulcanization isotherms which were taken on a moving die rheometer MDR 2000 (Alpha Technologies Ltd, Burnaby, BC, Canada) at 150 °C. The vulcanization was carried out in a hydraulic electric press with the size of the plates 400x400mm at a pressure of 10 MPa. The obtained composites had dimensions 60x80x1.5mm, from which were cut out tensile test samples in the form of dumbbells.*

7.3.3.4   CHARACTERIZATION OF THE COMPOSITES

7.3.3.4.1   Mechanical Properties

The mechanical properties of the studied vulcanizates comprising modified (AC.1) and non-modified (AC.0) filler are summarized in Table 7.11:

TABLE 7.11   Mechanical Properties of the Studied Vulcanizates.

	Type of filler	
	Non-modified	Modified
Shore A hardness	48	47
M_{100} (MPa)	0.6	0.9
M_{300} (MPa)	1.9	2.4
σ (MPa)	3.2	5.1
E_{rel} (%)	520	650
E_{res} (%)	15	15
Density (g/cm³)	1.13	1.15
Heat aging		
K_{σ} (%)	−12	−5
K_{g} (%)	−15	−7

As seen from Table 7.11, the modification increases the moduli 100 and 300%, the tensile strength and the elongation at break. The heat aging resistance also improves.

7.3.3.4.2 *Bound Rubber and Cross-link Density*

The amount of bound rubber after the modification of the filler also increases (Figure 7.35), what we take as evidence of enhanced "elastomer–filler" interaction.

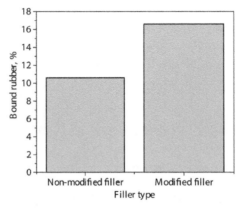

FIGURE 7.35   Effect of the non-modified and modified filler upon the amount of bound rubber (loading 50 phr).

A proof of the reinforcing effect of certain filler is its improved swelling resistance of the vulcanizates comprising it (Figure 7.36).

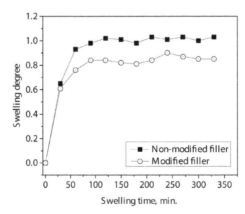

FIGURE 7.36   Effect of non-modified and modified filler on the swelling degree of the vulcanizates comprising them (loading—50 phr).

The improvement in the reinforcing effect of the modified filler in terms of the mechanical properties of the vulcanizates compared with those filled with the substrate is related to both the abovementioned changes in the texture after the modification, as well as to the presence of zinc oxide phase and its nano-sized state.

Direct correlation between the texture parameters of the modified filler and properties of vulcanizates at this stage has not been found. The results from the comparison of the samples containing non-modified and modified carbon-based fillers indicate that the influence of texture parameters exist and it is not confined to the specific surface. On the other hand, the statement that micropores (submicro- and supermicro-) that have significant contribution in shaping and in the values of specific surface area by BET, worsen the physicochemical and mechanical properties of vulcanizates because their main characteristic parameter (half-width) is relatively small and does not let filling their volume (e.g., with adsorbed molecules of the accelerator or with elastomer macromolecules[59]) has not been fully justified. In our case, viewing the improved enhancing effect of the modified filler quite obvious, where the volume of submicropores increases (Table 7.10) and that of super-micropores remains the same for the two materials.

In our opinion, volume and size distribution of mesopores are the texture parameters most important for the enhancing effect of the fillers and for the mechanical properties of the vulcanizates studied. They provide restriction to nanoscale particle size of the modifying agent (ZnO), and hence contribute to improving the opportunities for "elastomer–filler" contacts in various areas of nanoporous texture. Overall, this leads to increased "elastomer-filler" interaction, which is confirmed by the increased amount of bound rubber. Since the real mesopores have narrow parts, which can easily be blocked (e.g., by adsorbed molecules of the accelerator), an optimal average radius and volume of the mesopores are needed to use as much of their surface area for reacting with the molecules of the elastomers. Obviously, in this case, it is achieved for the modified filler (Table 7.10). Finally, such an average radius of mesopores contributes to the so-called mobile adsorption favors ability of elastomer molecules to move over the surface of the filler from a sorption center to another. This, according to Danneberg,[60] is important for the reinforcing effect of fillers. Therefore, the proportion of the specific surface area of the fillers (non-modified and modified) available for reaction with elastomer molecules is determined by various factors, including average radius and volume of the mesopores. The modifying agent is a kind of a regulator thereof, bringing them closer to the optimal values.

The zinc oxide phase deposited over the modified filler has been studied by Auger electron spectroscopy (AES). Due to the relative variability of Zn in the values Zn (2p) photoelectronic peak O (*1s*) were used to characterize the zinc phase $ZnL_3M_{4,5}M_{4,5}$. The dominant presence of ZnO (k.e. 990.3 eV) has been established (Table 7.12). Meanwhile, upon modification a shift of the values of about 1.5 eV (Table 7.12) towards the higher bounding energy O (1s) has been observed. According to some authors,[61] this shift is an evidence of a possible presence of hydroxyl groups in ZnO phase. The ability of such groups enhances the interaction of ZnO phase with elastomer macromolecules, which could explain the higher content of bound rubber and the better reinforcing effect of the modified filler.

TABLE 7.12 Kinetic Energy for $ZnL_3M_{4,5}M_{4,5}$ and Bounding Energy for O (*1s*) Peaks of Modified Filler AC.1 (eV).

Sample	Zn $L_3M_{4,5}M_{4,5}$, (eV) (ZnO)	O (1s), (eV) (ZnO)
AC.1	990.3	531.9
AC.0	–	530.3

7.3.3.4.3 Volume Resistivity and Thermal Conductivity

The changes in the "elastomer–filler" system having occurred after the modification and their interpretation confirm fully the observed changes in the degree of swelling and in bound rubber amount (already mentioned). The changes in such structure sensitive properties such as volume resistivity and coefficient of thermal conductivity of the tested samples (Figures 7.37 and 7.38) evidence the same.

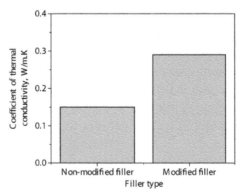

FIGURE 7.37 Effect of filler modification upon the coefficient of thermal conductivity of the vulcanizates comprising it.

A small increase has been observed in the thermal conductivity of the vulcanizates comprising modified filler, which is due to its more compact structure (Table 7.11).

The volume and surface resistivity of the vulcanizate comprising modified filler decrease slightly, probably due to the modifying agent and/or its ions that possess conductive properties better than those of the active carbon and elastomer matrix. The results on the effect of activated carbon modified with zinc oxide are published in.[59]

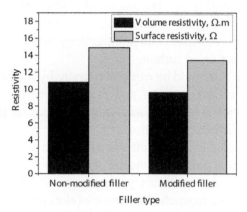

FIGURE 7.38   Effect of filler modification upon the specific electrical resistivity (volume in Ω.m, surface in Ω.

7.4   CONCLUSION

A method for in situ deposition of nano-sized magnetite in the texture of activated carbon up to 5.4% has been developed. Thus, two-phase hybrid fillers have been synthesized for the first time. The fillers have been characterized in detail by X-ray diffraction and photoelectron spectroscopy establishing the influence of the magnetite layer on the texture characteristics. The effect that the deposited magnetite has on the vulcanization process and density of vulcanization network of NR-based composites has been studied. The studies also spread over the impact of hybrid filler magnetite content on the composites mechanical properties, their electro- and thermal conductivity, dynamic characteristics. An attempt to evaluate the composites homogeneity has also been made.

It has been established that magnetite amount changes the texture characteristics of activated carbon, whereas a higher magnetite amount increases the volume and average diameter of mesopores while that of micropores decreases. It could be stated that the magnetite phase is deposited mainly over the external and less in the internal surface of activated carbon. When using organic medium for magnetite synthesis its penetration into the pores and deposition over the external surface is greater than in the case when aqueous medium is used for the purpose. The increasing amount of deposited magnetite hinders the vulcanization process and loosens the crosslink network. The relative elongation at break and residual elongation increase, while Shore A hardness and the modulus at 100% deformation decrease. The storage modulus values decrease while that of the tangent of mechanical loss angle get higher. Mobility of macromolecules also enhances. The dielectric constant and the tangent of dielectric loss angle also get higher. The thermal conductivity coefficient and homogeneity increase at higher magnetite amounts in the activated carbon.

The observed effects could be explained firstly by the impact that deposited magnetite has on the structure and properties of the dual phase filler (texture characteristics, adsorption activity) as well as by the change in the intensity of elastomer-filler interactions.

Studies have been carried out on the effect that the fillers synthesized have on the microwave properties, the real and imaginary part of the permittivity and on the real and imaginary part of permeability; as well as on the tangent of dielectric and magnetic loss angle of the composites based on NR.

It has been found that the total shielding effectiveness of the composite materials filled with activated carbon modified with magnetite is due mainly to the mechanism of reflection of electromagnetic power (reflective shielding effectiveness). However, in the 5–6 GHz range it is the mechanism of absorption (absorptive shielding effectiveness) that has a crucial role for all studied composites.

It has been found that the filler comprising externally about 5% of magnetite phase has the highest total shielding effectiveness amongst the dual phase fillers studied. The amount of magnetite phase above that concentration as well as the internal distribution (inside the volume of the finest pores) does not contribute to the improvement of the microwave characteristics probably on account of a steric hindrance. The absorption of electromagnetic waves is much more frequency dependent than the reflection.

The microwave properties of the tested composites are, to a greater extent, due to their dielectric and magnetic characteristics, and mostly to the dielectric and magnetic loss. The dielectric properties (the values of real and imaginary part of permittivity) are associated with the polarization and relaxation phenomena. The magnetic properties are determined by magnetic moments on the tetrahedral site occupied by ferric species, which are ferromagnetically aligned, while magnetic moments on octahedral sites occupied by ferrous and ferric species are antiferromagnetic, yielding a ferromagnetic behavior. Hence, the crystallochemical nature, the crystallographic structure, and the ratio between the carbon and magnetite phases have an apparent role in the properties of the studied fillers. This opens the opportunity to tailor the microwave properties of composites containing those fillers.

The introduction of the magnetic phase contributes to the improvement of the microwave characteristics of the composites because the co-existence of the dielectric and magnetic losses expands the frequency range in which the composites are of good microwave properties, determined in some cases by the high dielectric, and in others by the high magnetic losses.

The impregnation method known to technologists for the preparation of catalysts and sorbents has been adapted and applied to the carbon-containing fillers for elastomers to create a second phase therein. Such an application of the method has not been described in the literature.

The properties of the vulcanizates containing the modified fillers obtained by impregnation deposition of a second zinc oxide phase over activated carbon have been thoroughly investigated. It has been found that the modified fillers improve several important operational characteristics of the vulcanizates containing them much better than non-modified activated carbon.

The possible applications of composites containing activated carbon as a functional filler are related mainly to their microwave properties, especially to those of the dual phase fillers, whose one phase is of high dielectric losses and the other with high magnetic losses. Microwave absorbers based on them may be successfully employed in certain wavebands.

7.5   ACKNOWLEDGMENT

The work is the part of a project, funded by King Abdulaziz University, Saudi Arabia under Grant number MB/11/12/436. The authors acknowledge the technical and financial support. The authors would like also to thank to Prof. Ahmed A. Al-Ghamdi, Prof. Omar A. Al-Hartomy, Prof. Falleh R. Al-Solamy, and Assoc. Prof. Radostin Nikolov for their useful and long-term cooperation.

KEYWORDS

- **activated carbon**
- **modification of activated carbon**
- **hybrid dual-phase fillers**
- **in situ synthesized magnetite**

- **impregnation technology**
- **modification with zinc oxide**
- **microwave properties**
- **dielectric properties**
- **natural rubber**

REFERENCES

1. von Kienle, H.; Bäder, E. Activkohle und ihre industrielle Anwendung, Ferdinand EnkeVerlag: Stuttgart, 1980.
2. Rodriguez-Reinoso, F. The Role of Carbon Materials in Heterogeneous Catalysis. *Carbon* **1998**, *36*, 159–175.
3. http://www.shaurenvalour.com/granular-activated-carbon-2857823.html.
4. Burevski, D. The Application of the Dubinin-Astakhov Equation to the Characterization of Microporous Carbons. *Colloid Polym. Sci.* **1982,** *260*, 623–627.
5. Villarroel-Rocha, J.; Barrera, D.; Garcia Blanco, A. A.; Roca Jalil, M. E.; Sapag, K. Importance of the αs-Plot Method in the Characterization of Nanoporous Materials. *Adsorpt. Sci. Technol.* **2013**, *31*, 165–183.
6. Noh, J. S.; Schwarz, J. A. Estimation of Surface Ionization Constants for Amphoteric Solids. *J. Colloid Interface Sci.* **1990**, *139*, 139–148.
7. Mark, J. E.; Erman, B; Eirich, F. R., Eds.; *The Science and Technology of Rubber*, 3rd ed.; Elsevier Academic Press: Amsterdam; 2005.
8. Jana, P. B.; Mallick, A. K.; De, S. K. Effects of Sample Thickness and Fiber Aspect Ratio on EMI Shielding Effectiveness of Carbon Fiber Filled Polychloroprene Composites in the X-Band Frequency Range. *IEEE T Electromagn. C.* **1992**, *34*, 478–481.
9. Paul, C. R. *Introduction to Electromagnetic Compatibility*, 2nd Ed.; John Wiley & Sons, Inc.: Hoboken, 2006.
10. Ott, H. W. *Electromagnetic Compatibility Engineering*; John Wiley & Sons, Inc.: Hoboken, 2009.
11. Hernandez, B. Effect of Graphitic Carbon Nanomodifiers on the Electromagnetic Shielding Effectiveness of Linear Low Density Polyethylene Nanocomposites, PhD Thesis, Clemson University, December 2013. http://tigerprints.clemson.edu/all_dissertations.
12. Fenical, G. *The Basic Principles of Shielding*. Basics; 2014. http://incompliancemag.com/article/the-basic-principles-of-shielding/#ixzz4HZWIjRLG.
13. Sadasivuni, K. K.; Ponnamma, D.; Kim, J.; Thomas, S. *Graphene-Based Polymer Nanocomposites in Electronics*; Springer International Publishing: Switzerland, 2015.
14. Park, S.-H.; Theilmann, P. T.; Asbeck, P. M.; Bandaru, P. R. Enhanced Electromagnetic Interference Shielding Through the Use of Functionalized Carbon–Nanotube–Reactive Polymer Composites. *IEEE Trans. Nanotechnol.* **2010**, *9*, 464–469.

15. Hong, Y. K.; Lee, C. Y.; Jeong, C. K.; Lee, D. E.; Kim, K.; Joo, J. Method and Apparatus to Measure Electromagnetic Interference Shielding Efficiency and its Shielding Characteristics in Broadband Frequency Range. *Rev. Sci. Instrum.* **2003**, *74*, 1096–1102.
16. Wieckowski, T. W.; Janukiewicz, J. M. Methods for Evaluating the Shielding Effectiveness of Textiles. *Fibres Text. East. Eur.* **2006**, *14*, 18–22.
17. Chen, L. F.; Ong, C. K.; Neo, C. P.; Varadan, V. V.; Varadat, V. K. *Microwave Electronics Measurement and Material Characterization*; John Wiley & Sons Ltd.: Chichester, 2004.
18. Hunton, J. K.; Poulter, H. C.; Reis, C. S. High-Directivity Coaxial Directional Couplers and Reflectometers. *Hewlett Packard J.* **1955**, *7(2)*, 1–4.
19. Riley, R. B. A New Frequency Counter Plug-In Unit for Direct Frequency Measurements to 510 MC. *Hewlett Packard J.* **1961**, *12(5)*, 1–6.
20. Rouquerol, J.; Avnir, D.; Fairbridge, C. W.; Everett, D. H.; Haynes, J. M.; Pernicone, N.; Ramsay, J. D. F.; Sing, K. S. W.; Unger, K. K. Recommendations for the characterization of porous solids IUPAC. *Pure. Appl. Chem.* **1994**, *66*, 1739–1758.
21. Microwave Absorber: ECCOSORB®, Technical bulletin, Emerson & Cuming Microwave Products, a unit of Laird Technologies, http://www.eccosorb.com/products-eccosorb-high-loss-absorbers.htm.
22. ECCOSTOCK: Low Loss Dielectrics, Technical bulletin, Emerson & Cuming Microwave Products, a unit of Laird Technologies, http://www.eccosorb.com/products-eccostock-low-loss-dielectrics.htm.
23. ECCOSHIELD: EMI/RFI, Technical bulletin, Emerson & Cuming Microwave Products, a unit of Laird Technologies, http://www.eccosorb.com/products-eccoshield-emi-rfi.htm.
24. Shtarkova, R.; Dishovsky, N. Elastomer-Based Microwave Absorbing Materials. *J. Elastom. Plast.* **2009**, *41*, 163–174.
25. Dishovsky, N.; Grigorova, M. On the Correlation Between Electromagnetic Waves Absorption and Electrical Conductivity of Carbon Black Filled Polyethylenes. *Mater. Res. Bull.* **2000**, *35*, 403–409.
26. Zhang, C. S.; Ni, Q. Q.; Fu, S. Y.; Kurashiki, K. Electromagnetic Interference Shielding Effect of Nanocomposites with Carbon Nanotube and Shape Memory Polymer. *Compos. Sci. Technol.* **2007**, *67*, 2973–2980.
27. Attharangsan, S.; Ismail, H.; Baka, M. A.; Ismail, J. Carbon Black (CB)/Rice Husk Powder (RHP) Hybrid Filler-Filled Natural Rubber Composites: Effect of CB/RHP Ratio on Property of the Composites. *Polym. Plast. Technol. Eng.* **2012**, *51*, 655–662.
28. Ling, Q.; Sun, J.; Zhao, Q.; Zhou, Q. Effects of Carbon Black Content on Microwave Absorbing and Mechanical Properties of Linear Low Density Polyethylene/Ethylene–Octene Copolymer/Calcium Carbonate Composites. *Polym. Plast. Technol. Eng.* **2011**, *50*, 89–94.
29. Das, T. K.; Prusty, S. Review on Conducting Polymers and Their Applications. *Polym. Plast. Tech. Eng.* **2012**, *51*, 1487–1500.
30. Chandran, S. A.; Narayanankutty, S. K.; Mohanan, P. Microwave Characteristics of Polyaniline Based Short Fiber Reinforced Chloroprene Rubber Composites. *Polym. Plas. Technol. Eng.* **2011**, *50*, 453–458.
31. Weidenfeller, B.; Hofer, M.; Schilling, F. Thermal and Electrical Properties of Magnetite Filled Polymers. *Compos. Part. A– Appl.* **2002**, *S 33*, 1041–1053.

32. Mangnus, R. Processing and Properties of Magnetite – Rubber Blends. *Kaut. Gummi. Kunstst.* **2003**, *56*, 322–329.
33. Karyakin, Y. V.; Angelov, I. I. Pure Chemical Substances, Moscow, Chemistry, 1974 (in Russ.).
34. Nephedov, V. I. Handbook of X-ray Photoelectron Spectroscopy of Chemical Compounds, Moscow, Chemistry, 1984 (in Russ.).
35. Rossin, J. A. XPS Surface Studies of Activated Carbon. *Carbon* **1989**, *27*, 611–613.
36. Mark, J. E.; Erman, B.; Roland C., Eds.; *The Science and Technology of Rubber*, 4th ed.; Elsevier Academic Press; Amsterdam, 2013.
37. Bruck, S. Extension of the Flory–Rhener Theory of Swelling to an Anisotropic Polymer Systems. *J. Res. Nat. Bur. Stand. A Phys. Chem.* **1961**, *65A*, 485–487.
38. Al-Ghamdi, A. A.; Al-Hartomy, O. A.; Al-Solamy, F. R.; Dishovsky, N.; Zaimova, D.; Malinova, P.; Nihtianova, D. Preparation and Characterization of Natural Rubber Composites Comprising Conductive Carbon Black/Magnetite Hybrid Fillers Obtained by Impregnation Technology. *Polym. Plast. Technol. Eng.* Published online: 01 Mar 2016. http://dx.doi.org/10.1080/03602559.2015.1132461 (accessed 01 Mar 2016)
39. Lasance, C. The Thermal Conductivity of Rubbers/Elastomers. *Electronics Cooling,.* http://www.electronics-cooling.com/ (accessed Nov 2001).
40. Kuwagaki, H.; Meduro, T.; Tatami, J.; Komeda, K. An Improvement of Thermal Conductivity of Activated Carbon by Adding Graphite. *J. Mater. Sci.* **2003**, *38*, 3279–3284.
41. Park, N. W.; Lee, W. Y.; Kim, J. A.; Song, K.; Lim, H.; Kim, W. D.; Yoon, S. G.; Lee, S. K. Reduced Temperature-Dependent Thermal Conductivity of Magnetite Thin Films by Controlling Film Thickness. *Nano. Res. Lett.* **2014**, *9*, 96–104.
42. Lu, W.; Chung, D. D. L. Preparation of Conductive Carbons with High Surface Area. *Carbon* **2001**, *39*, 39–44.
43. Lee, B. Magnetite (Fe_3O_4): Properties, Synthesis, and Applications. *Lehigh Rev.* **2007**, *15*, (Paper 5). http://preserve.lehigh.edu/cas-lehighreview-vol-15/5.
44. Botros, S. H.; Moustafa, A. F.; Ibrahim, S. A. Homogeneous Styrene Butadiene/Acrylonitrile Butadiene Rubber Blends. *Polym. Plast. Technol. Eng.* **2006**, *45*, 503–512.
45. Botros, S. H.; Tawfic, S. Improvement of Homogeneity of SBR/CR Rubber Blends with SBR-g-AA Grafted Rubber. *Polym. Plast. Technol. Eng.* **2006**, *45*, 829–837.
46. Kucerova, J.; Ruziak, I.; Moskova, Z.; Kopa, I.; Krecmer, N.; Mokrysova, M.; Rusnakova, S. The Evaluation of Rubber Blends Homogeneity from Measurements of DC and AC *Electrical Conductivity*, 5th Research/Expert Conference with International Participation, Quality 2007, Neum, B&H, 06–09 June 2007, Proceedings, pp. 417–421, 2007.
47. O'Farreell, C. P.; Gerspacher M.; Nikiel, L. Carbon Black Dispersion by Electrical Measurements. *Kaut. Gummi. Kunstst.* **2000**, *53*, 701–710.
48. Ansar, M.; Atiq, S.; Alamgirand, K.; Nadeem, S. Frequency and Temperature Dependent Response of Fe_3O_4 nanocrsytallites. *J. Sci. Res.* **2014**, *6*, 399–406. http://www.banglajol.info/index.php/JSR.
49. Cullity, B. D. Introduction to Magnetic Materials; Wesley Reading: New York, 1972.
50. Maxwell, J. C. A Treatise on Electricity and Magnetism; Clarendon Press: Oxford, 1982.
51. Wagner, K. W. Zur Theorie der unvollkommenen Dielektrika. *Ann. Phys.* **1913**, *345*, 817–855. http://dx.doi.org/10.1002/andp.19133450502.
52. Rahman, M. M., Halder, P. K.; Ahmed, F.; Hossain, T.; Rahaman, M. Effect of Ca-substitution on the Magnetic and Dielectric Properties of Mn–Zn Ferrites. *J. Sci. Res.* **2012**, *4(2)*, 297–306.

53. Mu, G.; Chen, N.; Pan, X.; Shen, H.; Gu., X. Preparation and Microwave Absorption Properties of Barium Ferrite Nanorods. *Mater. Lett.* **2008**, *62*, 840–842.

54. Iwauchi, K. Dielectric Properties of Fine Particles of Fe_3O_4 and Some Ferrites. *Jpn. J. Appl. Phys.* **1971**, *10*, 1520–1528.

55. Maharramov, A. M.; Ramazanov, M.; Alizade, R.; Asilbeili, P. Structure and Dielectric Properties of Nanocomposites on the Basis of Polyethylene with Fe_3O_4 Nanoparticles. *Digest J. Nanomater. Biostruct.* **2013**, *8*, 1447–1454.

56. Debye, P. Polar Molecules, Chemical Catalogue Co., Inc., New York, 1929. *J. Soc. Chem. Industry* **1929**, *48*, 1036–1037.

57. Sambyal, P.; Singh, A. P.; Verna, M.; Gupta, A.; Singh, B. P.; Dhawan, S. K. Designing of MWCNT/ferrofluid/flyash Multiphase Composite as Safeguard for Electromagnetic Radiation. *Adv. Mater. Lett.* **2015**, *6*, 585–591.

58. Mishra, M.; Singh, A. P.; Sambyal, P.; Teotia, S.; Dhawan, S. K. Facile Synthesis of Phenolic Resin Sheets Consisting Expanded Graphite/γ-Fe_2O_3/SiO_2 Composite and its Enhanced Electromagnetic Interference Shielding Properties. *Indian J. Pure Appl. Phys.* **2014**, *52*, 478–485.

59. Malinova, P.; Nickolov, R.; Dishovski N.; Lakov, L. Modification of Carbon-Containing Fillers for Elastomers. *Kautsch. Gummi Kunstst.* **2004**, *57*, 443–445.

60. Danneberg, E. M. The Effects of Surface Chemical Interactions on the Properties of Filler-Reinforced Rubbers. *Rubber Chem. Technol.* **1975**, *48*, 410–444.

61. Нефедов В.; В. Черепин, В. Физические методы исследования поверхности твердых тел, Москва, Наука, 1983.

CHAPTER 8

ELASTOMER-BASED COMPOSITE MATERIALS COMPRISING FERRITES

CONTENTS

ABSTRACT

Regardless of the long lasting extensive theoretical and experimental work on creating effective microwave absorbers, there is an ongoing interest in obtaining such absorbers with tailored properties suitable for a variety of applications in the respective frequency range. Our study is aimed at studying some chemical, crystal, crystallographic, and structural characteristics of absorption-active fillers and their influence on the interaction of thus reinforced elastomeric composite materials with microwaves.

We have selected different kinds of ferrites for our research: magnetite, barium, and strontium-substituted and non-substituted hexaferrites having a high field of intrinsic magnetic anisotropy. The presence of a filler with high field of intrinsic magnetic anisotropy in a composite leads to qualitative changes in the frequency spectra of the real and imaginary parts of permittivity and permeability. Another result is significantly improved interaction of the composite with electromagnetic waves. The natural ferromagnetic resonance (NFMR) is an extremely important feature of used ferrite fillers, which manifests itself in the decimeter and the centimeter band of electromagnetic waves spectrum. Depending on their crystal composition, ferrites having various NFMR in predetermined areas of the frequency spectrum can be selected, so that interaction and absorption of microwaves is efficient in a certain frequency range.

In order to obtain broadband microwave absorbers, we have also studied some combinations of the used ferrites with carbon-containing fillers of high dielectric losses, such as graphite, carbon black, and activated carbon.

Various types of rubber have been used as a polymer matrix such as natural, acrylonitrile butadiene, chloroprene, butadiene, and styrene-butadiene rubber. It has been found that when the polymer matrix contains polar functional groups and bonds capable of altering its magnetic susceptibility, its chemical nature has a statistically significant impact on the degree of the composites interaction with electromagnetic waves.

The analysis of the obtained experimental results allows the conclusion that acrylonitrile butadiene rubber suits best as a polymeric matrix for composites required to interact effectively with microwaves. This, to a great extent, is due to the nitrile group present in its macromolecule.

We have developed and optimized voluminous and thin-film microwave absorbers, such with a concentration gradient, inclusive, and determined their microwave characteristics.

8.1 INTRODUCTION

Interdisciplinary research areas emerging at the boundaries of classic well-known research fields have been most characteristic features of recent-decade science. An exemplary case is the research on polymeric composite materials responding actively to the influence of ultra-high frequency (UHF) magnetic fields. The scope of the research involves polymer materials science, semiconductive electronics, and theoretical physics. Active response herein means electromagnetic energy absorbed by the volume of the composite material and its transformation into another type of energy, for example, heat. The aforementioned materials find application in three main spheres:

- Environment protection—applications related to the efforts of mankind to protect the environment from the influence of UHF electromagnetic magnetic fields (UHEMF), produced by industrial and other sources such as power plants, high-voltage transformers, radio, FM and TV transmitters, electronics, computing, and radiolocation equipment,[1,2] that is, fighting electromagnetic pollution
- Technical—applications related to elimination of parasite signals and noises in radio and TV equipment, improvement in the diagram of antenna direction, and reliability of telecommunications[3,4]
- Army—applications related to creating camouflage devices against modern surveillance equipment operating on radiolocation principles form land, air, and space.[5-8] Anti-radiolocation camouflage is one of the main methods for minimizing the effectiveness of radiolocation surveillance equipment in the total complex of radiation safety activities. Its aim is decreasing the radiolocation contrast between the camouflaged object and the background. This is possible by lowering the difference between their reflective abilities. The latter is connected with the effectiveness of their reflective surface.[5,9,10] On the other hand, one of the main means of reducing the effective reflective surface of objects is using camouflage of radio-absorbing covers[11-16] most of which are elastomer based.

No matter what their application is, polymeric composite materials with an active response to influence of electromagnetic fields, in most cases, should meet the following requirements: maximum absorbance and minimum reflection of electromagnetic waves (therefore, often those materials are termed as "microwave absorbers"), wide working band, good mechanical parameters, excellent weather resistance (sometimes resistance to aggressive media), good thermal properties, sustainable exploitation stability, and cause no technological problems.

The main principles for developing rubber-based microwave-absorbing composites involve[11]:

i)   Finding a suitable rubber matrix and suitable conductive filler or a system of fillers;
ii) Ensuring interaction between the composite and microwaves in the entire volume of the composite;
iii) Choice of an appropriate rubber matrix to insulate the filler particles completely from each other;
iv) Using fillers with high dielectric and magnetic loss values;
v)   Affording a multilayer gradient structure of the composite with respect to the filler concentration.

Considerable research should be performed to find out the effects of various chemical, crystallochemical, and physicochemical factors and processes on composites properties, so that the questions concerning the composition–structure–properties relationships can be answered, and thus, composites with tailored properties, meeting all requirements, can be developed. This has been confirmed by the survey on the publications dealing with polymeric elastomers.

Relative permittivity (ε_r) and permeability (μ_r) are universal parameters of particular importance for the description of the level at which the electromagnetic wave from the UHF range interacts with a polymeric composite material.[17,18] Depending on the values of these two parameters, the composite materials exhibit different capability to absorb electromagnetic waves. In an alternating field, ε_r and μ_r have a complex character and could be represented as follows:

$$\varepsilon_r = \varepsilon_r' - j\varepsilon_r''; \quad \mu_r = \mu_r' - j\mu_r'', \tag{8.1}$$

where, ε_r' and μ_r' are the real and parts of permittivity and permeability, while ε_r'' and μ_r'' are the imaginary parts, characterizing the ability of the material to absorb electromagnetic waves.[17]

When $\varepsilon_r'' \ll \varepsilon_r'$ and $\mu_r'' \ll \mu_r'$ (i.e., $\varepsilon_r'' \to 0$ and $\mu_r'' \to 0$), no absorption is observed. Most often such is the case with neat polymers. Most of them are dielectric substances of weak magnetism. Therefore, it is impossible to obtain a material actively responsive to electromagnetic waves based on a neat polymer. The polymeric composites comprising nonconductive fillers, such as silica, glass fibers, and so forth, could be included in this class.[17] If the imaginary parts of permittivity and permeability are not negligible, then the composites have the capacity of absorbing electromagnetic waves. The interaction of the electromagnetic wave from the UHF range with the composites is directly related to two other characteristic parameters which directly depend on the above two—namely, the tangents of dielectric and magnetic losses angles:

$$\tan \delta_\varepsilon = \varepsilon_r'' / \varepsilon_r'; \quad \tan \delta_\mu = \mu_r'' / \mu_r' \tag{8.2}$$

They are also related to the so-called wavenumber K:

$$K = K' - jK'' \tag{8.3}$$

The imaginary part of K is K″ and is called attenuation coefficient. Most often it is expressed as a logarithm of the ratio between the transmitted power (P_T) and incident power (P_I). However, it is known that when electromagnetic waves fall over a material besides being absorbed, a part of them is reflected at the interface of the media wherein the electromagnetic waves are spread. Therefore, when designing polymer composites for absorbers it is important to be aware of the parameters characterizing the degree of electromagnetic reflection. In the case, the so-called wave resistance Z and the reflection coefficient are of particular importance:

$$Z = \sqrt{\frac{\mu_r \cdot \mu_0}{\varepsilon_r \cdot \varepsilon_0}}, \tag{8.4}$$

where μ_0, ε_0 are the values of the permeability and permittivity of the media, respectively. The wave resistance of the media wherein electromagnetic waves are spread should be equal so that no reflection occurs.

Information about the electromagnetic reflection is obtained from the so-called reflection coefficient.

$$\Gamma = z_2 - z_1 / z_2 + z_1, \tag{8.5}$$

where z_1 and z_2 are the wave resistances of medium 1 and medium 2. When the wave resistance of the two media is equal, there is a full accordance between the media, and no reflection is observed. Quite often the standing wave ratio (SWR) is used as a parameter characterizing the reflection at the composite material–free space interface:

$$SWR = \frac{1+|\Gamma_o|}{1-|\Gamma_o|}$$ (8.6)

It is typical of all parameters, listed so far, to be frequency dependent. The dependence in the general case cannot be described by a uniform regularity.[17]

Till now practice has shown that only absorbers based on polymer composite materials can have values of the parameters meeting the requirements of significant practical applications.[19,20] As a rule, those materials are polymeric dielectric matrixes wherein the specific fillers actively absorbing UHF electromagnetic waves are distributed discretely. The fillers may be magnetic or nonmagnetic by nature.

The microstructure of the composite material (the filler particles size, the interparticle distance, their specific surface area, their phase volume, and especially the character of their distribution) is of particular importance for the effectiveness of such a system. The discrete distribution of the filler particles (i.e., their complete insulation by an optimally thick polymer shell) is a significant prerequisite for the normal functioning of the absorber.

According to Kovneristyi et al.,[17] the *first condition* for effective absorption of UHF energy by materials of the type "conductive filler dispersed in a dielectric matrix" is as follows:

Maximum absorption is achieved only in the cases when the dielectric phase (usually a polymer) insulates completely the conductive particles distributed evenly in the entire matrix volume.

The condition is valid for a wide temperature range, frequency range, and angles of the UHF radiation. The quantity of electromagnetic energy absorbed by the composite material and transformed into heat depends on the imaginary parts of its permittivity and permeability.

If the absorbed UHF energy is in a greater amount, then besides the absorption conditions, the conditions of thermal energy transfer by the microvolumes should be also taken into account. The stability of the properties of the composite material is related closely to the latter factor. The conditions of heat transfer depend on the size and concentration of filler particles. The nature and structural specifics of the dielectric matrix are also important. One should keep in mind the fact that the filler particles absorb much more energy than the dielectric matrix, the heat evolved in them is greater, and hence the heat flows of irradiated energy are directed from the filler particles to the dielectric matrix. Therefore, the nature of the dielectric matrix and the stability of its particles are important for the effective functioning of the absorber.

The *second condition* for effective absorption[17] is as follows:

Absorbing composite materials should comprise components possessing high stability of their structure, composition, and properties when being influenced by electromagnetic waves, heat, irradiation, corrosion, and so forth.

The nonconventional, strictly specific requirements that composite polymer materials absorbing UHF electromagnetic waves should meet has limited the range of suitable components—both polymers and fillers. On the other hand, meeting those requirements is a must for the effective functioning of the materials as absorbers. The list of rubber types includes: silicon rubber, fluorine-containing rubber, acrylonitrile-butadiene rubber, chloroprene rubber, urethane rubber, natural rubber, styrene-butadiene rubber and that of fillers includes traditionally used carbonyl iron, nickel cobalt powder, some types of ferrites, various carbon forms (carbon black, graphite, carbon fibers, fullerenes, carbon nanotubes, graphenes, etc.). Hybrid fillers possessing both high dielectric and high magnetic losses have been used in the recent years.[21–23] The summary of the conclusions drawn in several studies in the field[17,24,25] shows that solving the reversal of the task has a great practical value, that is, to develop an absorbent material with tailored values of electromagnetic waves absorption and reflection using selected ingredients with known values of the above parameters. A main problem in creating microwave absorbers is the maximum reduction of the reflection of electromagnetic waves at the air–absorber interface. The reflection is inevitable and occurs simultaneously with the absorption processes in the absorber. It occurs due to the mismatched impedances of the medium wherein the electromagnetic waves are spread (air) and of the absorbent material. The greater the mismatch in impedances, the greater is the reflection coefficient and SWR, which are the main parameters determining the effectiveness of each microwave absorber.

Low reflection can be achieved by a maximum match of the wave resistance of the composite material and that of the medium, that is, one should ensure initial resistance of the absorber equals to the wave resistance of the air medium. Although there are various routes to matching the impedance of the radiation absorbent materials with that of free space, their practical realization is always problematic, if the impedances mismatch is great. Moreover, difficulties increase when creating an absorber effective in a wide-frequency range.

Regarding the above aspect, the usage of metal sheets as absorbers, in particular, is impossible as one should match the extremely low-input impedance of the metal with the impedance of the free space for which values are the highest. This inflicts the usage of composite materials comprising metal, metal oxide, and semiconductive fillers dispersed finely in the dielectric polymeric matrix. In this process, the particles of the dispersed phase (absorption filler) are shelled and insulated by the polymeric material. Multilayered structure based on the gradient concentration principle for the fillers in the particular layers regarding the direction free space–electromagnetically shielded object have been used to decrease the reflection. Structures having geometrical *inhomogeneities on their external surface have also been used for the same purpose.*[11–16,19]

The analysis of the available periodical literature and patents shows that, there are scarce data about:

- Using hexaferrites of high intrinsic anisotropy field as fillers in polymer composites designed for effective absorption of UHF electromagnetic waves
- Using combinations of isotropic and highly anisotropic magnetic fillers for the same purposes
- Application of a magnetic field while preparing coatings for electromagnetic absorption, so that the arrangement of the fillers could be optimized and the microwave properties of the prepared materials could be improved.
- Using combinations of fillers for microwave absorbers, especially such of the type "an isotropic magnetic filler, highly anisotropic magnetic filler, and nonmagnetic filler of high dielectric losses."
- The effect that a number of factors, such as microstructures, crystallochemical, and so forth have upon the interaction of composite materials with UHF electromagnetic waves.

This chapter aims at studying how some chemical, crystallochemical, and crystallographic characteristics of the absorption-active filler affect the interaction between the elastomeric materials and UHF electromagnetic waves, so that microwave absorbers of good properties could be prepared.

Various types of ferrites, for example, barium and strontium substitute and non-substituted ones have been chosen for our investigations. Some combinations of ferrite and carbon-containing fillers have been studied as well.

8.2 CHEMICAL COMPOSITION AND CRYSTAL CHEMISTRY OF FERRITES

The ferrites we used with nominal composition $FeO.Fe_2O_3$, $BaFe_{12}O_{19}$, $SrFe_{12}O_{19}$ are complex oxide compounds whose one part is an oxide of three-valent iron and the other one is an oxide of a different metal. The ferrite composition could be presented by the general formula:

$$\left(Me_2^k + O_k^{2-}\right)_m \left(Fe_2^{3+} + O_3^{2-}\right)_n \tag{8.7}$$

where: Me is the characterizing metal which can be iron, nickel, zinc, barium, strontium, magnesium, and so forth; k is the valence of the metal; m and n are the numbers which could have a certain value.

There are known ferrites of different structures.[26–29] But, those of our greatest interests were:

1. Ferrospinels: These have the structure of the natural mineral spinel $MgO.Al_2O_3$ —a class to which ferrite belongs.
2. Hexaferrites: These are ferrites having a hexagonal structure of the $MeO.6Fe_2O_3$ type, where Me = Pb, Ba, Sr—a class to which barium and strontium ferrites belong.

The above ferrites are polycrystal substances wherein the bond forces, as it is the case of most materials of that type, are determined by electrostatic interactions between the positive metal ions and the negative oxygen ions. In some cases, however, when the size of metal cations (e.g., those of barium) is large enough, they can substitute the oxygen ions at the lattice vertexes. Although, often the cations have radii many folds smaller than the anions so they merge into the space between them and may look as located in the center of a polygon the vertexes of which are occupied by anions. In the oxides of transition metals, which are of major practical importance, the most common coordination between the cations and anions is the tetrahedral, octahedral, and dodecahedral one as shown in Figure 8.1.

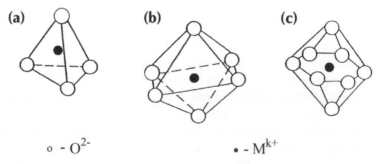

(a)   **(b)**   **(c)**

o - O^{2-} $\bullet$ - M^{k+}

FIGURE 8.1 Main types of oxygen surroundings in ferrites: (a) tetrahedral, (b) octahedral, and (c) dodecahedral.

8.3   SPECIFIC PROPERTIES OF HEXAFERRITES

8.3.1   ANISOTROPY

Specific characteristic of hexaferrites determining our interest to them in this study is the availability of an effective field of intrinsic magnetic anisotropy.

The magnetic energy of hexaferrites depends strongly on the magnetic texture of the material. In this connection, oxide magnets of the type are isotropic and anisotropic. In case of anisotropic magnets in a specific direction, magnetic energy is considerably larger than in other directions. According to the definition of Tikadzumi,[30] magnetic anisotropy is energetically favorable orientation of spontaneous magnetization of the ferromagnet in the direction of some characteristic of its crystallographic axes. Actually, this phenomenon changes the intrinsic energy of ferrite depending on the orientation of the magnetization vector in the crystal.

Materials of this type are magnetized more easily in certain directions than in the others. The latter fact was very important for our research because some of the studied composite materials containing hexaferrites (mostly thin-layered ones) were formed by applying an external magnetic field while casting an absorbing film. In order to achieve saturated magnetization of the sample when being magnetized along the preferred direction, it is sufficient to apply even a very weak external magnetic field.[30]

The hexaferrites we used have axis of slight magnetization which direction is along their hexagonal axis "c".

8.3.2   ELECTROMAGNETIC SPECTRUM. FERROMAGNETIC RESONANCE

The use of ferrous materials in a reaction with UHF electromagnetic fields is related to a change in the μ_r' and μ_r'' parts of their complex permeability (Eq. 8.1), which are generally high frequency dependent.[31]

Different areas are formed in the frequency dependence of the real and imaginary part of the complex permeability. The most important for the purposes of our study is the so-called UHF field range. It is the ferromagnetic resonance (FMR) area which is directly connected to the behavior of ferrite filler, hence to the entire composite material in its interaction with UHF electromagnetic fields. The nature of FMR is determined by the spin precession of the magnetic moments of electrons which are under the simultaneous effect of constant and alternating field.[31,32]

If the electron is under the influence of a constant external magnetic field with intensity Ho, arising forces tend to orient the plane of the electronic orbit perpendicularly to the field. The vector of the magnetic moment is oriented in the direction of this field and begins to rotate around it with a certain frequency. In the real case, after some time, the vector of the magnetic moment overlaps with the constant magnet field and the motion of its end point is spiral.[31]

When the above-described system is subjected to UHF alternating field H with a circular frequency applied in the plane perpendicular to the permanent magnetic field, the losses in the ferrite can be compensated at a frequency close to the precession frequency of the system. This excites the sustained precession and FMR is observed. Here, ferrite absorbs a part of the energy of the high-frequency field and thus interacts with it. This is the so-called natural ferromagnetic resonance (NFMR), while the absorbed energy is transformed into heat.[31]

The NFMR occurs in the decimeter and centimeter range of UHFs. This parameter is a crucially important feature of ferrite fillers used in our investigations and was the second main reason for our research interest in hexaferrites.

Depending on their crystallochemical composition, one can choose ferrite fillers having NFMR in different desired ranges of the frequency spectrum, so as the interaction to be effective, and microwave absorption to be in a predetermined frequency range, respectively.

8.4 EXPERIMENT

8.4.1 SAMPLES

8.4.1.1 RUBBERS

Acrylonitrile–butadiene rubber, SKN-40 (Russia).

Content of bound acrylonitrile—39 ± 1%, Mooney viscosity ML (1 + 4) 373K—65±7, density—1000 kg/m³.

Chloroprene rubber, Baypren 210 (Bayer, Germany)—medium crystallization ability, Mooney viscosity ML (1 + 4) 373K—40–45, density—1230 kg/m³.

Butadiene rubber, Buna CB 10 (Bunawerke Huls GmbH)—content of cis-1,4 units—96%, Mooney viscosity ML (1 + 4) 373K—45–50, density—910 kg/m^3.

Styrene–butadiene rubber, Bulex 1500 (Neftohim, Bulgaria)—content of bound styrene—about 23%, content of cis-1,4 butadiene units—about 80%, Mooney viscosity ML (1 + 4) 373K—38–53, density—940 kg/m^3.

8.4.1.2   ABSORPTION-ACTIVE FILLERS

8.4.1.2.1   Isotropic Fillers

Magnetite (Fe_3O_4) of cubic structure, isotropic (anisotropy constant K= 8 × 10^{-3} J/m^3), particles size about 1 μm.

8.4.1.2.2   Carbon Containing Fillers

Graphite (made in Russia)

Carbon black—furnace carbon black N 776, N 550, N 330, N 220 (made in Russia) and acetylene black P1250 (made in Germany).

Activated carbon K13 (made in Russia)—carbon absorbent obtained by carbonization of coal and charcoal, particle diameter—3–5 nm, specific surface area—400 m^2/g.

8.4.1.2.3   High Anisotropic Magnetic Fillers

Cobalt–titanium-substituted barium hexaferrite with a general formula $BaCo_{0.85}Ti_{0.85}Mn_{0.1}Fe_{10.2}O_{19}$ with NFMR at a frequency of 12 GHz and intensity of the field of intrinsic anisotropy Ha = 4.4kOe, anisotropy constant—0.42 J/m3.

Cobalt–titanium-substituted barium hexaferrite with a general formula $BaCo_{0.75}Ti_{0.75}Mn_{0.1}Fe_{10.4}O_{19}$ with NFMR at a frequency of 16 GHz and intensity of the field of intrinsic anisotropy Ha = 8kOe.

Strontium hexaferrite ($SrFe_{12}O_{19}$) with NFMR at 40 GHz and intensity of the field of intrinsic anisotropy Ha = 18kOe.

The particle size of all ferrites was about 1 μm according to data from the electronic microscopy studies.

8.4.2 METHODS

8.4.2.1 ABSORPTION AND REFLECTION OF ELECTROMAGNETIC WAVES OF ULTRA-HIGH FREQUENCY

The absorption and reflection of elastomer composites were tested by two methods: wave guiding and in free space (reflectometric).

First method (wave guiding): The 10 × 20-mm sample was filling the entire cross section of a waveguide for the desired frequency range. Absorption Lt, and the value of the SWR were measured. The transmission and reflection were calculated from them. Based on their values the part of the absorbed power in the overall capacity balance was calculated also.

Second method (measurements in the free space): The sample of a significantly larger size (297 × 210 mm) was placed frontally between two horn antennas. The induced attenuation Lt was measured with regard to the attenuation without a sample and the reversible loss Lr, by which the transmission of electromagnetic waves, reflection, and the reflection and absorption are calculated.

For more complete evaluation assessment of the effectiveness of absorbing elastomer compositions and coatings based on them the percentage of contribution of the particular power was evaluated on the basis of power balance:

$$P_I = P_R + P_A + P_T \tag{8.8}$$

where P_R, P_T, and P_A are the reflected, transmitted and absorbed power. From the above expression one can easily find out the following absorption ratio as a function of reflection and the transmission powers

$$A_R = 1 - P_R - P_T \tag{8.9}$$

and can even calculate the percentage contribution of each characteristic and power in the total balance.

8.4.2.2 FREQUENCY MEASUREMENTS OF COMPLEX PERMEABILITY AND PERMITTIVITY

The measurements were performed using a computerized vector impedance analyzer (Hewllet-Packard, model 8548). The parameters of a circuit, including a segment of a coaxial line containing a sample of the tested absorbing elastomer compositions, were determined.

The vector circuit analyzer performed measurements in the references plane A-A and B-B, and so forth of S-parameters (S_{11} reflection coefficient of the "air–absorbent material" interface and S_{21} coefficient of transmission through the material) at set frequency values. The results of these measurements were used to calculate the real and imaginary parts of the complex permeability and permittivity.

8.4.2.3   OPTICAL MICROSCOPY STUDIES

The optical microscopy studies were performed on a polarizing microscope (MIN8). The films investigated were obtained by casting rubber solutions containing absorption-active fillers on a highly transparent glass support. The thickness of the films was about 100 nm. In some cases, an external magnetic field of Ho = 1 kOe was applied in the process of film formation. The direction of the force lines was perpendicular to the plane of the film.

8.5   RESULTS AND DISCUSSION

8.5.1   SYNTHESIS AND CHARACTERIZATION OF THE FERRITES USED

We used an opportunity for changing the physical parameters of ferrites and especially the frequency range of their NFMR, so that the crystal structure and intensity of the field of their intrinsic magnetic anisotropy could be amended by introduction of specific additives into the material. The existing three types of gaps which can host the ions of the additive are the theoretical prerequisite to realize the idea. On the other hand, the ferrite lattice is very elastic and the gaps can also host cations having a radius larger than the theoretically calculated ones. This allows the substitution of Fe^{3+} ions, aimed at introducing additives into the hexaferrite, so that some of its properties could be changed purposefully; in particular, the frequency range of its NFMR and the field of its intrinsic magnetic anisotropy, which determine the frequency range in which ferrite is used. Those are the two key characteristics of the materials in our research.

A combination of Co^{2+} and Ti^{4+} substituting a certain amount of Fe^{3+} was introduced into the substrate barium hexaferrite. We established a drastic narrowing of the frequency range of the NFMR and a decrease of the intrinsic anisotropy field intensity (H_a) with the increasing degree of Fe^{3+} substitution (Figure 8.2(a), (b)), that is, the possibility to change the two key characteristics of barium hexaferrite affecting considerably the properties of the entire composite via changing the crystallochemical structure of hexaferrite.

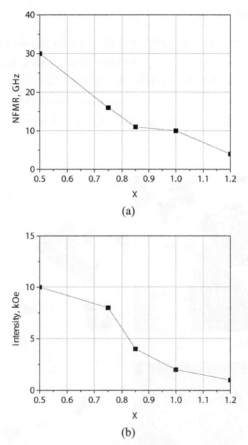

(a)

(b)

FIGURE 8.2 (a) Dependence of natural ferromagnetic resonance frequency, and (b) the intensity of intrinsic field anisotropy on the degree of Fe^{3+} substitution (X) in barium hexaferrites with cobalt and titanium ones.[33]

The cobalt–titanium combination was preferred to that of scandium and indium due to economic reasons. The degree of Fe^{3+} substitution was determined in the desired frequency range by the particular requirement for NFMR set by prospective applications of the developed composite materials. We had an idea to include magnetic fillers of high anisotropy in the investigations carried out. Those fillers differ considerably in the frequency range of their NFMR and in their intrinsic field anisotropy. Therefore, we also needed materials for which values of the latter characteristic are much higher than those of barium hexaferrites. So, we chose strontium hexaferrites. The investigations revealed that substituted barium and strontium hexaferrites had the following characteristic nominal compositions, frequency ranges of NFMR, and intensity of the field of intrinsic anisotropy:

$$BaCo_{0.85}Ti_{0.85}Mn_{0.1}Fe_{10.2}O_{19} - NFMR = 12 \text{ GHz, } Ha = 4.4kOe$$
$$BaCo_{0.75}Ti_{0.75}Mn_{0.1}Fe_{10.4}O_{19} - NFMR = 16 \text{ GHz, } Ha = 8kOe$$
$$SrFe_{12}O_{19} - NFMR = 40 \text{ GHz, } Ha = 18kOe$$

The particle shape and size of the ferrite fillers used was determined by electron microscopy (Figure 8.3).

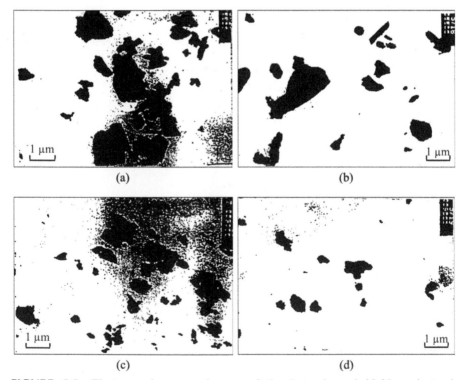

FIGURE 8.3 Electron microscopy images of the isotropic and highly anisotropic magnetite fillers used: (a) Fe_3O_4 (magnetite), (b) $BaCo_{0.85}Ti_{0.85}Mn_{0.1}Fe_{10.2}O_{19}$, (c) $BaCo_{0.75}Ti_{0.75}Mn_{0.1}Fe_{10.4}O_{19}$, (d) $SrFe_{12}O_{19}$.

8.5.2 VOLUMINOUS MICROWAVE ABSORBERS BASED ON ELASTOMER COMPOSITES

8.5.2.1 ELASTOMER COMPOSITES COMPRISING FILLERS OF HIGH MAGNETIC LOSSES

According to literature and patents, the most used fillers of high magnetic losses are carbonyl iron and some ferrites which have undergone special treatment aimed at increasing Fe^{2+}, and successively increasing their tan $\delta_\mu = \mu_r''/\mu_r'$.[34–44] We supposed natural magnetite could be of interest as such a filler. It possesses high magnetic losses and permeability due to its high Fe^{2+} amount and electron exchange between Fe^{2+} and Fe^{3+}. The experiments performed on 2-mm thick monolayered absorbent sheets prepared form styrene–butadiene composites showed that in the ranges between 8–12 and 12–18 GHz the electromagnetic wave absorbance is strongly dependent on magnetite content in the composites (Figure 8.4).[45,46]

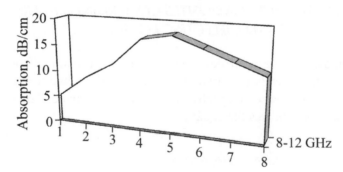

FIGURE 8.4 Dependence of electromagnetic wave absorption on the magnetite/rubber ratio in an elastomeric composite (X—the magnetite/rubber ratio).

As the Figure 8.4 shows, the absorption maximum appears at 400–600 parts per hundred of rubber (phr) magnetite amount in the composite. At a higher filler concentration, the probability of a direct contact between the particles enhances. This is the likely reason for the poorer properties of the absorbers, while at lower filler concentrations (Figure 8.4) the magnetic losses are low and fail to guarantee the needed degree of electromagnetic wave absorbance. It is also seen that magnetite is more suitable filler for the 8–12 GHz range. Beyond that range the absorbers are of worsened properties, as has been proven by the frequency spectrum of the tangent of magnetic losses angle of the elastomeric composites (Figure 8.5).[47]

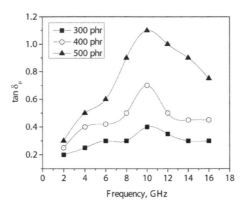

FIGURE 8.5 Frequency dependence of the tangent of magnetic losses angle of an elastomer composite comprising magnetite at 300, 400, and 500 phr (X—magnetite/rubber = 3:4:5). (Reprinted with permission from Dishovski, N.; Petkov, A.; Nedkov, I.; Razkazov, I. Hexaferrite Contribution to Microwave Absorbers Characteristics. *IEEE Trans. Magn.* **1994,** *30,* 969–971. © 1994 IEEE.)

8.5.2.2 ELASTOMER COMPOSITES COMPRISING FILLERS OF HIGH MAGNETIC AND DIELECTRIC LOSSES

Such composites have been developed because the band of the working frequency of the already prepared elastomer absorbers had to be expanded and their efficiency at the respective frequencies improved. The theoretical basis of the approach was the equation:

$$\tan \delta = \tan \delta_\varepsilon + \tan \delta_\mu = \frac{\varepsilon_r^{''}}{\varepsilon_r^{'}} + \frac{\mu_r^{''}}{\mu_r^{'}} \tag{8.10}$$

where:

tan δ_ε is tangent of dielectric losses;

tan δ_μ is tangent of magnetic losses;

ε_r', μ_r' is real parts of relative permittivity and permeability;

ε_r'', μ_r'' is imaginary parts of relative permittivity and permeability.

The necessity of simultaneous presence of fillers possessing high ε_r'' and μ_r'' values in the respective frequency range is obvious.

The intention to prepare a material which absorbs the electromagnetic waves effectively and reflects them weakly in the 2–10 GHz range led to the development and optimization of a composite comprising 450 phr magnetite, 20 phr graphite, and 100 phr elastomer. The introduction of small amounts of graphite was evoked by the low reflection of electromagnetic waves and by the need to improve the effectiveness of the absorber in the 2–10 GHz range, wherein magnetite is less effective than graphite. On the other hand, the composite should be made easier by substituting a part of magnetite with graphite (Table 8.1).

The level of electromagnetic wave reflection was corrected by preparing a multilayered structure with an ascending concentration of the fillers in the direction from free space to shielded object as a means to achieve a smooth transition between the impedances of the absorber and the air. The total thickness of the absorber was 12 mm.

Concentration of the absorbing fillers in the particular layers (in phr) was as follows:

- I^{st} layer (outmost): magnetite at 100 phr
- IInd layer: magnetite at 100 phr and graphite at 10 phr
- IIIrd layer: magnetite at 200 phr and graphite at 20 phr
- IVth (inner): magnetite at 450 phr and graphite at 20 phr

The profound study on the formation of a composite absorbing UHF electromagnetic waves effectively and reflecting them weakly, revealed the proper choice of elastomer matrix to be a decisive factor. The investigations we carried out showed that the chemical nature of the elastomer had a statistically significant effect (at the same concentrations of absorbing filler and equal other conditions) upon the interaction between the electromagnetic waves and the composite (Figure 8.6).

TABLE 8.1 Comparisons of the Microwave Properties of Mono- and Multilayered Composites.

	Frequency (GHz)	ε_r'	μ_r'	Tan δ_ε	Tan δ_μ	Absorption (dB)
Monolayered composite comprising 500 phr magnetite and 100 phr elastomer[48]	2	7.01	1.94	0.036	0.40	2.5
	4	6.02	1.80	0.031	0.50	6.2
	6	6.02	1.64	0.030	0.62	11
	8	5.85	1.65	0.030	0.70	12
	10	4.82	1.55	0.026	0.80	16
Multilayered composite comprising magnetite (100–450 phr) and graphite (10–20 phr)[49]	2	11.03	1.98	0.140	0.42	4.3
	4	9.75	1.78	0.120	0.49	8.1
	6	7.02	1.65	0.083	0.68	14.5
	8	5.35	1.60	0.044	0.70	17
	10	5.03	1.57	0.035	0.95	21

The density of the first composition was 2720 kg/m³, and of the second was 2540 kg/m³.

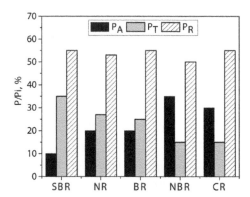

FIGURE 8.6 Effect of the chemical nature of the rubber matrix used upon the power balance at 9.4 GHz.

The dispersion analysis (level of significance $\alpha = 0.05$) have shown that the effect is statistically significant at the expressed absorbed and transmitted power (within the limits of 20%) and is close to rejecting the hypothesis of a statistical correctness at the reflected power, that is, the nature of the elastomer matrix influenced mainly the absorbed and transmitted power. The elastomers containing a highly polar functional groups or bonds (nitrile, chloroprene) in their macromolecule determine a higher absorption rate and a lower transmission degree of electromagnetic waves, and better interaction and better microwave characteristics of the composition, respectively. The polarity appears to be more influential factor than the ability of the polymer to crystallize. If there are no polar functional groups in the macromolecule, then the ability of the elastomer to crystallize significantly affects its absorbing properties. Elastomers of the same nature and crystallization ability possess quite close microwave properties.

The analysis of the obtained experimental results allow the conclusion that butadiene nitrile rubber is the most suitable polymer matrix in structures meant to interact effectively with UHF electromagnetic waves. This is mainly due to nitrile groups present in its macromolecules.

Using butadiene–nitrile rubber in the role of elastomeric matrix is imposed by other exploitation reasons, such as resistance to gasoline and motor oils, good confection stickiness, tendency to form healthy adhesive seams with glues based on polar elastomers, good compatibility with coatings, and absorption of electromagnetic waves in the near infrared region.

Considering these facts, we developed a four-layered coating based on buta-diene–acrylonitrile composites filled with magnetite and conductive carbon black (our experiments showed the latter facilitated better microwave performance than graphite). The thickness of the layers was optimized and the concentration of absorption active fillers therein (in phr) in the direction free space to the shielded object was as follows[50,51]:

- I[st] layer (outermost): thickness 1.5 mm; carbon black P1250—25 phr; magnetite—50 phr
- IInd layer: thickness 1.5 mm; carbon black P1250—50 phr; magnetite—100 phr
- IIIrd layer: thickness 1.0 mm; carbon black P1250—100 phr; magnetite—150 phr
- IVth layer (inner): thickness 3.0 mm; carbon black P1250—25 phr; magnetite—50 phr

The total thickness of the coating was 7 mm.

The coating is distinguished by good microwave characteristics in a wide-frequency range (8–18 GHz), and it owns and retains pronounced minima regarding the reflection of electromagnetic waves in two different subranges: 9–10 GHz (X-band) and 17–18 GHz (Ku-band). The coating retains its good microwave characteristics (especially the reflective ones) when a massive metal piece is firmly attached behind it.

We established a particularly interesting fact that with minimal thickness change of the fourth coating layer, we could shift the reflection minima to a wide-frequency range (Figure 8.7).

Absorption and reflection characteristics of the composite were determined by three methods: wave guiding with a matching load, and with a matched metal behind the sample, and by the method in the free space.

As a result, we obtained a fairly complete picture (Table 8.2) of power distribution (in accordance with the power balance) at the interaction between the instant power of the UHF electromagnetic wave and the developed multilayered composite.

The good properties of the developed multilayered composite in terms of electromagnetic waves reflection and absorption are obvious. Particularly impressive are the very low reflected power values.

Additional studies were conducted using a sweep generator which showed that throughout the studied range of UHF spectrum (8–18 GHz) the reflected energy was less than 10%. The composition and structure of the created anti-radiolocation coating are patented.[50]

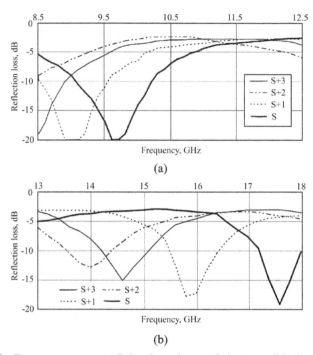

FIGURE 8.7 Frequency spectra of the dependence of the reversible losses, Lr on the thickness in (a) X-band and (b) Ku-band, (S—standard thickness of the IVth layer, 3 mm).

TABLE 8.2   Characteristics of the Developed Composite at 9.4 GHz.

Wave guiding with a matched load				Wave guiding with a matched metal behind the sample			In the free space		
Lt (dB)	SWR	P_T (%)	P_R (%)	P_A (%)	SWR	P_R (%)	P_A (%)	Lr (dB)	P_R (%)	P_A (%)
7.5	3.8	18	34	48	1.4	2.8	97.2	14.2	3.8	96.2

8.5.2.3   ELASTOMER COMPOSITES COMPRISING FILLERS WITH HIGH FIELD OF INTRINSIC MAGNETIC ANISOTROPY

Our research was motivated by the following theoretical formulation:

If an isotropic magnetic filler is introduced into a polymer matrix, the frequency dependence of the magnetic spectrum of the coating based on such a composite and its absorption of electromagnetic waves could be described satisfactorily within the model of weakly-bound spectra.[52] According to this model, each particle of a magnetic filler can be viewed as "gyromagnetic oscillator", being a center which under the influence of UHF electromagnetic waves absorbs, accumulates, and emits energy in a certain time interval. If, however, a highly anisotropic hard magnetic filler is introduced into the system simultaneously with an isotropic magnetic material, which is under the influence of the magnetic anisotropy field of its particles, the magnetic vectors of the adjacent isotropic magnetic particles will orient toward the direction of the magnetic field influencing them at some polar angle and azimuth angle. In other words, the effect of the anisotropy magnetic field of the hard magnetic filler has, to a certain extent, the effect of the applied external magnetic field regarding the isotropic particles. As a result, the homogeneity of the absorbing structure becomes higher, and a resonance absorption appears which can be expressed by the following integral dependence valid when an external magnetic field (H_0) is applied in the process of forming an absorbing structure:

$$\chi^*(\omega, H) = p \int_0^\infty \chi^* \alpha p \left(\omega, H_\alpha, H_{\alpha p}\right) \varphi\left(H_{\alpha p}\right) dH_{\alpha p} \qquad (8.11)$$

where $\chi''\alpha p$ (ω, H_α, $H_{\alpha p}$) is the absorption of each particle in terms of orientation, $H_{\alpha p}$ is the intensity of the field of intrinsic magnetic anisotropy; H_α is the average value of the field for a group of particles, $\varphi(H_{\alpha p})$ is a δ function characterizing the probability distribution of the field, that is, $\varphi(H_{\alpha p}) = \delta$ ($H_{\alpha p} - H_\alpha$); ω is the angular frequency of electromagnetic radiation.[53]

The main conclusion that can be drawn from the above theoretical postulate and the corresponding mathematical expression is that the absorption capacity of a composite material comprising a combination of magnetic fillers, one of which is of high magnetic intrinsic field anisotropy, will strongly depend on the anisotropic properties of each particle, mostly on the properties of the anisotropic magnetic field. Maximum absorption of the structure will be in the frequency range of the NFMR of the filler with high anisotropy. On the other hand, the vectors of the magnetic moments of the isotropic magnetic particles can be expected to orientate under the influence of the magnetic field of the highly anisotropic particles. This is the prerequisite for creating a long-range magnetic order in the composite material structure. As a result of the latter fact, one can expect a higher degree of interaction between the coating and the electromagnetic radiation, and effective microwave properties. We developed a series of formulations for absorbing coatings containing an isotropic magnetic filler (magnetite) and a magnetic filler with a high intrinsic anisotropy magnetic field (cobalt–titanium–substituted barium hexaferrites of the general formula $Ba(CoTi)_xFe_{12-2x}O_{19}$). The preliminary experiments have shown that the degree of substitution of iron cations in barium hexaferrites with cobalt–titanium ones, under the same other conditions, has a strong influence on the microwave characteristics of the absorber in the studied 8–22 GHz range, especially in the subrange of 8–12GHz (Figure 8.8).

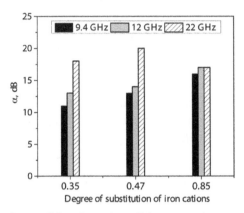

FIGURE 8.8   Dependence of the absorption of the composite material on the degree of substitution of iron cations with cobalt–titanium in the structure of barium hexaferrites.

The observed effect is due to the influence that the degree of substitution of iron cations has upon the intensity of the intrinsic magnetic anisotropy field of the filler and on the frequency range of its NFMR already discussed. It is evident that with the increasing degree of substitution, the absorption of the elastomeric composition is improved mainly in the range of 8–12 GHz. This is due to the narrowing of the frequency range of the NFMR of ferrite in the direction of $x = 0.35$ to $x = 0.85$ and its overlapping with the frequency range of our investigations (where $x = 0.85$, NFMR is at a frequency of 12 GHz; Figure 8.2(a)). The graphic representations of electromagnetic waves absorption as a function of the mass ratio of magnetite and hexaferrite showed a great difference regarding this parameter in the field of NFMR of ferrite and beyond it (Figure 8.9). The mass ratio of magnetite and hexaferrite was selected as a result of an optimization procedure performed on the basis of an experimental design (an orthogonal multifactor compositional plan of the experiment is used).

It is of crucial importance that NFMR range of the filler with high field of intrinsic magnetic anisotropy is within the desired frequency range wherein we have been searching for maximum absorption. The possibility to adjust it by changing the structure of hexaferrites via substituting iron cations in the crystal lattice at a certain degree is also evident. To find the essence of the problem, research has been performed which has shown that the nature, extent, and the level of interaction between the composite and the electromagnetic waves in the presence of magnetic fillers depends on their impact on the imaginary part of the complex permeability. The effect of the fillers used has been investigated (Figure 8.10).

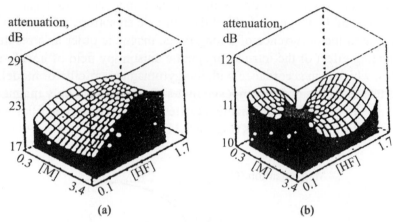

(a)   (b)

FIGURE 8.9 Spatial representation of the dependence of microwave absorption on the magnetite: hexaferrite mass ratio in an elastomer composite: (a) at 12 GHZ—in the NFMR range of hexaferrite having a general formula: $BaCo_{0.85}Ti_{0.85}Mn_{0.1}Fe_{10.2}O_{19}$, (b) at 18 GHz—beyond that range (the amounts of magnetite and hexaferrite are encoded).

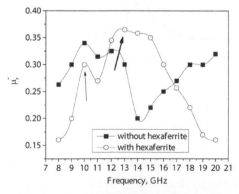

FIGURE 8.10 Effect of the hexaferrite on the frequency spectrum of the imaginary part of magnetite permeability μ_r'' in the 8–20 GHz range.

The analysis of the obtained frequency dependencies shows that the addition of highly anisotropic magnetic filler transforms the nature of the spectrum of magnetic losses from relaxation into a resonance one. This is confirmed by the significant increase that has been observed in the values of the imaginary part of permeability of the composite in the NFMR range of hexaferrite (in the 12–15 GHz range, indicated by a bold arrow). A less pronounced maximum is seen at about 10 GHz (marked in the figure with a regular arrow). In our opinion, this is due to the presence of magnetite particles in the system and orientation of their magnetic moments under the influence of the magnetic field of the intrinsic anisotropy of hexaferrite, that is, due to the occurrence of a long-range magnetic order as a result of the orienting effect of the intrinsic magnetic anisotropy field of hexaferrite particles, which is in accordance with the gyromagnetic oscillator model.[53]

Hysteresis curves of the composite materials comprising only magnetite or only hexaferrite or the magnetite–hexaferrite combination as filler were taken to confirm our hypothesis (Figure 8.11).

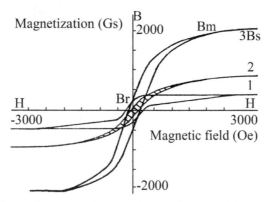

FIGURE 8.11 Hysteresis curves of composite materials containing: magnetite (curve 1), hexaferrite (curve 2), and the combination of magnetite and hexaferrite (curve 3) as filler.[54]

It is obvious that the hysteresis properties of the composites change abruptly when magnetite and hexaferrite both are present in the elastomer matrix. We assume that the changes in Br (residual magnetic induction), Bm (maximum magnetic induction), and in Bs (saturation induction) as well as the total increase in the hysteresis loss are due to the formation of elements of long magnetic range order (orientation) in the matrix of the composite under the influence of the introduced highly anisotropic hexaferrite. This applies fully to the changes in Bm–Bs segment of the curve. Interesting changes have been observed in the frequency spectra of the real and the imaginary part of the complex permittivity of the composite (Figure 8.12) that occur under the influence of the highly anisotropic filler.[55]

The analysis of the figure shows marked change in the dielectric proper-
ties and, above all, significant decrease in dielectric loss tan $\delta_\varepsilon = \varepsilon_r''/\varepsilon_r'$ of
the elastomer composite. As the elastomeric matrix possesses the dielectric
properties of the composite in the case investigated, we believe the changes
occurring in the presence of a highly anisotropic filler are primarily due
to changes in the supramolecular structure of elastomeric matrix and the
oriented distribution of nitrile butadiene rubber macromolecules over the
surface of hexaferrite particles under the impact of the field of intrinsic
magnetic anisotropy. The reason for this effect is proven[56] by a dependence
of the magnetic susceptibility of the elastomeric macromolecule on some
of the functional groups and bonds it contains (double bonds, polar groups,
etc.), sensitive to magnetic fields forces. The change in the magnetic susceptibility is considered to cause the residual magnetization, which is the real
reason for the orientation of elastomeric macromolecules due to the field of
intrinsic magnetic anisotropy of the filler. In fact, this interfacial phenomenon gives rise to structural micro-heterogeneities resulting in differences
in the mutual arrangement of macromolecules in the surface layer and the
transition layers, located at a different distance from the surface (in some the
influence of the field of intrinsic anisotropy is still felt) and in the volume
of the elastomeric matrix. Micro-heterogeneities also occur at a supramolecular level resulting into differences in the packaging nature of supramolecular structures in the surface layers and volume.

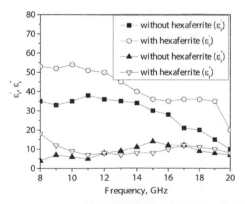

FIGURE 8.12   Frequency spectra of the real and imaginary parts of permittivity.

The observed significant lowering of the dielectric losses of the hexa-
ferrite-filled composite is caused by the formation of elements of dominant
orientation as well as by the certain orderness degree of the polymer–high
anisotropic filler interface, under the influence of the magnetic field of the

latter and of sensitive functional groups and bonds in the elastomer macro-molecule (the nitrile group has the strongest impact in the case). Similar effects are not observed when only magnetite is present in the composite. The thesis has been supported by evidence obtained from the optical micros-copy studies of thin-layered absorbers of similar compositions (described later) as well as by studies of the equilibrium degree of swelling of the vulcanizates, that is, absorbers based on polar and nonpolar elastomers. It turned out that there is no physical polymer–filler interaction in the case when using fillers of high field of intrinsic magnetic anisotropy and nonpolar elastomers (natural rubber) as a matrix. The respective orientation of the macromolecules on the first surface of the second led to a sharp increase in the swelling degree (Figure 8.13, curve 1). In absorber-based polar matrices (nitrile rubber) when high anisotropic filler at up to 100 phr is introduced, the equilibrium degree of swelling changes negligibly. This proves the enhanced physical interaction between the particles of the high anisotropic magnetic filler and the macromolecules of polar matrix. Here, it results into orientation of macromolecules under the influence of magnetic lines in the field of intrinsic magnetic anisotropy (Figure 8.13, curve 2). This is another evidence of the different nature of interfacial phenomena and their resultant micro-heterogeneities when fillers having high field of intrinsic magnetic anisotropy and polar elastomeric matrices are present simultaneously in the compositions.

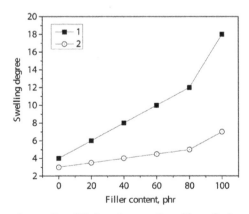

FIGURE 8.13   Dependence of equilibrium degree of swelling of microwave absorbers based on natural (1) and nitrile (2) rubber on the content of a filler (cobalt–titanium-substituted barium hexaferrite) having high field of intrinsic magnetic anisotropy.

In fact, the combination of an isotropic and highly anisotropic compo-nent leads to a significant improvement in the absorption of the coating in a wide frequency range. The formulation of the composite which the coating is made of is patented.[57]

The effect that the thickness of the absorbent composite has on its absorption effectiveness has been tested for the particular investigated system of fillers with magnetic nature (Figure 8.14).[58]

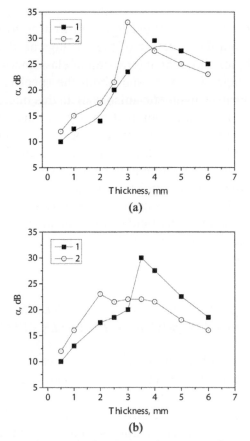

FIGURE 8.14 Dependence of electromagnetic waves absorption on the absorber thickness (monolayered even vulcanizate): (a) at 9.4 GHz, (b) at 22 GHz (1—ratio of rubber:hexaferrite:magnetite = 1:1.5:0.6; 2—ratio of rubber:hexaferrite:magnetite = 1:1.5:3.0).

The studies on our system of magnetic fillers revealed that maximum absorption is achieved at a 2–4 mm thickness, depending on the hexaferrite:magnetite mass ratio. An increase in the content of the magnetite in the studied frequency range leads to shifting the maximum absorption to lesser thickness.[59,60]

8.5.3 THIN-LAYER MICROWAVE ABSORBERS BASED ON ELASTOMER COMPOSITES

Development and optimization of formulations for polymer composites intended for production of thin-layered coatings, of effective absorption and low reflection of UHF electromagnetic waves, is another major area in our research. The ability to observe the samples directly by optical microscopy in normal and polarized light allowed us to obtain significant information about the structure of the coatings needed to prove our worked-out hypotheses about the interaction of elastomeric compositions with UHF electromagnetic waves and about the specific role of fillers with high field of intrinsic magnetic anisotropy in this interaction. Special attention has been paid to the combination of two magnetic fillers, that is, isotropic and anisotropic fillers, since with this combination we had achieved the highest level of the desired microwave characteristics. This proved to be particularly interesting in the case of voluminous microwave absorbers in both fundamental and purely applied aspect. By applying external magnetic field while preparing the absorbent coating, we tried to get an idea of the impact that the field of intrinsic magnetic anisotropy of hexaferrite has on the magnetite particles and on the macromolecules of the elastomeric matrix.

8.5.3.1 PARTICLES DISTRIBUTION IN THE FILLERS OF DIFFERENT MAGNETIC NATURE COMPRISING THIN-LAYER ABSORBERS

Our optical microscopy observations showed that when using an isotropic filler, such as magnetite, without an external magnetic field (or filler with a high field of intrinsic magnetic anisotropy), the thin-layered absorbing structure lacked any organization in the distribution of the magnetic filler particles during its formation process. The particles were grouped into chaotically located agglomerates of different shape and size. The latter depended on the concentration of the filler in the polymer solution from which the film was prepared. Although the filler particles had sharp edges and smooth walls, the agglomerates of the isotropic magnetite had an almost spherical or oval shape (Figure 8.15). Under identical other conditions noteworthy were the effects of the chemical nature of the elastomer matrix used and that of the solvent.

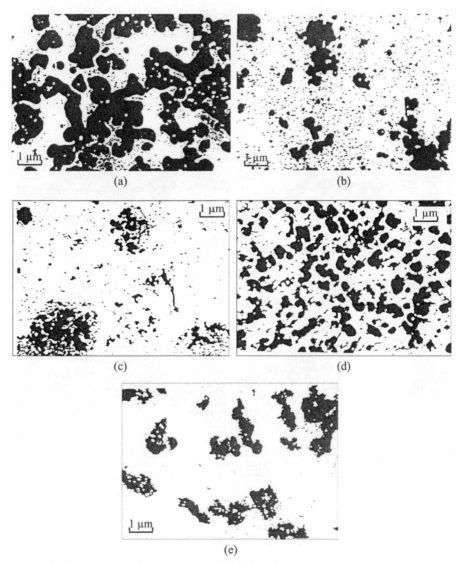

FIGURE 8.15   Effect of the chemical nature of the elastomer matrix and of the solvent on the shape and size of aggregates formed by magnetite particles without external magnetic field (magnification 45×): (a) NR—nonpolar, crystallizing, (b) Chloroprene rubber—polar, crystallizing, (c) NBR—polar, non-crystallizing, (d) SBR—nonpolar, non-crystallizing (solvent benzene, 5% solution), (e) NBR (solvent acetone, 5% solution). NR: natural rubber, NBR: nitrile–butadiene rubber, SBR: styrene–butadiene rubber.

The differentiation is evident. The fact that, elastomeric matrices— polar or nonpolar—being of a similar nature have a similar effect on the distribution of the filler particles, is considered normal. When the matrix is of the same nature, the solvent also has noticeable influence.

It was worth finding out whether in the case of high anisotropic magnetic fillers without external magnetic field the particles would be distributed randomly.

We carried out microscopic observations on films containing high aniso-tropic fillers (Figure 8.16).

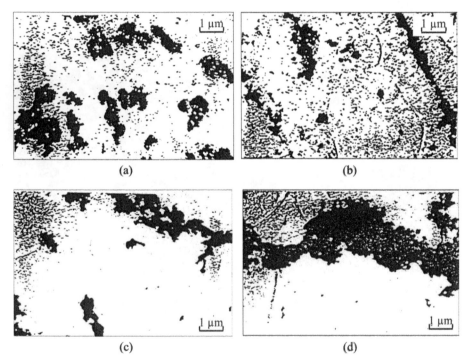

FIGURE 8.16 Optical microscopy images of absorbing films (nitrile rubber; solvent acetone; magnification 210×; without external magnetic field) containing: (a) magnetite (reference), (b) $BaCo_{0.85}Ti_{0.85}Mn_{0.1}Fe_{10.2}O_{19}$, (c) $BaCo_{0.75}Ti_{0.75}Mn_{0.1}Fe_{10.4}O_{19}$, (d) $SrFe_{12}O_{19}$.

Magnetite confirmed the already made observations concerning the lack of any organization in the distribution of particles and their grouping into chaotically arranged agglomerates of any shape and size. The same could also be argued about ferrite of the general formula: $BaCo_{0.85}Ti_{0.85}Mn_{0.1}Fe_{10.2}O_{19}$.

However, there is a noticeable trend toward transforming the random irregular shape of chaotically arranged formations into a chain-like one (Figure 8.16(b)). With the increase in the intensity of the field of the internal magnetic anisotropy in $BaCo_{0.75}Ti_{0.75}Mn_{0.1}Fe_{10.4}O_{19}$, especially in $SrFe_{12}O_{19}$, this trend gradually increases and the particles of these fillers are arranged in a disorderly formations, but their shape is no longer irregular (as in magnetite). It is predominantly chain-like (Figure 8.16(c), (d)), and size of agglomerates increases in the indicated direction.

In the films containing magnetic fillers with a high field of intrinsic anisotropy, one notices elements of organization in the arrangement of the filler particles.

8.5.3.2 CHANGES OCCURRING IN THE DISTRIBUTION OF FILLERS OF DIFFERENT MAGNETIC NATURE CAUSED BY AN EXTERNAL MAGNETIC FIELD

The issue of changes occurring in the particle arrangement organization of isotropic and anisotropic magnetic fillers at application of an external magnetic field is also very important. Studies carried out in this aspect have shown that in all cases the application of an external magnetic field triggers organization in their particles arrangement. However, it strongly depends on the magnetic nature of the respective filler. In case of isotropic magnetite, the observed formations are of approximately the same shape and size of the cluster type distributed evenly throughout the matrix. They consist of a central part (core) and branches protruding from it (Figure 8.17). With lowering the concentration of magnetite in the polymeric solution, the central part of the cluster decreases and the size of branches increases (Figure 8.17(a)–(c)); the latter are of very well pronounced dendrite-like structure, regardless of the chemical nature of the elastomer matrix (Figure 8.17(d), (e)).

It has been found that highly anisotropic fillers, in the presence of an external magnetic field, also facilitate organization in the arrangement of their particles. Evenly distributed fibril structures throughout the volume oriented in the direction of the magnetic field forces have been observed. With decreasing the anisotropic filler concentration the particles size shrinks, but their shape does not change (Figure 8.18).

When anisotropic fillers are used, dendrite-like structures have not been observed in contrast to the case when isotropic ones are used (Figure 8.17(d), (e)). There is a tendency to increasing the fibril width in the transition in the $BaCo_{0.85}STi_{0.85}Mn_{0.1}Fe_{10.2}O_{19} \rightarrow BaCo_{0.75}Ti_{0.75}Mn_{0.1}Fe_{10.4}O_{19}$ direction (i.e. at decreasing the degree of iron cations substitution under the same other conditions).

The above discussion allows the main conclusion that the external magnetic field applied in the process of forming a thin-layer structure gives

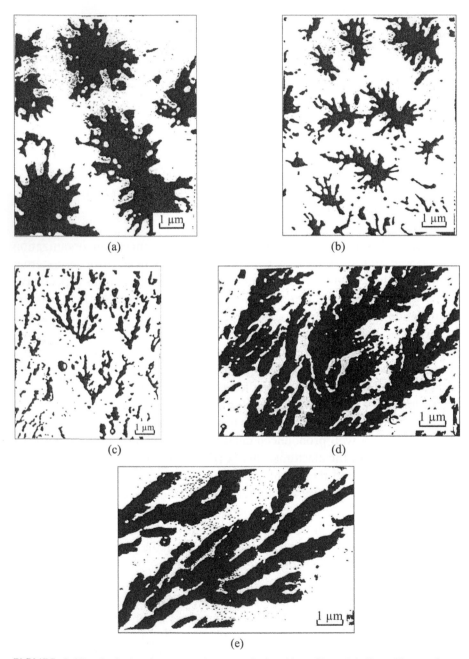

FIGURE 8.17 Optical microscopy images of absorbing films (nitrile rubber; solvent acetone; magnification: a, b, c—45×; d, e—210×) at different solvent concentrations: (a) 18 wt. %; (b) 9 wt. %; (c) 5 wt. %.

rise to organization in the arrangement of the filler particles and clusters them into agglomerates evenly distributed throughout the volume of the elastomeric matrix. The magnetic nature of the filler has the decisive impact on the shape, type, and location of these agglomerates. Isotropic formations prevail in isotropic magnetic fillers (magnetite). Highly anisotropic formations, oriented in the direction of the magnetic field forces, prevail in anisotropic fillers.

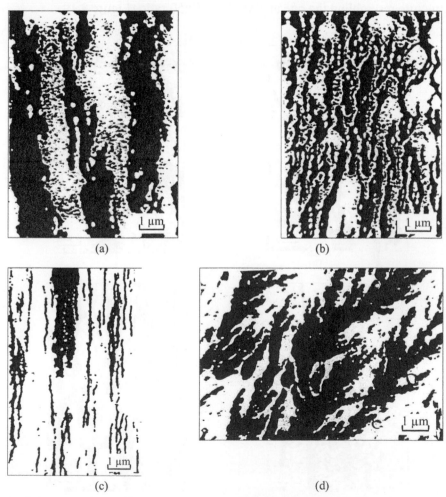

(a) (b)

(c) (d)

FIGURE 8.18 Optical microscopy images of films comprising $BaCo_{0.75}Ti_{0.75}Mn_{0.1}Fe_{10.4}O_{19}$ at different concentrations in the elastomer solution (nitrile rubber-acetone; magnification 45×): (a) 18 wt. %, (b) 9 wt. %, (c, d) 5 wt.%.

Of particular interest has been the study of the distribution of the filler particles in the elastomer matrix when filled with both magnetite and hexaferrite and the effect of applied external magnetic field. It has been found that in this case, the type of the observed structures highly depends on the mass ratio of the fillers. Due to their very different magnetic nature, the structural homogeneity of the film is observed only at a M:HF = 50:50 mass ratio. In other cases, we observe separate zones with predominating particle distribution typical either of magnetite or of hexaferrite, and a transition zone with a structure resembling the one at a M:HF = 50:50 ratio (Figure 8.19).

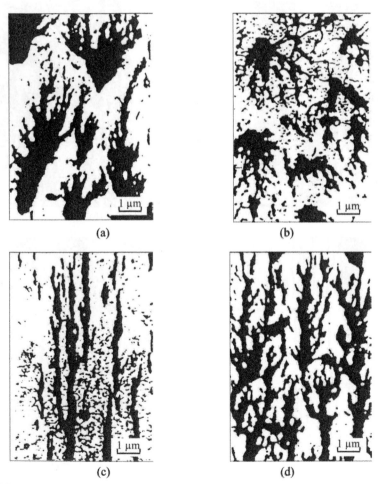

FIGURE 8.19 Optical microscopy images of absorbing films comprising magnetite and hexaferrite at different mass ratios, formed in the presence of a magnetic field (magnification 48×): (a) at a 50:50 ratio, (b–d) different zones at a M:HF = 75:25 mass ratio.

The observed peculiarities of the structures in the transitional areas as well as the magnetite:hexaferrite = 50:50 mass ratio (fibrils with branches protruding out of them with a dendritic structure) give us the reason to assume that in such cases the composition structures include particles of both fillers—fibril hexaferrite particles forming the core and magnetite ones forming the branches (Figure 8.19(a)–(d)).

8.5.3.3   INTERACTION BETWEEN THIN-LAYER COATINGS AND HIGH-FREQUENCY ELECTROMAGNETIC WAVES

One of the stages in our work was to trace the structural changes in an elastomeric film containing magnetic fillers occurring under the influence of an external magnetic field and of the field of intrinsic magnetic anisotropy of fillers, and how those changes affect the parameters characterizing its interaction with electromagnetic waves.

Without doubt, the occurrence of organization in the arrangement of the filler particles (no matter if they are isotropic or anisotropic, and regardless of interfacial phenomena caused by them) significantly improves the interaction of the film with electromagnetic waves, its microwave characteristics as a result of the applied external magnetic field during its formation (Figure 8.20).

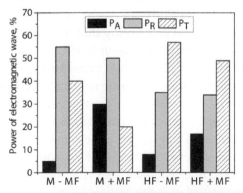

FIGURE 8.20   Microwave characteristics (at 9.4 GHz) of coatings comprising magnetite (M) and substituted barium hexaferrite (HF) with a general formula $BaCo_{0.85}STi_{0.85}Mn_{0.1}Fe_{10.2}O_{19}$ formed in the presence of an external magnetic field.

The percentage of the transmitted power at the expense of absorbed power decreases significantly, due to the improved level of interaction of the film with electromagnetic waves as a result of the uniform and non-chaotic distribution of the filler particles in its volume under the influence of the external magnetic field. The improvement is also due to the higher probability, in a statistical sense, of an electromagnetic wave to encounter absorber active particles of the filler in its path, and not to pass through the radio transparent elastomeric material.

The main conclusion that can be drawn is that the established organization in the arrangement of the filler particles in the elastomer matrix and the achieved macrostructural uniformity, as a result of applied external magnetic field, improves the level of materials interaction with electromagnetic waves.

Applying the gyromagnetic oscillator model to the structures studied,[52] one can expect a strong dependence of absorption properties on the field of magnetic anisotropy of the filler used and on the resonance absorption in the NFMR frequency range of the low anisotropic filler. This has been confirmed when analyzing the frequency dependence of the real and imaginary part of the complex permeability of coatings based on acrylonitrile–butadiene rubber containing magnetite and barium hexaferrite $BaCo_{0.85}STi_{0.85}Mn_{0.1}Fe_{10.2}O_{19}$ at a M:HF = 50:50 mass ratio (Figure 8.21(a)).

The change in the character of the spectrum of the imaginary part of permeability in the direction of resonance behavior when applying an external magnetic field ($H_o > 0$) is evident. We assume that the second (less pronounced) peak in the frequency dependence of μ'' is due to the orientation of the vectors of the magnetic moments of magnetite particles. In accordance to these frequency maxima, the best microwave characteristics of the coatings have been observed. It is also noteworthy that the applied external magnetic field in the process of coatings formation leads to a noticeable decrease of permittivity values (Figure 8.21(b)). We consider the observed changes in the frequency dependence of the magnetic spectra to be a confirmation of the "gyromagnetic oscillator" model.[52] The presence of a high anisotropic filler improves the long-range magnetic order, which is the reason for the resonance behavior observed. The changes in permittivity occur due to changes in the supramolecular structure of the polar elastomeric matrix at the elastomer–high anisotropic filler interface, that is, to oriented arrangement of the polymeric macromolecules over the particles surface of the high anisotropic filler occurring under the influence of their magnetic field.

The studies show the perceptiveness of the absorbing coatings containing a combination of two magnetic fillers-isotropic and anisotropic. One of these has a high field of intrinsic magnetic anisotropy and a crucial role in the character of frequency spectrum of magnetic losses of the coating, especially in the range of NFMR. On the other hand, the intrinsic magnetic anisotropy field affects the elastomer macromolecules. It has been established

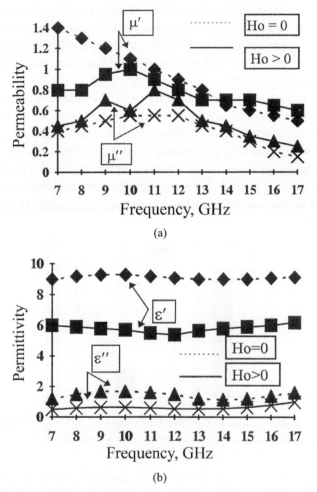

FIGURE 8.21 Frequency spectra of (a) permeability and (b) permittivity of thin absorbing coatings containing magnetite and barium hexaferrite.

that elastomers containing polar groups, double and triple bonds have an exactly fixed value of their magnetic susceptibility, which for molecules without own magnetic moment has two components—diamagnetic and paramagnetic. Under the influence of an external magnetic field or the influence of an intrinsic magnetic anisotropy field of the anisotropic fillers used, the magnetic susceptibility of macromolecules containing groups or bonds sensitive to such influence changes, what gives rise to residual magnetization. The effect is particularly pronounced in polar elastomers (nitrile, chloroprene), whose high paramagnetic macromolecules can be easily oriented by the magnetic field forces, particularly those over the particles surface of the high anisotropic filler. These interfacial phenomena explain the micro-heterogeneities arising as a result of the changes in the structure-sensitive properties of composites, and the extent of interaction with UHF electro-magnetic waves, inclusive.

We investigated, in a merely applied aspect, the influence of coating thick-ness on its absorption and reflection characteristics, as well as the effective-ness (absorption and reflectance) of the multilayered coatings constructed on a principle of gradient filler concentration in the different layers.

It has been found that with increasing the coating thickness in a techno-logically feasible range (0.5–2.0 mm), the effectiveness evaluated according to the power balance at 9.4 GHz has improved significantly (Figure 8.22).

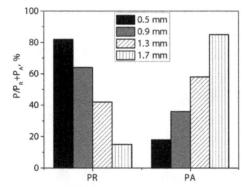

FIGURE 8.22 Power balance at 9.4 GHz for coatings of different thicknesses obtained from solutions of elastomeric compositions coated over a metal support.

As seen, a 1.5–2.0-mm thick coating is capable of ensuring reflection of electromagnetic waves around 15%; meanwhile its absorption is within the 85% limit.

A way to further improve the effectiveness of these coatings is their construction on a gradient principle: the only homogeneous layer 1.7 mm thick is replaced by three sublayers of the same total thickness, the absorp-tion-active fillers concentration decreases from the innermost toward the free space. Our research has demonstrated that this reduces the reflection considerably.

The formulations of solutions of elastomeric compositions effectively absorbing and weakly reflecting the UHF electromagnetic waves, which we have developed and optimized, are patented.[61–66]

8.6 CONCLUSION

The influence of some chemical, crystal, crystallographic, and structural factors on the rate of interaction of the elastomer composites and UHF electromagnetic waves has been investigated with selected representatives of different elastomer groups and ferrites as absorbing active fillers. It has been found that both polymeric matrix and fillers discreetly dispersed therein have their own contribution to the degree of interaction with UHF electromagnetic waves.

The chemical nature of the polymer matrix has a statistically significant impact on the degree of interaction of the composites with electromagnetic waves, when the former contains polar functional groups and bonds capable of altering their magnetic susceptibility under the influence of a magnetic field—external or of the intrinsic magnetic anisotropy of the fillers introduced.

The crystallochemical nature and the crystallographic structure of ferrites have a crucial role in the extent of interaction of a composite with UHF electromagnetic fields. Particularly significant and specific in importance for improving the interaction is the presence of a strong field of intrinsic magnetic anisotropy and an NFMR in a certain frequency range.

When the composite comprises a filler with high intrinsic magnetic field anisotropy, one observes qualitative changes in the frequency spectra of the real and imaginary parts of permittivity and permeability, which improve the interaction of the composite with electromagnetic waves to a large extent.

Fillers with high field of intrinsic magnetic anisotropy have a strong impact on the structural micro-heterogeneity of the composite and micro-heterogeneities at a supramolecular level. The influence of the magnetic field of intrinsic anisotropy, in our opinion, is limited to formation of elements of the long-range magnetic order in the composite when an isotropic magnetic filler is present. It also contributes to the arrangement of polymer macromolecules over the particles surface of the filler due to its field of intrinsic anisotropy. Here, the functional groups and bonds in the polymer also have their impact.

It has been found that the combination of two magnetic fillers, one isotropic and the other with high field of intrinsic magnetic anisotropy, as well as the appropriately selected frequency range of the NFMR are the promising parameters to create composite structures that possess optimum micro-heterogeneity and interact effectively with electromagnetic waves.

Changing the crystal structure of the fillers with high field of intrinsic magnetic anisotropy by introducing ions of certain metals into their crystal lattice can purposefully manage and alter the interfacial phenomena and the resulting micro-heterogeneities.

8.7 ACKNOWLEDGMENT

The authors express their gratitude to:

- Prof. Ivan Nedkov, DSc. from the Institute of Electronics, Bulgarian Academy of Sciences for the synthesis and characterization of barium and strontium hexaferrites (substituted and non-substituted), and for measuring the complex permittivity and permeability.
- Assoc. Prof. Boris Vitchev, Ph.D. and Assoc. Prof. Kosta Kostov, Ph.D. from the Institute of Electronics, Bulgarian Academy of Sciences for measuring the microwave properties of the composite materials.
- Dr. Rashko Dimitrov from the University of Chemical Technology and Metallurgy for performing the optical microscopy studies.

KEYWORDS

- **elastomers**
- **ferrites**
- **magnetite**
- **substituted barium hexaferrites**
- **strontium ferrites**
- **composites**
- **interaction with electromagnetic waves**
- **microwave absorbers**
- **specific characteristics**

REFERENCES

1. Habash, R. W. Y. *Electromagnetic Fields and Radiation: Human Bio-effects and Safety*; CRC Press, Marcel Dekker, Inc.: New York, USA, 2001.

2. Adams, C. N. *Electromagnetic Health: Making Sense of the Research and Practical Solutions for Electromagnetic Fields (EMF) and Radio Frequencies (RF) Logical Books*; Wilmington Delaware: USA, 2012.
3. Chung, D. D. L. Electromagnetic Interference Shielding Effectiveness of Carbon Materials, Review. *Carbon* **2001**, *39*, 279–285.
4. Gnecco, L. T. *Design of Shielded Enclosures: Cost-Effective Methods to Prevent EMI*; Newnes, Butterworth-Heinemann: USA, 2000.
5. Micheli, D.; Pastore, R.; Gradoni, G.; Primiani, V. M.; Moglie, F.; Marchetti, M. Reduction of Satellite Electromagnetic Scattering by Carbon Nanostructures Multilayers. *Acta Astronaut.* **2013**, *88*, 61–73.
6. Wei, S.; Li, R.; Chen, L.; Yao, J. Research of fiber radar absorbing materials. *Adv. Mater. Res.* **2013**, *602–604*, 835–838.
7. Saville, P. Review of Radar Absorbing Materials, Defence R&D Canada—Atlantic, Technical Memorandum DRDC Atlantic TM 2005-003, 2005. http://dtic.mil/dtic/tr/fulltext/u2/a436262.pdf.
8. Vinoy, K. J.; Jha, R. M. *Radar Absorbing Materials: From theory to Design and Characterization*; Kluwer Academic Publishers:Boston, 1996.
9. Knott, E. F.; Shaeffer, J. F.; Tuley, M. T. *Radar Cross Section*, 2nd ed.; Artech House: Norwood, 1993.
10. Mottahed, B. D.; Manoochehri, S. A Review of Research in Materials, Modeling and Simulation, Design Factors, Testing and Measurements Related to Electromagnetic Interference Shielding. *Polym.- Plast. Technol. Eng.* **1995**, *34*, 271–346.
11. Dishovsky, N. Rubber based composites with active behavior to microwaves. *J. Univ. Chem. Technol. Metall.* **2009**, *44*, 115–122.
12. Dobrovenskii, V. V.; Zasovin, E. A.; Mirovitskii, D. I.; Cherepanov, A. K. Microwave-Absorbing Composites with Absorbing-Grating Layers. *Usp. Sovrem. Radioelektron.* **2000**, *2*, 61–66.
13. Savi, P.; Miscuglio, M.; Giorcelli,M.; Tagliaferro, A. Analysis of Microwave Absorbing Properties of Epoxy MWCNT Composites. *Prog. Electromagnetics Res. Lett.* **2014**, *44*, 63–69.
14. Shneiderman, Y. A. Microwave Absorbing Materials. *Zarubezh Radioelektron* **1975**, *2*, 93–113.
15. Singh, V. K.; Shukla, A.; Patra, M. K.; Saini, L.; Jani, R. K.; Vadera, S. R.; Kumar, N. Microwave Absorbing Properties of a Thermally Reduced Graphene Oxide/Nitrile Butadiene Rubber Composite. *Carbon* **2012**, *50*, 2202–2208.
16. Micheli, D.; Pastore, R.; Apollo, C.; Marchetti, M.; Gradoni, G.; Primiani, V. M.; Moglie, F. Broadband Electromagnetic Absorbers Using Carbon Nanostructure-Based Composites. *IEEE Trans. Microwave Theory.* **2011**, *59*, 2633–2646.
17. Kovneristyi, Y. K.; Lazareva, I. Y.; Ravaev, A. A. *Materialy, Pogloshchayushchie SVCh-Izluchenie (Microwave-Absorbing Materials).*, Nauka: Moscow, 1982.
18. Kim, S. S., Kim, S. T.; Yoon, Y. C.; Lee, K. S. Magnetic, dielectric, and microwave absorbing properties of iron particles dispersed in rubber matrix in gigahertz frequencies. *J. Appl. Phys.* **2005**, *97*(10), F905.
19. Danlee,Y.; Huynen, I.; Bailly, C. Thin smart multilayer microwave absorber based on hybrid structure of polymer and carbon nanotubes. *Appl. Phys. Lett* **2012**, *100*, 213105.
20. Srikanth, V. V. S. S.; Raju, K. C. J. *Graphene/Polymer Nanocomposites as Microwave Absorbers in Graphene-Based Polymer Nanocomposites in Electronics*; Springer International Publishing: Burlin, 2015; pp 307–343.

21. Al-Ghamdi, A. A.; Al-Hartomy, O. A.; Al-Solamy, F.; Dishovsky, N.; Malinova, P.; Shtarkova, R. Microwave properties of natural rubber based composites containing carbon black-magnetite hybrid fillers. *Sci. Eng. Compos. Mater.* **2016**, (revision sent-05.04.2016).

22. Al-Ghamdi, A. A.; Al-Hartomy, O. A.; Al-Solamy, F. R.; Dishovsky, N.; Malinova, P.; Atanasova, G.; Atanasov, N. Conductive Carbon Black/Magnetite Hybrid Fillers in Microwave Absorbing Composites Based on Natural Rubber. *Compo.s Part. B- Eng.* **2016**, *96*, 231–241.

23. Al-Ghamdi, A. A.; Al-Hartomy, O. A.; Al-Solamy, F. R.; Dishovsky, N.; Nickolov, R.; Atanasov, N.; Ruskova, K. Effect of Activated Carbon/in Situ Synthesized Magnetite Hybrid Fillers on the Microwave Properties of Natural Rubber Composites. *Advanced Materials Proceedings*, in press.

24. Thomas, S.; Stephen, R. *Rubber Nanocomposites: Preparation, Properties and Applications*; Wiley: Singapore, 2010.

25. Morgan, P. *Carbon Fiber and Their Composites*; Taylor &Francis Group: Boca Raton, FL, USA, 2005.

26. Ферриты и магнитодиэлектрики: справочник/под общ. ред. Н.Д. Горбунова и Г.А. Матвеева. М.: Сов. радио, (1968).

27. Ферриты: физ. свойства и практ. применения/Я. Смит, Х. Вейн; пер. с англ. Т.А. Елкиной, под ред. Ю.П. Ирхина и И.Е. Старцевой.—М.: Издательство иностранной литературы, (1962).

28. Ю. Ситидзе, Х. Сато, Ферриты, М.: Мир, (1964).

29. Ферриты: строение, свойства, технология производства/Л.И. Рабкин, С.А. Соскин, Б.Ш. Эпштейн.—Л.: Энергия, (1968).

30. Тикадзуми С. Физика ферромагнетизма Магнитные свойства вещества, М. Мир (1983).

31. Радо, Д., Рапт, Р., Эмерсон, В. Ферромагнитный резонанс, Сборник, М. Издателство иностр. лит. (1952).

32. Журавлев Г.И. Химия и технология ферритов, Ленинград: Издательство "Химия", 1970.

33. Nedkov, Iv., A. Petkov, V. Karpov, Microwave Absorption in Sc- and CoTi- Substituted Ba Hexaferrite Powders. *IEEE Trans. Magn.* **1990**, *26*, 1483–1484.

34. Sun, Y.; Zhou, X.; Liu, Y.; Zhao, G.; Jiang, Y. Effect of Magnetic Nanoparticles on the Properties of Magnetic Rubber. *Mater. Res. Bull.* **2010**, *45*, 878–881.

35. Guo, Z.; Lei, K.; Li, Y.; Ng, W. H.; Prikhodko, S.; Hahn, H. T. Fabrication and characterization of iron oxide nanoparticles reinforced vinyl-ester resin nanocomposites. *Compos. Sci. Technol.* **2008**, *68*, 1513–1520.

36. Wang, Y.; Edwards, E.; Hooper, I.; Clow, N.; Grant, P. S. Scalable Polymer-Based Ferrite Composites with Matching Permeability and Permittivity for High-Frequency Applications. *Appl. Phys.* **2015**, A120, 609–614.

37. Wang, Y.; Grant, P. S. NiZn Ferrite/Fe Hybrid Epoxy-Based Composites: Extending Magnetic Properties to High Frequency. *Appl. Phys.* **2014**, A117, 477–483.

38. Kong, L. B.; Li, Z. W.; Lin, G. Q.; Gan, Y-B. Magneto-Dielectric Properties of Mg–Cu–Co Ferrite Ceramics: I. Densification Behavior and Microstructure Development. *J. Am. Ceram. Soc.* **2007**, *90*, 3106–3112.

39. Souriou, D.; Mattei, J. L.; Chevalier, A.; Queffelec, P. Influential Parameters on Electromagnetic Properties of Nickel-Zinc Ferrites for Antenna Miniaturization. *J. Appl. Phys.* **2010**, *107*, 09A518—09A518.

40. Biernholtz, H.; Kenig, S.; Dodiuk H. Dielectric, Magnetic and Mechanical Properties of Ferrite Composites. *Polym. Adv. Technol.* **1992**, *3*, 125–131.
41. Yang, H.; Lin, Y.; Zhu, J.; Wang, F. Electromagnetic Properties and Mechanical Properties of Ni0.8Zn0.2Fe2O4/polyolefin Elastomer Composites for High-Frequency Applications. *J. Appl. Polym. Sci.* **2009**, *114*, 3510–3514.
42. Tan, D. Q.; Cao, Y.; Irwin, P. C. Electromagnetic Interference Shielding Polymer Composites and Methods of Manufacture. U.S. Patent US20,090,101,873, Oct 19 2010.
43. Fuchs, A.; Gordaninejad, F.; Hitchcock, G.; Elkins, J.; Zhang, Q. Tunable Magneto-Rheological Elastomers and Processes for Their Manufacture. U.S. Patent US 20,050,116,194 A1, Jun 02 2005.
44. Singh, P.; Babbar, V. K.; Razdan, A.; Srivastava, S. L.; Puri, R. K. Complex Permeability and Permittivity, and Microwave Absorption Studies of Ca (CoTi) x Fe 12– 2x O 19 hexaferrite composites in X-Band Microwave Frequencies. *Mat. Sci. Eng. B-Solid* **1999**, *67*, 132–138.
45. Dishovky, N.; Nedkov, I.; Petkov, A.; Razkazov, I. *Microwave Properties of Vulcanizates Containing Oxide Materials and Traditional Elastomer Fillers, XII International Conference on Microwave Ferrites, 19–23. 09.1993*; Bulgaria, Conference Proceedings, 1, 125, 1993.
46. Nedkov, I.; Milenova, L.; Dishovsky, N. Microwave Polymer-Ferroxide Film Absorbers. *IEEE Trans. Magn.* **1994**, *30*(6), 4545–4547.
47. Dishovski, N.; Petkov, A.; Nedkov, I.; Razkazov, I. Hexaferrite Contribution to Microwave Absorbers Characteristics. *IEEE Trans. Magn.* **1994**, *30*, 969–971.
48. Dishovsky N.; Mladenov Iv, Iliev, Composition for producing of microwave absorbers. *Bul. Patent* 75,443, 1987.
49. Mladenov Iv, Peshleevsky R., Andreev S., Dishovsky N., Material for antiradar camouflage. Bul Patent 213, 1987.
50. Dishovsky N. et al., Camouflage coating. Bul. Patent 63,212B1, 1998.
51. Dishovsky, N.; Kostov, K.; Vichev, B. *Rubber Based Multilayer Microwave Absorber, XIII* International Conference on Microwave Ferrites, Prahova Valley, Romania, Sept 23–27,1996, Conference Proceedings, 174; 1996.
52. Mikhailovsky, L. K.; Kitaytsev, A. A.; Koledintseva, M. Y. *Advances of Gyromagnetic Electronics for EMC Problems,* Proc. Int. IEEE Symp. Electromag. Compat., Washington, DC, Aug 21–25, 2000, 2, 773–778, 2000.
53. Nedkov, I.; Milenova, L.; Dishovsky, N. Microwave Polymer-Ferroxide Film Absorbers. *IEEE Trans. Magn.* **1994**, *30*, 4545–4547.
54. Dishovsky, N. The Substituted Barium Hexaferrites—Fillers with Possibilities for Structural Changes, EUROFILLERS 97, Manchester (UK), Septe 8th–11th, 429, 1997.
55. Dishovsky, N.; Starkova, R.; Dodov, N.; Vichev, B.; Kostov, K. *Rubber Based Microwave Absorbers,* International Rubber Conference, Paris, May 12–14, 1998, 381, 1998.
56. Alexeev, A.; Kornev, E.; Magnetic elastomers, Moscow, Khimia, in Russian, 1987.
57. Peshleevsky, R; Dishovsky, N.; Nedkov, Iv, Elastomer composition, Bul. Pat, N BG 146 Y1, 1992.
58. Dishovsky, N.; Nedkov, I. Microwave Absorbing Rubber Composites Containing Hard Magnetic Fillers, EUROFILLERS'95, Mulhouse, France, Sept 11–14, 1995, 507, 1995.
59. Dishovsky, N.; Kostov, K.; Vichev, B. Dual—Band Microwave Multilayer Rubber Absorber, In Chapter Microwave Physics and Techniques *(NATO Adv Sci Inst Series)* **1997**, *33*, 425–430.

60. Dishovsky, N.; Nedkov, I.; R Dimitrov, R. Magnetic Field Induced Microwave Absorption in Filled Elastomer Films, EUROFILLERS'95, Mulhouse, France, Sept 11–14, 1995, 503, 1995.
61. Andreev, S.; Peshleevsky, R.; Dishovsky, N. Elastomer coating, Bul. Pat. N 51810, 1991.
62. Andreev, S.; Peshleevsky, R.; Dishovsky, N. Camouflage coating. Bul. Pat., N BG145Y1, 1992.
63. Dishovsky N. et al. Camouflage Coatings and Device. Bul. Pat. N 63612, 2001.
64. Dishovsky, N. Composite for Absorbing and Shielding of Electromagnetic Irradiations. Bul. Patent. 66,453 B1/2014.
65. Dishovsky, N. Composite for Electromagnetic Wave Shielding. Bul. Patent 66,452 B1/2014.
66. Dishovsky, N. Electroconductive Composite. Bul. Patent 66,445 B1/2014.

ELASTOMER-BASED COMPOSITE MATERIALS COMPRISING CARBIDE AND BORIDE CERAMICS

CONTENTS

ABSTRACT

Polymer composites with conductive properties have been of researchers'
interest due to the great opportunities they provide for a variety of applications. Usually, they are insulators, but with suitable fillers, their electrical
conductivity can be varied within a wide range. There is a theoretical possibility to create polymer composite materials with semiconductor properties,
comparable with those of inorganic semiconductors, by mixing the polymer
matrix with suitable fillers. The difficulties are due to impurities and imperfections existing in the structure of the macromolecular network, which
because of the transport of charged particles is uncontrollable.

We used SiC, B_4C, and TiB_2 as functional fillers in elastomer composites
based on natural and acrylonitrile butadiene rubber. The dielectric (dielectric
constant and dielectric loss angle tangent) properties of the obtained vulcanizates and the changes in them caused by different factors (fillers concentrations, pressure, bending, temperature, etc.) have been determined.

The mechanism of charge transfer has been clarified via the parameters
characterizing it (activation energy, mobility, content, and type of the charge
carriers) as well as the impact that the composites structure has on their electrical and mechanical properties. The change in the electrical and tensile properties of the composites resulting from the thermal aging has been monitored.

The results from our investigations have shown that the composites could
be included in electric circuits and transform nonelectric variables into electric. We have determined experimentally the ranges in which the composites
studied can act as high frequency fillers, thermal sensors, sensors of pressure
and great bending, shields in nuclear industry, materials with high coefficient of thermoconductivity, and so forth.

9.1   INTRODUCTION

Polymers and polymer composites with power transmission properties are
subject to complex research as materials with greater opportunities for
diverse applications.[1-7] Usually, they are isolators,[8] but in a number of cases
with suitable fillers, their electric conductivity may be varied within a wide
range, as the matrix and the filler have a specific effect on the electrical
conductivity of the composites.[9-11] When mixed with fillers having a high
electrical conductivity (carbon black, graphite, metal powders), the composites prepared are with resistivity close to that of metals. The properties of
these composites change depending on the concentration of the filler. At

very low concentrations, the resistivity is close to that of the polymer matrix. It gradually decreases with the increasing filler quantity, and at a given critical (percolation) threshold, resistance sharply decreases by several orders of magnitude.[12–18]

Polymer composite materials with semiconductor properties comparable with those of inorganic semiconductors have not been prepared yet. There is a theoretical possibility to obtain such materials, as the overlapping of π-bonds and the distance between the macromolecules in polymers can reach a sufficiently small value (closeness) favoring the transition of the charge carriers. The difficulties are due to the unavoidable impurities and structural imperfections of the macromolecular network, because of which the transport of charged particles is uncontrollable.[19] This drawback, analogous to achieving a high conductivity could be compensated by mixing the polymer matrix with suitable fillers that could be found by carrying out a considerable number of experiments. An eventual success will be a great achievement in polymer materials science due to the extremely simple technology and small funds required to create polymer composites with typical semiconductor properties. We decided to use in this study carbide and boride ceramics /SiC, B_4C, TiB_2/ as functional fillers aiming to prepare composites possessing the above features.

Our objective was to study the electrical properties of elastomer composites containing those fillers and to evaluate the opportunities for their specific applications. To reach this goal, we had to

1. Determine the changes in the electrical properties of the composites under the influence of various factors (reinforcing, pressure, temperature, oxidation processes).
2. Clarify the mechanism of charge carriers' transfer by defining characteristic parameters (activation energy, mobility, concentration, and type of charge carriers).
3. Identify the role that the composites structure has upon the electrical and mechanical properties.
4. Define the most suitable areas of application of the composite.

There are formulations of composites based on various elastomers filled with metal ceramics, always in combination with conductive powders (carbon black, graphite, metal powders) described in the patent literature. Pressure sensors with good sensitivity, but with a strong nonlinearity in certain pressures intervals, are made of such composite.[13–18] The authors[15] recommend using TiC, B_4C, WC, BN, and AlN as a combination with carbon black, pointing SiC and Si_9N_4 as preferable fillers. The purpose of fillers combination is to slow the sharp resistance decrease with increasing pressure, what is typical only of carbon black filled vulcanizates. The authors state an increase in resistance in the presence of SiC, whereas the increase is smaller if the carbide is in the form of fibers. Most probably, the fibers facilitate better arrangement of carbon black particles, what guarantees conductivity higher than in the case of powder SiC.

However, there is no thorough study on the electric properties of composites containing carbide ceramics. That is valid for many other properties of theirs.

9.2 EXPERIMENT

9.2.1 SAMPLES

9.2.1.1   RUBBERS

9.2.1.1.1 Acrylonitrile–Butadiene Rubber

Acrylonitrile butadiene rubber (NBR) with 33–35% bound acrylonitrile (Perbunan N) produced by LANXESS was used.

9.2.1.1.2 Natural Rubber

The natural rubber (NR) used in the study was a Standard Malaysian Rubber 10 (SMR 10) type.

The used NBR *that is polar no crystallizing and nonpolar crystallizing* NR *have been chosen due to the great difference in their properties, especially the electric ones (the volume resistivity of the polar rubbers is* 10^8–10^9 Ω.m, whereas that of the nonpolar ones—10^{13}–10^{14} Ω.m.

9.2.1.2   FILLERS

9.2.1.2.1   Carbides

9.2.1.2.1.1   Silicon carbide (SiC)

Pure silicon carbide is an isolator. Depending on the nature of the impurities therein SiC may have semiconductor properties, it is characterized by n- or p-type conductivity.[20, 21] The silicon carbide used herein was provided by Wako Chemical Company Tokyo, Japan and had an average particle size of about 1 μm.

9.2.1.2.1.2   Boron carbide (B_4C)

Analogously to SiC, boron carbide is extremely hard (~10 according to Mohs scale) and has semiconductor properties.[20, 21] The boron carbide was provided by Wako Chemical Company Tokyo, Japan and has an average particle size of about 2 μm.

9.2.1.2.1.3   Titanium diboride (TiB_2)

Titanium diboride is known for its high conductivity at low temperatures. When TiB_2 (a conductive filler) is used to reinforce the composites, their conductivity acquires a semiconductive course.[22] Wako Chemical Company Tokyo, Japan also provided TiB_2 the particle size of which varied about 2–4 μm.

9.2.1.3   RUBBER COMPOSITES

Two series of composites were prepared with each of the above fillers: NBR-based, which were compound with SiC, B_4C, or TiB_2 at various concentrations and NR-based, compound with the respective fillers at the same concentrations. The compositions of the used rubber compounds in phr are presented in Tables 9.1 and 9.2.

TABLE 9.1   Formulations of Acrylonitrile Butadiene Rubber (NBR)-Based Composites Filled with SiC, B_4C, and TiB_2.

	Parts per hundred rubber					
NBR	100	100	100	100	100	100
ZnO	5	5	5	5	5	5
Stearic acid	2	2	2	2	2	2
SiC/B_4C/TiB_2	0	10	15	25	35	45
IPPD	1.5	1.5	1.5	1.5	1.5	1.5
ZDEC	1	1	1	1	1	1
MBT	0.5	0.5	0.5	0.5	0.5	0.5
Sulfur	2	2	2	2	2	2

TABLE 9.2 Formulations of Natural Rubber (NR)-Based Composites Filled with SiC, B$_4$C, and TiB$_2$.

	Parts per hundred rubber					
NR	100	100	100	100	100	100
ZnO	4	4	4	4	4	4
Stearic acid	2	2	2	2	2	2
SiC/B$_4$C/TiB$_2$	0	10	15	25	35	45
IPPD	1.5	1.5	1.5	1.5	1.5	1.5
MBT	3	3	3	3	3	3
Sulfur	1	1	1	1	1	1

The samples were prepared on an open mixer by controlling the compounding temperature and time. The rubber compounds were vulcanized on a hydraulic press at 160°C (NBR-based composites) and at 155°C (NR-based composites) under 12 MPa for 10 min. The optimum vulcanization was determined on a Monsanto oscillating disk rheometer. The samples were plates of 90 × 60 × 2 mm.

9.2.2 METHODS

9.2.2.1 CHARACTERIZATION OF CONDUCTIVITY PECULIARITIES OF THE POLYMER COMPOSITES

The following parameters were determined

1. Volume resistivity—ρ_v, Ω.m.
2. Volume conductivity—$\sigma_v = 1/\rho_v$, Ω^{-1}.m^{-1}.

9.2.2.1.1 Determining the Volume Resistivity

The volume resistivity for flat samples in a uniform electric field is calculated by the formula[23]

$$\rho_v = R_v \cdot S / h \tag{9.1}$$

where R_v is the measured Ohmic electrical resistance, Ω; S is the area of the measurement electrode, m^2; and h is thickness of the sample, m.

The volume electric resistivity of the studied composites was measured on a Teraline III *teraohmmeter (Germany)*.

The change in the volume resistivity upon aging (72 h at 100°C for NBR-based composites and 72 h at 70°C for NBR-based composites) was determined by calculating the aging coefficient (K_ρ, %), obtained regarding ρ_v of the vulcanizates according to the equation:

$$K\rho = \frac{\rho_{V1} - \rho_{V2}}{\rho_{V1}} \cdot 100\%,$$ (9.2)

where ρ_{V1} is the volume resistivity prior to aging; ρ_{V2} isthe volume resistivity upon aging.

9.2.2.1.2 Determining the Volume Resistivity as a Function of the Applied Pressure

Pressure of 10, 15, 20, 25, and 35 kPa was used. The measurements were conducted at room temperature with increasing and decreasing the pressure to determine the presence or absence of hysteresis.

Measurements of ρ_v were carried out at high pressure (0–10 MPa). The samples were placed in a steel sleeve between the two electrodes, on which, by means of a load cell, pressure was exerted, and the change in resistivity was monitored on a teraohmmeter.Determining the Compression Coefficient of the Volume Resistivity

The change in the volume resistivity could be presented with some approximation as a linear function of the change in pressure:[24]

$$\Delta\rho = PC\rho_V \cdot \rho_0 \cdot (P - P_0),$$ (9.3)

where ρ_0 is the resistivity without pressure; $PC\rho_V$ is proportion coefficient, called compression coefficient of the volume resistivity, is the change in resistivity by 1 Ω at increasing the pressure with one unit.

At $P_0 = 0$, where P_0 is the initial pressure, follows

$$\Delta\rho = PC\rho_V\rho_0 P$$ (9.4)

$$\rho_1 = \rho_0 \pm \Delta\rho$$ (9.5)

$$\rho_1 = \rho_0 \pm PC\rho_V\rho_0 P_1 \text{ at a pressure } P_1$$ (9.6)

$$\rho_2 = \rho_0 \pm PC\rho_V \rho_0 P_2 \ at \ a \ pressure \ P_2 \qquad (9.7)$$

$$\rho_1 = \rho_0 \left(1 \pm PC\rho_V P_1\right) \qquad (9.8)$$

$$\rho_2 = \rho_0 \left(1 \pm PC\rho_V P_2\right) \qquad (9.9)$$

The equation has the sign "+", when the resistivity increases, and the sign "–", when it decreases.

If $\rho_1 = \rho_0 \left(1 + PC\rho_V P_1\right)$ and $\rho_2 = \rho_0 \left(1 + PC\rho_V P_2\right)$, after right-array division to eliminate ρ_0, as $\rho_2 > \rho_1$, the obtained ratio is the following:

$$\frac{\rho_2}{\rho_1} = \frac{1 + PC\rho_V P_2}{1 + PC\rho_V P_1}, \ then \qquad (9.10)$$

$$\rho_2 (1 + PC\rho_V P_1) = \rho_1 (1 + PC\rho_V P_2) \qquad (9.11)$$

$$\rho_2 + PC\rho_V \rho_2 P_1 = \rho_1 + PC\rho_V \rho_1 P_2 \qquad (9.12)$$

$$\rho_2 - \rho_1 = PC\rho_V \left(\rho_1 P_2 - \rho_2 P_1\right) \qquad (9.13)$$

$$PC\rho_V = \frac{\rho_2 - \rho_1}{\rho_1 P_2 - \rho_2 P_1} \qquad (9.14)$$

If $\rho_1 = \rho_0 \cdot (1 - PC\rho_V \cdot P_1)$ and $\rho_2 = \rho_0 \cdot (1 - PC\rho_V \cdot P_2)$, after right-array division to eliminate ρ_0, as $\rho_1 > \rho_2$, we obtain the following:

$$\frac{\rho_1}{\rho_2} = \frac{1 - PC\rho_V P_1}{1 - PC\rho_V P_2}, \ then \qquad (9.15)$$

$$PC\rho_V = \frac{\rho_1 - \rho_2}{\rho_1 P_2 - \rho_2 P_1} \qquad (9.16)$$

The formula allows calculating the compression coefficient by measuring the resistivity at two different pressure values.

9.2.2.1.4 Determining the Volume Resistivity as a Function of Bending

Viewing a potential usage of the vulcanizates studied as tension trans-ducers in pressure sensors, we carried out an experiment following our own technique. The experiment simulated the work conditions of a tension transducer of mechanical parameters into electrical.

The elementary tension transducers are 79 × 20 × 2 mm large belts of the vulcanizates studied. Copper folio was stuck with conductive glue to the upper surface of the tension elements. The tension transducer was stuck with the same glue to a 190 × 25 × 0.37 mm belt, simulating a bending support. The copper folio and the steel belt were connected to a teraohmmeter with wires. The belt was bent, and by a limiting rake, it was kept at certain radii of the bow thus obtained. The radii are set by patterns overlapping with the steel belt bow. The initial radius of the experiments performed was 100 mm. The bending degree (in %) was calculated by the following formula:[25]

$$\delta = \frac{r_0 - r}{r_0} \cdot 100\%,\qquad(9.17)$$

where δ is the bending, r_0 and r are the initial and changing radii at a different bending degree.

9.2.2.1.5 Determining the Volume Resistivity as a Function of Temperature

The vulcanizate was placed between the electrodes of an air thermostat. The resistivity measurement was performed at constant current at 100 V and pressure of 10 kPa. The scheme of the set for determining the volume resistivity as a function of the changing temperature is presented in Figure 9.1.

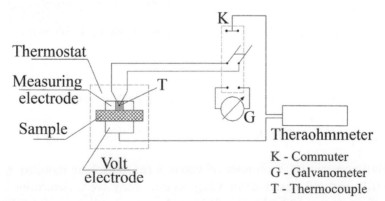

FIGURE 9.1 Scheme of the set for determining ρ_v with changing the temperature. (Reprinted from Dishovsky, N.; El—Tantawy, F.; Dimitrov, R. Effect of Bi-containing Superconducting Ceramic on the Volume Resistivity of Butyl Rubber Composites. *Polym. Test.* **2004.** *23,* 69–75.© 2004 with permission from Elsevier.)

9.2.2.1.6 Determining the Temperature Coefficient of Volume Resistivity

The temperature coefficient of resistivity () is calculated according to the next formula:[26]

$$\alpha = \rho_{T2} - \rho_{T1} / (\rho_{T1}T_2 - \rho_{T2}T_1),$$ (9.18)

where ρ_{T1} is volume resistivity at temperature T_1; ρ_{T2} is volume resistivity at temperature T_2.

9.2.2.1.7 Activation Energy

The activation energy is determined at a zone transition according the formula[27]

$$\sigma_V = \sigma_0 e^{-Eaz/2kT}$$ (9.19)

and for the hopping mechanism according to the formula:

$$\sigma_V \sqrt{T} = \sigma_0 e^{-Eah/2kT}$$ (9.20)

9.2.2.1.8 Determining the Volume Resistivity in a High Electric Field

A RLC bridge (3535 Z Hitester, Hioki, Japan), with a relative error of ±2% was used.

9.2.2.2 DETERMINING THE TYPE OF CONDUCTIVITY

9.2.2.2.1 Effect of Hall

The Hall constant (R_H), number of carriers (n), and their mobility (μ_m) in the NBR- and NR-based composites under study were determined on a Hall effect sensor 102, Hitachi, Tokyo, Japan according to the following equations:[28]

$$R_H = \frac{V_H h}{I_H B}; \ n = \frac{B I_H}{q h V_H}; \ \mu_m = \frac{\sigma_V}{ne},$$
(9.21)

where B is the induction of applied magnetic field, q—the charger per carrier, h—sample thickness, I_H and V_H are the current and voltage, obtained when determining the Hall effect, e—the elementary charge (charge of the electron), and σ_V—conductivity. The measurements were carried out at a relative error of ±3%.

9.2.2.2.2 Thermocouple Method

The thermocouple method has been used to determine the type of conductivity (conductivity of the negative or positive electric charges) in semiconductors and elastomer composites. The thermocouple is presented schematically in Figure 9.2.

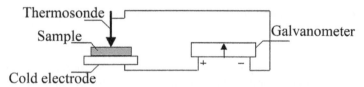

FIGURE 9.2   A set for determining the type of conductivity according to the thermocouple method.

When the probe (heated to about 80–100°C) touches the surface of the sample, the charges prevailing in number "run away" because of the high temperature. If the latter are negative, the probe is charged positively, and the lower electrode (the cold one)—negatively. In that case, the arrow of the galvanometer (it is a "zero" type instrument where "0" is in the middle of the scale) moves right (to the + terminal). When the conductivity is of positively charged electrons, the deflection of the pointer is in the opposite direction.

9.2.2.3 VOLT–AMPERE CHARACTERISTICS

There is a particular voltage versus circuit dependence for each material comprising electric charges that is not subjected to thermal, mechanical, and so forth effects. The dependence is called volt–ampere characteristic of a conductor. The volt–ampere characteristics are measured according to the scheme presented in Figure 9.3.

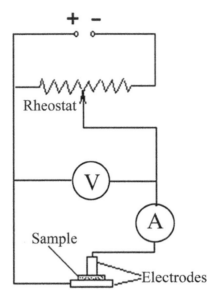

FIGURE 9.3 Scheme of the experimental set for measuring volt–ampere characteristics of rubber samples.

If the volt–ampere dependence is linear, the conductivity of the material is ohmic, that is, Ohm law applies to it. These characteristics are typical of metals. Semiconductors, polymers, and polymeric composites have nonlinear volt–ampere characteristics. Ohm law applies to them. The charge transport in those materials is realized by a mechanism controlled by quantum mechanics.

9.2.2.4   STRUCTURE AND PROPERTIES CHARACTERIZATION OF COMPOSITES

9.2.2.4.1   Estimating the Content of Bound Rubber

The method of C. M. Blow[29] was used to estimate the content of bound rubber, expanding its applicability and defining the experimental steps.

9.2.2.4.2   Microscopic Studies on the Composites

The microscopic studies on the composites were performed on a MIN-9 (Russia) microscope with an immersion lens.

The samples were prepared by cutting the nonvulcanized specimen from the respective rubber compound and placing it between two folio plates (polyethylene terephthalate). Then, the specimen was pressed at 70°C and zero distance between the press plates. The thin films obtained were thick enough to detect the details of our interest. Micrographs under nonpolarized light were taken at a magnification of 1500×.

9.2.2.4.3   Determining the Tensile Stress-Strain Properties

The tensile stress-strain properties were determined according to Reference.[30] The change in those properties upon samples aging was monitored: 72 h at 100°C for NBR-based composites and 72 h at 70°C for NR-based composites. The aging coefficient (K_{cr}, %) was determined according to the tensile strength of the unfilled NBR- and NR-based composites and of those filled at 45 phr (TiB$_2$, B$_4$C, SiC), according to the formula:[31]

$$K_\sigma = \frac{\sigma_1 - \sigma_2}{\sigma_1} \cdot 100, \% \qquad (9.22)$$

where σ_1 is the tensile strength of the vulcanizate prior to its aging; σ_2—the tensile strength of the vulcanizate upon its aging.

9.3   RESULTS AND DISCUSSION

9.3.1   CONDUCTIVITY PECULIARITIES OF NBR- AND NR-BASED COMPOSITES COMPRISING SIC, B$_4$C, AND TIB$_2$ AS A FILLER

9.3.1.1   CHANGES IN THE VOLUME RESISTIVITY DEPENDING ON FILLERS CONCENTRATION

The dependences of volume resistivity on fillers (SiC, B$_4$C, and TiB$_2$) concentration for NBR- and NR-based composites at 25°C and 10 kPa are presented in Figures 9.4 and 9.5.

As seen from Figure 9.4, volume resistivity increases with increasing SiC and B$_4$C concentration, whereas at a higher TiB$_2$ concentration, it decreases. The dualism in the values of volume resistivity for one and the same matrix is caused by the difference in the own resistivity of the matrix and filler. It is 1.82×10^8 Ω.m for NBR, 2.3×10^9 Ω.m for SiC, and 2.1×10^8 Ω.m for B$_4$C. Having a high-volume resistivity, SiC and B$_4$C act as barriers to the charge transfer and redirect the charges mainly through the matrix. At a higher filler concentration, the elastomer layers between the particles get thinner what hinders additionally the charge transfer through the composite, hence the resistivity increases.

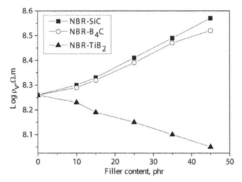

FIGURE 9.4   The volume resistivity as a function of fillers concentration for composites NBR-SiC, NBR-B$_4$C, and NBR-TiB$_2$.

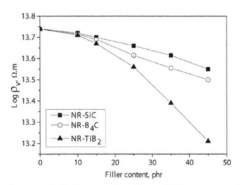

FIGURE 9.5   The volume resistivity as a function of fillers concentration for composites NR-SiC, NR-B$_4$C, and NR-TiB$_2$.

Having a lower volume resistivity ($\rho_v = 1.26 \times 10^7$ Ω.m) with regard to the matrix TiB$_2$ becomes a "preferred" zone for charge transfer. In the case, the higher amount of filler particles favors the charge transfer, despite of the thinning membranes between them.

Such a duality has not been observed for polymeric composites filled with conductive fillers, as in all the cases, the materials (carbon black, graphite, metal powders) have significantly lower volume resistivity than the matrix. The charge carrier transfer in those composites is realized with the active participation of the fillers.[32]

The resistivity of NR-based composites decreases in the presence of the three fillers (SiC, B_4C, and TiB_2), which possess resistivity lower than that of the matrix ($\rho_v \approx 5.5 \times 10^{13}$ Ω.m). They are factors bringing an additional amount of electrons to the ion transport in the matrix, thus changing the resistivity of the composites in a descending order following their own resistivity (Figure 9.5). SiC and B_4C also have their own contribution to NBR-based composites, but owing to the low resistivity of the matrix, their stopper effect of electron transfer is suppressed. The electron amount, contributed by the carbides to the high ohmic NR-matrix, though a small one, causes an increase in the resistivity.

9.3.1.2 CHANGES IN THE VOLUME RESISTIVITY DEPENDING ON THE APPLIED PRESSURE

9.3.1.2.1 Low Pressure

The values of volume resistivity for composites NBR-SiC, NBR-B_4C, and NBR-TiB_2 as a function of pressure up to 35 kPa at 25°C are presented in Figures 9.6–9.8.

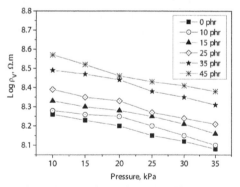

FIGURE 9.6 Pressure dependence of ρ_v for NBR-SiC composites. (Reprinted with permission from Todorova, Z.; El-Tantawy, F.; Dishovsky, N.; Dimitrov, R. Investigation of Conductivity Characteristics of Nitrile Butadiene Rubber Vulcanizates Filled with Semiconducting Carbide Ceramic. *J Appl. Polym. Sci.* **2007**, *103*, 2158–2165. © 2007 with permission from John Wiley.)

With the increasing pressure, the values of volume resistivity (ρ_v) for all three types of composites decrease. The curves in the plots for NBR-based composites comprising carbides (Figures 9.6 and 9.7) have a slope almost the same as that of the curve for an unfilled NBR composite what emphasizes the dominant role of the matrix. The distribution in an ascending order with regard to their filler content is due to the unblocked charge carriers resulting from the higher amount of filler particles. The descending order with increasing pressure as in the case of NBR-TiB$_2$ composites (Figure 9.8), whose curves order changes into descending at higher filler amounts, is due to thickening of the membranes what favors the charge transfer between the structure units.

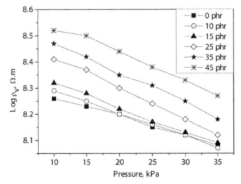

FIGURE 9.7 Pressure dependence of ρ_v for NBR-B$_4$C composites. (Reprinted with permission from Todorova, Z.; El-Tantawy, F.; Dishovsky, N.; Dimitrov, R. Investigation of Conductivity Characteristics of Nitrile Butadiene Rubber Vulcanizates Filled with Semiconducting Carbide Ceramic. *J Appl. Polym. Sci.* **2007**, *103*, 2158–2165. © 2007 with permission from John Wiley.)

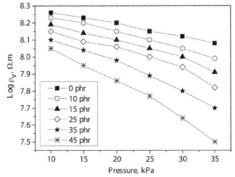

FIGURE 9.8 Pressure dependence of ρ_v for NBR-TiB$_2$. (Adapted from Todorova, Z.; El-Tantawy, F.; Dishovsky, N.; Dimitrov, R. Investigation of Some Electrical Properties of Elastomer Composites Filled with TiB2. *J. Elastom. Plast.* **2007**, *39*, 69–80. © 2007 with permission from SAGE Publications.)

In a more compact structure, these eased transfers could be realized, only if the charge carriers are electrons. If the charge carriers have been predominantly ions, then their transfer through a more compact structure would be hindered, and the resistivity would increase with the increasing pressure.[33,34]

The conductivity has been proven by the thermocouple method and Hall constant, which is negative. The determined values of the volume resistivity of composites NR-SiC, NR-B$_4$C, and NR-TiB$_2$ as a function of pressure (up to 35 kPa) at 25°C are presented in Figures 9.9–9.11.

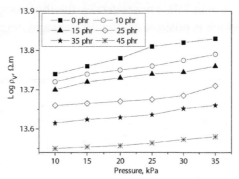

FIGURE 9.9 Pressure dependence of ρ_v for NR-SiC composites.

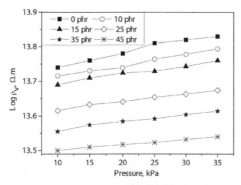

FIGURE 9.10 Pressure dependence of ρ_v for NR-B$_4$C composites.

As seen from Figures 9.9 and 9.10, ρ_v values increase at higher pressure. The result is attributed to the compacting of the matrix under the applied pressure that hinders the transfer of charge carriers (predominantly ions).

The values of ρ_v for composites with silicon carbide are higher due to its resistivity, which is higher than that of B$_4$C. As Figure 9.11 shows, the resistivity for NR-based composites filled with TiB$_2$ decreases at higher filler amounts and higher pressure due to thinning of the rubber layers between the conductive fine particles. The effect is more pronounced for the samples with a higher filler amount (35 and 45 phr) that enables building more compact electrically conductive structures of the filler particles under the set pressure, hence lowers significantly the resistivity. That evidences the change in conductivity from ionic into electronic for NR-based composites in the presence of TiB$_2$. It can be assumed that TiB$_2$ acts as an emitter of electrons in the process of charge carrier transfer. The mentioned ionic conductivity of NR-based composites filled with carbides and the electronic conductivity of the composites filled with TiB$_2$ are proven by the thermocouple method and by Hall constant, which is positive (in the former) and negative (in the latter case), respectively.

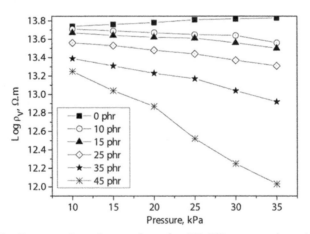

FIGURE 9.11 Pressure dependence of ρ_v for NR-TiB$_2$ composites. (Adapted from Todorova, Z.; El-Tantawy, F.; Dishovsky, N.; Dimitrov, R. Investigation of Some Electrical Properties of Elastomer Composites Filled with TiB2. *J. Elastom. Plast.* **2007**, *39*, 69–80. © 2007 with permission from SAGE Publications.)

The changes in values of ρ_v for both types of rubber at increasing and lowering the pressure have been also determined. As there is no difference in the resistivity values, it is not necessary to present the plots of the dependences.

The dependences are presented at a relative error of ± 4%, what defines the position of the plots and the linear mode of ρ_v versus pressure dependence for NBR- and NR-based composites. Having in mind the lack of hysteresis in pressure increase and decrease, the studied composites could find application as pressure sensors. The results in favor of such a possibility will be presented further on.

9.3.1.2.2 Compression Factor of Volume Resistivity (PCP$_V$)

The compression factor of the volume resistivity has been determined for all aforementioned composites.

The changes in $PC\rho_V$ for NBR- and NR-based composites as a function of filler concentration are presented in Figures 9.12 and 9.13.

The compression factor of the volume resistivity of NBR-based composites (Figure 9.12) is negative. That means resistivity decreases at a higher pressure. The "sensitivity" to compression influences increases with the increasing filler amount. The "sensitivity" of NBR-TiB$_2$ composites is the highest, whereas that for NBR-SiC composites—the lowest. As seen from Figure 9.13, NR-based composites filled with carbides have a positive compression factor, whereas NR-TiB$_2$—a negative one. In that case, the highest "sensitivity" to compression is also exhibited by composites filled with TiB$_2$. Composites NR-SiC and NR-B$_4$C have very close $PC\rho_V$ values. As they do not exceed the relative error of ±3%, that is not a reason to consider the dependency graphs as two particulate characteristics revealing that the "sensitivity" of the vulcanizates with boron carbide is higher than that of the composites with silicon carbide.

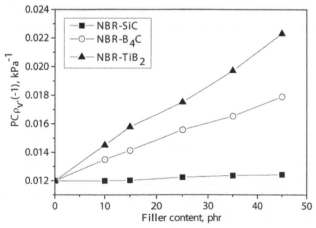

FIGURE 9.12 $PC\rho_V$ as a function of SiC, B$_4$C, and TiB$_2$ concentration in NBR-based composites.

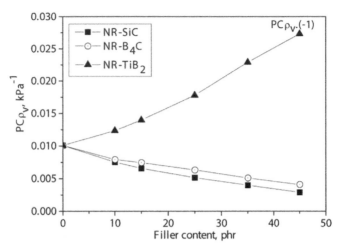

FIGURE 9.13 $PC\rho_v$ as a function of SiC, B$_4$C, and TiB$_2$ concentration in NR-based composites.

As the particles cannon get into contact at the chosen filler amount and relative low pressure (the pattern of the plots does not change), the observed change in the compression "sensitivity" of the composites with the increasing filler content is due to the coordinating role of the filler particles regarding the matrix macromolecules. During the compounding process, the filler particles penetrating between the macromolecule chains disentangle and arrange them parallelly. That structure of the matrix is more sophisticated than the globular one and under pressure favors the contacts between the macromolecules, which get denser, hence $PC\rho_V$ values increase at a higher concentration of the fillers.[35–39]

9.3.1.2.3 High Pressure

To clarify the effect that the structural changes in the studied composites have upon the composites conductivity, we have monitored their behavior under pressure varying 0–10 MPa, viewing the possibility of contact between the filler particles and reaching the percolation threshold. The changes in the volume resistivity of NBR-based composites filled with B$_4$C, SiC, and TiB$_2$ as a function of pressure are presented in Figures 9.14 and 9.15.

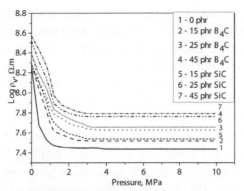

FIGURE 9.14   Pressure dependence of ρ_v for composites NBR-B$_4$C and NBR-SiC.

As seen from the figures, the volume resistivity decreases with the pressure increasing up to 2 MPa, then the values become constant. The decrease of ρ_v is explained by the dominating electron transport in NBR-based composites. The denser structure favors the electron transfer from a localized state to another. Those zones of localization in NBR-based composites are defined by the location of nitrile groups amongst which the exchange of electrons occurs via a hopping mechanism. Due to this peculiarity of the elastomer, composites resistivity increases at a higher concentration of the fillers, as the latter have a higher resistivity than the matrix. Hence, the electron transport through the composites is hindered. Such fillers are B$_4$C and SiC. As Figure 9.14 shows, the resistivity of the composites filled with B$_4$C and SiC increases at higher filler amounts, the curves are closer to the higher resistivity values, following both the increasing filler amounts and the values of their own resistivity (the curves for SiC-filled composites are higher than those for B$_4$C) at the same concentration.

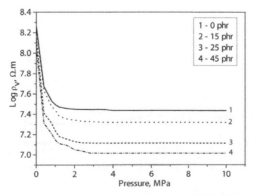

FIGURE 9.15   Pressure dependence of ρ_v for composites NBR-TiB$_2$.

Titanium diboride having volume resistivity lower than that of NBR favors the electron transport and at higher amounts causes a decrease of vulcanizates resistivity. The curves in Figure 9.15 are under each other according to the filler concentration.

The horizontal part of the curves defining the resistivity unchanging with the increasing pressure shows that the structure is compact enough and does not undergo changes. As the pattern of those curves is like the one of the unfilled vulcanizate, one cannot assume that the filler particles get in contact. Hence, the separating membranes built by the matrix bar the percolation effect. Most probably, the hydrostatic property of nonshrinking is involved. Therefore, the dimension characteristics of the structure are retained, so the resistivity remains the same in the pressure interval 2–10 MPa.

The changes in the volume resistivity as a function of pressure for composites NR-SiC, NR-B$_4$C, and NR-TiB$_2$ are presented in Figures 9.16–9.18.

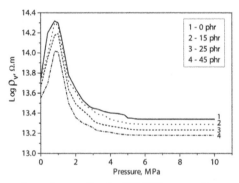

FIGURE 9.16   Pressure dependence of ρ_v for NR-SiC composites.

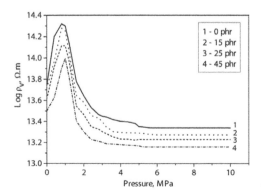

FIGURE 9.17   Pressure dependence of ρ_v for NR-B$_4$C composites.

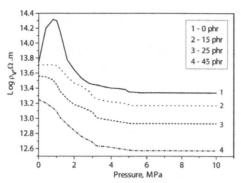

FIGURE 9.18 Pressure dependence of ρ_v for NR-TiB$_2$ composites.

As seen in Figures 9.16 and 9.17 from 0 to about 1 MPa, there is a pronounced increase in the values of volume resistivity. This change in resistivity is due to the mixed type of conductivity occurring by ions and electrons. Ionic conductivity is the dominant for this type of composites, what is evidenced by the increasing resistivity at the higher pressure and established by detected by the thermocouple transport positive charges. The increasing slope of the curves corresponds to ionic conductivity, which decreases with increasing the pressure as the denser structure hinders the transport of ions, yet dominant over the amount of electrons. The process continues until the maximum resistivity. From the maximum on (up to ~3 MPa), with the increasing compactness of matrix structure, the transport of electrons is facilitated, and the course of the curves turns from ascending into a descending one. Parris et al.[40] investigating NR-based vulcanizates filled with three types of carbon black have established increasing resistivity in the range 0–1 MPa. In their opinion, the resistivity is controlled by the macromolecular layers between the particles.

Figure 9.18 shows changes in the volume resistivity of NR-based composites filled with TiB$_2$ at various pressures. The decrease of ρ_v is explained by the electrical properties of TiB$_2$, which are more specific, if compared with those of carbides. This filler brings too much electrons, thus the effect of the ionic conductivity of the matrix becomes insignificant in comparison with the growing transport of electrons. For this reason, the increasing resistivity registered in composites with carbides was not observed, and the curves pass directly to decreasing values. Those results show that under the influence of pressure and type of filler, the conductivity can change from ionic to electron one.

The resistivity characteristics for all NR-based composites at a pressure higher than 3 MPa are parallel to the abscissa. The explanation of the result is identical with that concerning the course of the curves for NBR-based composites.

Figure 9.19 shows changes in the volume resistivity of NR-based composites filled with SiC, B₄C, and TiB₂ as a function pressure up to 50 MPa.

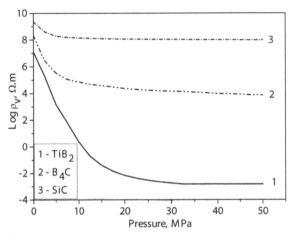

FIGURE 9.19   Pressure dependence of ρ_v for SiC, B₄C, and TiB₂.

As the figure shows, ρ_v values decrease at higher pressure due to the denser particles and thus formed larger contact area. At the studied pressure, the lowest value $\rho_v = 1.26 \times 10^7$ Ω.m is for TiB₂, which drops to $\rho_v = 1.48 \times 10^{-3}$ Ω.m at 30 MPa. The measurements up to 50 MPa have shown that the resistivity values remain constant at a pressure higher than 30 MPa. The curves become straight lines parallel to the abscissa as a result of agglomerating structure of the filler. The particles of SiC have almost not agglomerated (Figure 9.56) under the applied pressure. They have only become denser. The agglomeration of TiB₂ is the greatest (Figure 9.58). The compression process in that filler starts with agglomerates destruction followed by compression of the fragments. The structural changes occur in a wide pressure interval: 0–30 MPa. The characteristic of B₄C, which is more prone to agglomeration than SiC, is in the middle of the order. The order of the curves is according to the own resistivity of the fillers and to the probability of contacts between their particles; the contact area of TiB₂ particles being the largest.

The compression characteristics of the three fillers allow an important conclusion about the composites structure. It is stated in many studies[40–44] that two types of particle distribution occur in polymer compounds reinforced with powder fillers. The first type of distribution is of the particles separated from the polymer by a film, whereas the second type is of particles getting into a direct contact. The comparison of the results about the resistivity of the studied composites and fillers obtained under pressure reveals that particles in the compounds do not get into contact. If such a contact did occur, the changes in the vulcanizates resistivity would be much closer to the resistivity values of the fillers. As the graphic dependences show, the differences are not only in the values of the volume resistivity but also in the pattern of the curves and in the pressure at which the patterns of the curves changes. Hence, those, although indirect results and the microscopic studies allow stating that, matrix macromolecules and filler particles (which do not get into contact) associate during the compounding process.

9.3.2 TENSOEFFECT IN THE NBR-AND NR-BASED COMPOSITES FILLED WITH SIC, B_4C, AND TIB_2

The resistance (R) depending on the bending of the steel belt to which the specimens of NBR- and NR-based composites filled (SiC, B_4C, TiB_2) at 45 phr were glued. As seen, the tendency of changes in R values both for NBR-(Figure 9.20) and NR-based (Figure 9.21) composites is the same as the one for the varying volume resistivity as a function of pressure (up to 35 kPa), presented in Figures 9.6–9.11.

A creeping effect of the rubber layers parted by orientation of the structure occurs in the course of bending. That leads to some easing of electron transport and hindering of ion transfer, hence the lower and higher resistance values. The used conductive glue favors the contact between the specimen and the steel belt what is a prerequisite for obtaining lower resistance values.

As the results show, the studied specimens are prone to greater deformation without breaking unlike the conventional (inorganic) semiconductor materials that are brittle and fragile allowing deformation up to 1%.[24]

The experiment shows that the choice of an appropriate matrix and fillers allows manufacturing various convenient pressure and deformation sensors with a wide range of applications implementing a relatively simple technology.

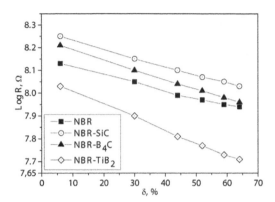

FIGURE 9.20 Resistance of NBR-based composites filled (SiC, B$_4$C, TiB$_2$) at 45 phr depending on the applied bending.

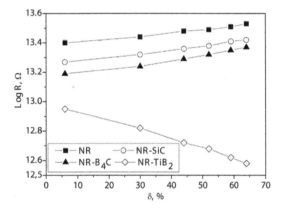

FIGURE 9.21 Resistance of NR-based composites filled (SiC, B$_4$C, TiB$_2$) at 45 phr depending on the applied bending.

9.3.3 TEMPERATURE DEPENDENCE OF VOLUME RESISTIVITY

The temperature dependencies of volume resistivity for composites NBR-SiC, NBR-B$_4$C, and NBR-TiB$_2$ are presented in Figures 9.22–9.24.

With the increasing temperature, the volume resistivity of all studied samples decreases. The rather close resistivity values as a function of temperature define an interval wherein the values for all composites change. Its limits are the curve for the unfilled sample and that for the sample filled at 45 phr.

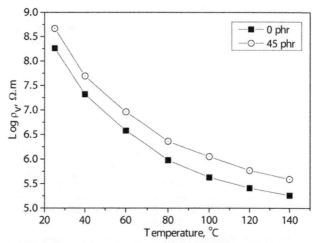

FIGURE 9.22 Temperature dependence of ρ_v for NBR-SiC composites. (Reprinted with permission from Todorova, Z.; El-Tantawy, F.; Dishovsky, N.; Dimitrov, R. Investigation of Conductivity Characteristics of Nitrile Butadiene Rubber Vulcanizates Filled with Semiconducting Carbide Ceramic. *J Appl. Polym. Sci.* **2007,** *103*, 2158–2165. © 2007 with permission from John Wiley.)

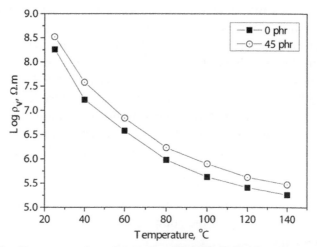

FIGURE 9.23 Temperature dependence of ρ_v for NBR-B$_4$C composites. (Reprinted with permission from Todorova, Z.; El-Tantawy, F.; Dishovsky, N.; Dimitrov, R. Investigation of Conductivity Characteristics of Nitrile Butadiene Rubber Vulcanizates Filled with Semiconducting Carbide Ceramic. *J Appl. Polym. Sci.* **2007,** *103*, 2158–2165. © 2007 with permission from John Wiley.)

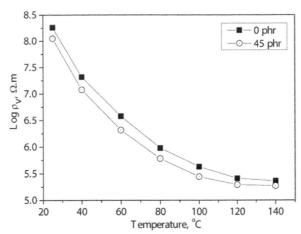

FIGURE 9.24 Temperature dependence of ρ_v for NBR-TiB$_2$ composites. (Adapted from Todorova, Z.; El-Tantawy, F.; Dishovsky, N.; Dimitrov, R. Investigation of Some Electrical Properties of Elastomer Composites Filled with TiB2. *J. Elastom. Plast.* **2007**, *39*, 69–80. © 2007 with permission from SAGE Publications.)

The resistivity of the composites monitored at rising and lowering the temperature yields the same values. Temperature hysteresis has not been observed for those compositions. As seen from the figures, the curves for the temperature dependence of the resistivity of composites NBR-SiC and NBR-B$_4$C are located above the one of the unfilled sample, whereas those for composites NBR-TiB$_2$—bellow it. In the case of NBR-based composites filled with carbides, the uniform matrix and the close resistivity values of SiC and B$_4$C cause a small difference in the temperature dependence of the resistivity for the two composite. As explained, this result is due to the properties of SiC and B$_4$C, which having a resistivity higher than that of NBR hinder the transport of electrons to an extent greater than electrons emission into the composite does. Having a lower resistivity TiB$_2$ is a source of a larger amount of electrons. With these properties, that filler weakens significantly the resistivity of the composites which drops below the one of the unfilled sample.

The temperature dependencies of volume resistivity for NR-based composites are presented in Figures 9.25–9.27.

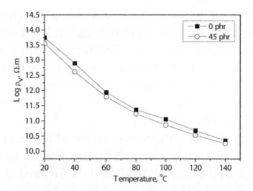

FIGURE 9.25 Temperature dependence of ρ_v for NR-SiC composites.

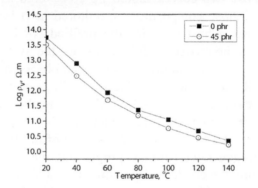

FIGURE 9.26 Temperature dependence of ρ_v for NR-B$_4$C composites.

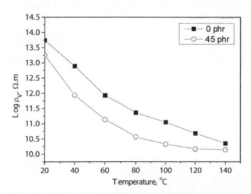

FIGURE 9.27 Temperature dependence of ρ_v for NR-TiB$_2$ composites. (Adapted from Todorova, Z.; El-Tantawy, F.; Dishovsky, N.; Dimitrov, R. Investigation of Some Electrical Properties of Elastomer Composites Filled with TiB2. *J. Elastom. Plast.* **2007,** *39*, 69–80. © 2007 with permission from SAGE Publications.)

With the increasing temperature, the volume resistivity of the three NR-based composites decreases. As in the case of NBR-based composites, the rather close resistivity values as a function of temperature also define an interval wherein the values for all composites with the chosen filler amounts change. The limits of the interval are the curve for the unfilled sample and that for the sample filled at 45 phr. As seen from the figures, the curves for the temperature dependences of ρ_v for NR-based composites are also quite close, due to the volume resistivity of SiC and B_4C, unlike the one of TiB_2. The difference between the curve for the unfilled composite and those for NR-based composites filled with SiC and B_4C (Figures 9.25 and 9.26) is of minimal value—0.46, whereas the absolute error is of ± 0.006. Hence, those two curves should be considered as separate ones. The resistivity of those compositions monitored at rising and lowering the temperature has a hysteresis for the unfilled NR-based vulcanizate (Figure 9.28). With the increasing filler amounts, the hysteresis decreases, and at high loads (45 phr), almost has not been observed. The hysteresis in the case of NR-SiC composites (Figure 9.29) is greater than the one for NR-B_4C composites (Figure 9.30), whereas in the case of NR-TiB_2 composites (Figure 9.31), there is no hysteresis.

The occurrence of a hysteresis in the values of unfilled vulcanizate is caused by the ions generated with the temperature increasing up to 140°C. At that temperature, the conductivity is the highest. At that point, temperature starts lowering with the highest amount of ions. Therefore, in the reverse course till reaching ambient temperature, the resistivity has lower values.

In unfilled composites, the elastomer amount emitting ions is lower. The number of ions decreases with the increasing filler concentration. Meanwhile, the electrons amounts brought into the composites by the fillers are in the following order: SiC, B_4C, and TiB_2. Each filler proportionally to the brought in electrons masks the ionic conductivity and lowers the resistivity hysteresis, which in the case of TiB_2 is lost completely.

The deviations in the results from the temperature dependence of resistivity are within the range of ± 6%.

The temperature dependencies of volume resistivity for the studied fillers as determined at 100 V are presented in Figure 9.32.

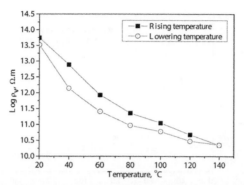

FIGURE 9.28 Changes in ρ_v values with rising and lowering the temperature of unfilled NR-based composite. (Reprinted from Todorova, Z.; El-Tantawy, F.; Dishovsky, N.; Dimitrov, R. Investigation of Some Electrical Properties of Elastomer Composites Filled with TiB2. *J. Elastom. Plast.* **2007**, *39*, 69–80. © 2007 with permission from SAGE Publications.)

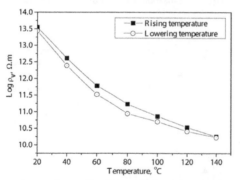

FIGURE 9.29 Changes in ρ_v values with rising and lowering the temperature of a NR-based composite filled with SiC at 45 phr.

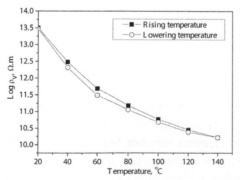

FIGURE 9.30 Changes in ρ_v values with rising and lowering the temperature of a NR-based composite filled with B_4C at 45 phr.

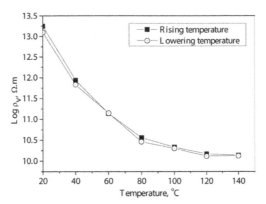

FIGURE 9.31 Changes in ρ_v values with rising and lowering the temperature of a NR-based composite filled with TiB$_2$ at 45 phr. (Reprinted from Todorova, Z.; El-Tantawy, F.; Dishovsky, N.; Dimitrov, R. Investigation of Some Electrical Properties of Elastomer Composites Filled with TiB2. *J. Elastom. Plast.* **2007**, *39*, 69–80. © 2007 with permission from SAGE Publications.)

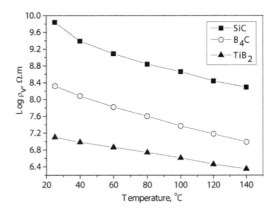

FIGURE 9.32 Temperature dependence of ρ_v for SiC, B$_4$C, and TiB$_2$.

The curves follow a descending order typical of semiconductors to which class also belong SiC, B$_4$C, and TiB$_2$. The curve for TiB$_2$ is at the lowest position, then comes one for B$_4$C, whereas the curve for SiC occupies the highest position. As seen from the equation $\sigma = e.n.\mu$, which relates the conductivity (σ) to electron number (n), mobility (μ) and charge of electrons (e), the number of electrons and their mobility in TiB$_2$ is the greatest; therefore, the filler has the best effect upon the improved conductivity of the composites comprising it.

9.3.4   TEMPERATURE COEFFICIENT OF VOLUME RESISTIVITY

The temperature coefficient of volume resistivity (α_t) is one of the important characteristic of the electrical properties of elastomer composites.

According to the temperature dependencies of resistivity for the composites based on the two types of rubber, two values for α_t defined by the straight segments of the curves in the intervals 20–80°C and 80–140°C have been obtained (Figures 9.22–9.27).

The obtained α_t values for NBR-based composites are the following:

- unfilled NBR composite

$$\alpha_{20-80} = -1.25 \times 10^{-2} {}^{\circ}C^{-1}; \alpha_{80-140} = -6.49 \times 10^{-3} {}^{\circ}C^{-1};$$

- NBR-SiC composite

$$\alpha_{20-80} = -1.24 \times 10^{-2} {}^{\circ}C^{-1}; \alpha_{80-140} = -6.56 \times 10^{-3} {}^{\circ}C^{-1};$$

- NBR-B$_4$C composite

$$\alpha_{20-80} = -1.25 \times 10^{-2} {}^{\circ}C^{-1}; \alpha_{80-140} = -6.52 \times 10^{-3} {}^{\circ}C^{-1};$$

- NBR-TiB$_2$ composite

$$\alpha_{20-80} = -1.25 \times 10^{-2} {}^{\circ}C^{-1}; \alpha_{80-140} = -6.11 \times 10^{-3} {}^{\circ}C^{-1}.$$

The obtained α values for NR-based composites are the following:

- unfilled NR composite

$$\alpha_{20-80} = -1.25 \times 10^{-2} {}^{\circ}C^{-1}; \alpha_{80-140} = -6.80 \times 10^{-3} {}^{\circ}C^{-1};$$

- NR-SiC composite

$$\alpha_{20-80} = -1.25 \times 10^{-2} {}^{\circ}C^{-1}; \alpha_{80-140} = -6.81 \times 10^{-3} {}^{\circ}C^{-1};$$

- NR-B$_4$C composite

$$\alpha_{20-80} = -1.25 \times 10^{-2} {}^{\circ}C^{-1}; \alpha_{80-140} = -6.80 \times 10^{-3} {}^{\circ}C^{-1};$$

- NR-TiB$_2$ composite

$$\alpha_{20-80} = -1.25 \times 10^{-2} {}^{\circ}C^{-1}; \alpha_{80-140} = -6.09 \times 10^{-3} {}^{\circ}C^{-1}.$$

The obtained α values for SiC, B_4C, and TiB_2 are the following:

$$SiC : a_{20-140} = -6.97 \times 10^{-3} {}^{\circ}C^{-1};$$
$$B_4C : a_{20-140} = -6.85 \times 10^{-3} {}^{\circ}C^{-1};$$
$$TiB_2 : a_{20-140} = -6.03 \times 10^{-3} {}^{\circ}C^{-1}.$$

The temperature coefficients of volume resistivity for the unfilled and filled composites in the 20–80°C intervals are the same due to the dominating effect of NBR and NR matrixes. In the second interval (80–140°C), the temperature coefficients for NRB-based composites filled with carbides are higher than those of the unfilled composite and of $NBR-TiB_2$. Therefore, those composites are more sensitive to temperature changes. Composites $NR-TiB_2$ have the lowest temperature coefficients of volume resistivity of all NR-based composites, whereas those of NR-SiC and $NR-B_4C$ are almost the same. The comparison of α_t values of the composites based on the two types of rubber shows that those of NR-based composites are higher because of the more pronounced effect of NR matrix. Filler TiB_2 is the least sensitive to changes in temperature, being the most conductive, then come B_4C and SiC. Regarding the composites—their α_t values are dominated by the properties of the elastomer matrix. The results obtained allow the conclusion that all compositions studied belong to the class of "negative thermistors."[35-39]

9.3.5   CHARGE CARRIERS TRANSFER

The changes in the resistivity of the studied composites dependent on thermal and mechanical treatment define them as materials possessing semiconductor properties. The chemical composition and macromolecular conformations of the matrix, the specifics of the fillers, the substances used in the vulcanization process, and so forth random adulterants are an intricate complex of conditions affecting the origin, type, and transport of charge carriers through the bulk of the vulcanizates studied. Despite of the theoretical interpretations of the conductivity mechanism, a complete theory explaining the electric properties of that type of composites has not been postulated. The exponential dependence of the electrical conductivity on temperature gives grounds to apply the *Brillouin zone* theory that explains the electrical properties of inorganic semiconductors. However, the low mobility of the charge carriers

<1 cm²/(V.s) does not meet the requirements of that theory and limits its application to polymer materials. The low mobility of the charge carriers is due to their localization in areas wherein they stay longer, than during the pauses in electrons movement and holes in the inorganic semiconductors. Due to this peculiarity of polymer composites, the theoretic works dealing with their electrical properties apply the theory of charge carrier hopping transport. According to that theory, the conductivity versus temperature is also an exponential function, but the activation energy is greater than the one in the transfer zones. The reason is in the greater distance between the electron localizations and the steric hindrance caused by the micro-Brown movement of macromolecules. The charge carriers activation energy values (E_{az}) in the transfer zone are: $E_{az}=0.62$ eV (SiC), $E_{az}= 0.53$ eV (B_4C) and $E_{az}=0.30$ eV (TiB_2). Those results correlate with the volume resistivity of the three fillers.

The charge carriers activation energy (E_{az}) at a zone transfer and activation energy at a hopping transport (E_{ah}) for NBR-based composites are presented in Figures 9.33 and 9.34, while those for NR-based composites—in Figures 9.35 and 9.36.

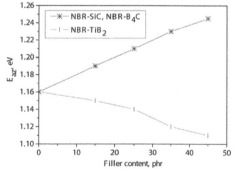

FIGURE 9.33 E_{az} as a function of filler amount for NBR-SiC, NBR-B_4C and NBR-TiB$_2$ composites.

Since the values for the respective energies for NBR-SiC and NBR-B$_4$C composites (Figures 9.33 and 9.34) and NR-SiC and NR-B$_4$C composites (Figures 9.35 and 9.36) are commeasurable with the measurement error (±3%), they are presented by a single curve.

As seen from Figures 9.33 and 9.34, E_{az} and E_{ah} values for composites NBR-SiC and NBR-B$_4$C increase at higher filler amounts, while those for NBR-TiB$_2$. The hopping transport has higher activation energy than the zone charge charier transport. That is also valid for NR-based composites (Figures

9.35 and 9.36), wherein the charge carrier transport proceeds at an activation energy higher than that of the process in NBR-based composites. The
values of activation energies of NBR matrix decrease with the increasing
filler amount.

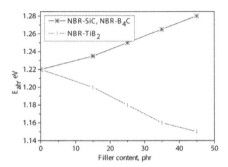

FIGURE 9.34 E_{ah} as a function of filler amount for NBR-SiC, NBR-B$_4$C and NBR-TiB$_2$
composites.

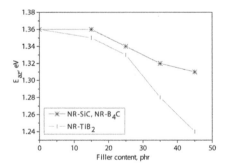

FIGURE 9.35 E_{az} as a function of filler amount for NR-SiC, NR-B$_4$C and NR-TiB$_2$
composites.

The fact that the elastomer matrixes have activation energies (0 phr—
curves in Figures 9.33–9.36) higher than those of SiC, B$_4$C, and TiB$_2$, means
the charge carrier transport through the matrixes is a hopping one, while that
through the filler—a zone one. At higher filler amounts the activation energy
of the zone transport in the very filler particles remains unchanged since
fillers structure and properties do not change. The values of E_{ah} for NBR-SiC
and NBR-B$_4$C increase due to the thinning elastomer membranes between
the filler particles. That decreases the number of locations for charge carriers
exchange (mainly –CN– groups) and enlarges the distance between them.
Thus the dominating E_{ah} values contribute to estimate higher E_{az} values.

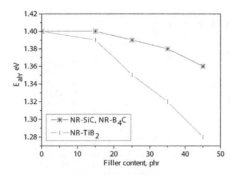

FIGURE 9.36 E_{ah} as a function of filler amount for NR-SiC, NR-B$_4$C and NR-TiB$_2$ composites.

The structure elastomer–filler of composites NBR-SiC and NBR-B$_4$C is such that the charge carriers transport proceeds at activation energy higher than that required for the transport in the structural units with a lower barrier. In our case such units are SiC and B$_4$C particles which have less influence on the composites conductivity than the matrix. If the conductivity of the fillers was dominant, total E_{az} would decreased with the increasing the filler amount, as shown by the plot of NBR-TiB$_2$ composites (Figure 9.33). The total energy of zone transport for NR-based composites (Figures 9.35 and 9.36) also decreases on account of the higher number of particles realizing the charge carrier transport. The values of E_{ah} also decrease because the thinning of elastomeric membranes between the particles favors charge carriers hopping. The values of E_{az} and E_{ah} for TiB$_2$ composites are determined by the own conductivity of the filler. The decrease of activation energy for NR-TiB$_2$ is greater than that for NBR-TiB$_2$ composites due to the difference in the volume resistivity of NR and TiB$_2$ which is greater than that between the resistivity of NBR and TiB$_2$. It should be noted that, E_{az} values for the three fillers are not reached neither by NBR-TiB$_2$ composites, nor by any of NR-based composites. That is another proof that filler particles do not get between each other regardless of the applied extreme treatment, high temperature, in the case.

Another result in support of the probability of dominating hopping transport (and influence of the matrix), are ρ_v values determined using alternating electric field. According to the authors,[45] the hopping transport of charge carriers, the resistance measured at alternating current should decrease with increasing frequency. Such is the result of the tested samples shown in Figures 9.37–9.42. Therefore, taking into account the difference in the

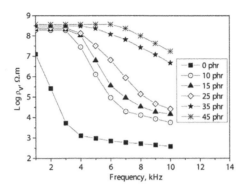

FIGURE 9.37   Frequency dependence of ρ_v for NBR-SiC composites.

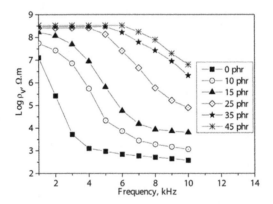

FIGURE 9.38   Frequency dependence of ρ_v for NBR-B$_4$C composites.

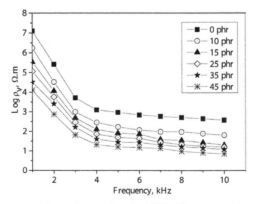

FIGURE 9.39   Frequency dependence of ρ_v for NBR-TiB$_2$ composites.

activation energies of the fillers and composites, most probably the charge carriers hopping transport through the matrix occurs. The former has a dominating impact on the electrical properties of the composites. The change in the activation energy with the increasing filler amount can be assumed as a macroeffect revealing the role of the filler particles as emitting or blocking carriers, depending on the type of matrix they are dispersed in. Titanium diboride can be considered a defined emitter since all composites comprising it have decreasing values of activation energy. Both carbides (SiC and B_4C) have more specific properties. In NBR matrix, they act as blocking traps of electrons and increase the activation energy. Under their influence, the activation energy decreases because the fillers exhibit properties of charge carriers (electrons) emitters. Perhaps, the duality in the fillers behavior is determined by the balance of the charge carriers in them and in the matrix. Where the charge carriers transported in the composite originating from the matrix are more than those from the filler, the latter acts as a blocking trap. If the filler is the one emitting charge carriers in a larger amount than the matrix, then it is defined as an emitter.[35-37]

9.3.6 CHANGES IN VOLUME RESISTIVITY IN THE HIGH FREQUENCY ELECTRIC FIELD

The changes in the volume resistivity as a function of the electric field frequency for NBR-based composites filled with SiC, B_4C, and TiB_2 as a function of the electric field frequency are presented in Figures 9.37–9.39.

As Figures 9.37 and 9.38 show, in the 1–6-kHz interval, the curves for the volume resistivity of NBR-based composites filled with carbides have plateaus and ascending positioning regarding the ordinate proportional to the filler amount. In this frequency range, the oscillators of unfilled NBR-composite, having a wide range of own frequencies and freedom of movement, start vibrating consistent with the increasing frequency. That to the greatest extent reduces its volume resistivity, in comparison to that of other composites. In the 6–10-kHz range yet smaller charged kinetic units enter all composites gradually in a resonance, what contributes to the reduction of volume resistivity in this frequency range. At higher filler amounts the number of oscillators decreases. They are blocked by the filler particles, their oscillations subside and the volume resistivity increases. These high-frequency characteristics of NBR-based composites filled with carbides reveal an opportunity for their applications as high-pass filters. As seen from Figure 9.39, for unfilled NBR, as well as for NBR-TiB_2 composites, with the increasing frequency the oscillators start vibrating successively with their decreasing size. The effect is most pronounced for TiB$_2$ at an amount of 45 phr, where the concentration of charge carriers (mostly electrons) is the highest and the volume resistivity drops to the greatest extent. Unlike SiC and B$_4$C, TiB$_2$ having a resistivity ($\rho_v = 1.26 \times 10^7$ Ω.m) lower than that of the matrix brings the resistivity values of its composite below those of the unfilled sample.

The changes in the volume resistivity as a function of the electric field frequency for NBR-based composites filled with SiC, B$_4$C, and TiB$_2$ are presented in Figures 9.40–9.42.

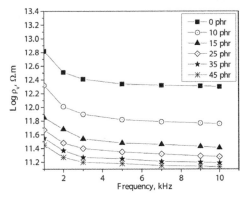

FIGURE 9.40   Frequency dependence of ρ_v for NR-SiC composites.

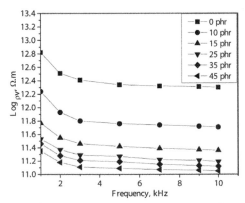

FIGURE 9.41   Frequency dependence of ρ_v for NR-B$_4$C composites.

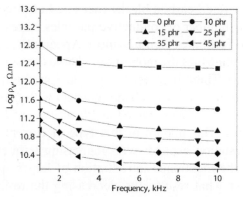

FIGURE 9.42 Frequency dependence of ρ_v for NR-TiB$_2$ composites.

As seen all three types of NR-based composites the volume resistivity decreases at higher filler amounts and increasing frequency of the electric field from 1–3 kHz. The greatest decrease is for NR-TiB$_2$ composites. They are marked by the lowest resistivity, because of the volume resistivity of TiB$_2$ which lower if compared with that of B$_4$C and SiC. Therefore, the oscillators of NR-TiB$_2$ composite, having a wide range of own frequencies and freedom of movement, start vibrating consistent with the increasing frequency and cause the greatest decrease of volume resistivity, if compared with that of the other composites.

The curves representing the high frequency characteristics of the composites do not have a plateau as the volume resistivity of the three fillers is lower that of NR matrix. The mechanism of charge carriers transport is the same as in NBR-based composites. The difference is that curves presenting ρ_v values for NR-based composites are below the one for the unfilled sample. The order follows the decrease of fillers resistivity: SiC, B$_4$C, and TiB$_2$. The influence that the electric field frequency has on resistivity shows that in composites, besides the free charge carriers, there are kinetic units of the dipole-segmental type associated with molecular chains having the freedom only to oscillate, without participating in the transport of carriers. Those chains form the so-called displacement currents. With increasing frequency free charge carriers (electrons and ions) of small mass in the electric field resonate at higher frequencies can under the influence of the alternating electric field pass not along the molecular chains, but also from a macromolecule to another. Those are the most likely reasons for the lower resistivity values measured at alternating current, if compared with the values obtained at direct current. The course of the curves is identical to that obtained for the impedance of an elastomer filled with increasing amounts of conductive filler (e.g., carbon black) measured at direct current. In this case, the increasing number of conductive particles increases the amount of charge carriers and facilitates their transport. Apparently, in the case of high impedance filled composites the mobility of charge carriers in the bulk of the material is restricted. Therefore, resistivity values are high at direct current. In a high-frequency electric field, charge carriers with "restricted freedom" of movement act as oscillators involved in "displacement currents" vibrating at their localizations. The amount of this type of current elements is very high which manifestation expands with increasing frequency. The process is very similar to the mentioned introduction of larger amounts of highly conductive particles what results into decreasing the resistivity by several orders of magnitude as in the case of the composites studied.

9.3.7 MOBILITY AND QUANTITY OF CHARGE CARRIERS

The quantity and mobility of charge carriers as a function of filler concentration for NBR-based composites filled with SiC, B_4C, and TiB_2 are presented in Figures 9.43 and 9.44. The two parameters were determined via the Hall effect.

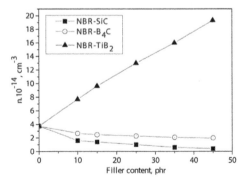

FIGURE 9.43 Quantity of charge carriers as a function of filler concentration for NBR-TiB_2, NBR-B_4C and NBR-SiC composites.

As the figures show, NBR-based composites filled with B_4C and SiC the quantity of charge carriers decreases while their mobility increases. On the contrary for NBR-TiB_2, composites the quantity of charge carriers decreases while their mobility increases. The changes in and n for NBR-based composites with the various filler are due to the difference in the quantity of charge carriers which determines the deference in the resistivity of the fillers (ρ_v for

$B_4C - 2.1 \times 10^8$ Ω.m and ρ_v for SiC $- 2.3 \times 10^9$ Ω.m) and matrix (ρ_v for NBRe $\approx 1.82 \times 10^8$ Ω.m) are higher than that for the matrix, while the resistivity for TiB_2 ($\rho_v = 1.26 \times 10^7$ Ω.m) is lower than that of NBR. The particles of TiB_2 favor the decrease of composites resistivity. As seen from Figures 9.45 and 9.46, the quantity of charge carriers in NR-based composites filled with SiC, B_4C, and TiB_2 at different concentrations increases with the increasing filler amount, while their mobility decreases.

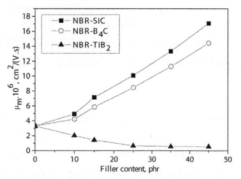

FIGURE 9.44 Mobility of charge carriers as a function of filler concentration for NBR-TiB_2, NBR-B_4C and NBR-SiC composites.

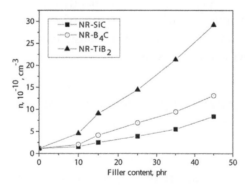

FIGURE 9.45 Quantity of charge carriers as a function of filler concentration for NR-TiB_2, NR-B_4C, and NR-SiC composites.

The highest n values are those for NR-TiB_2 composites while the lowest for NR-SiC, while the charge carrier mobility is the lowest for the composites filled with TiB_2 and the highest for the composites filled with SiC. The deference is again explained by the own volume resistivity of the fillers. All three fillers have ρ_v values lower than those of the NR ($\rho_v = 5.5 \times 10^{13}$ Ω.m).

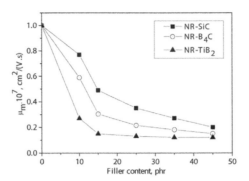

FIGURE 9.46 Mobility of charge carriers as a function of filler concentration for NR-TiB$_2$, NR-B$_4$C, and NR-SiC composites.

The results allow the conclusion that the conductivity depends on the resistivity value of the matrix and fillers. When the resistivity value of the matrix is higher than those of the fillers, the conductivity of the composites increases with the increasing filler amount. If the resistivity of the matrix is lower than those of the fillers, the conductivity of the composites decreases with the increasing filler amount. In all cases, the comparison of the resistivity measured regarding the respective charge carrier quantity and mobility leads to the conclusion that the conductivity depends on the mobility of charge carriers. The lower mobility of the charge carriers in polymer composites with increasing conductivity at a higher charge carriers quantity has been established by other authors as well.[23]

9.3.8   VOLT–AMPERE CHARACTERISTICS OF NBR-BASED COMPOSITES COMPRISING SIC, B$_4$C, AND TIB$_2$

The volt–ampere characteristics allow elucidating the specifics of conductivity mechanism and structure of the studied specimens. Concerning the composites of reinforced polymers, those characteristics prove the distribution of fillers particles in the polymeric matrix and the dissociation degree of their agglomerates and aggregates.

The volt–ampere characteristics of the fillers (SiC, B$_4$C, and TiB$_2$) and NBR-based composites thus filled have been determined. The results for SiC, B$_4$C, and TiB$_2$ are presented in Figure 9.47.

The arrangement of the curves representing the volt–ampere characteristics is in an ascending order according to the volume resistivity of the three fillers and show increase in the current with the increasing voltage. The curve representing the characteristic of SiC is in the lowest position, while the one for TiB$_2$—the highest. The linear character of the curves shows that the current versus voltage dependence obeys Ohm law and the contacts between the fillers particles are ohmic.

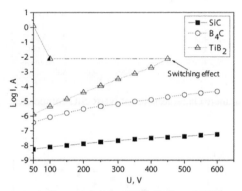

FIGURE 9.47 Volt–ampere characteristics of SiC, B$_4$C, and TiB$_2$.

The volt–ampere characteristic of TiB$_2$ reveals a peculiarity typical of some semiconductor materials. At 450 V, the resistivity of TiB$_2$ drops drastically and at the same the current decreases to 100 V. The spontaneous change in the conductivity is known as a "switching effect"—under the influence of external electric field electrons mobility accelerates to a degree at which they start colliding with other atoms of the medium, thereby ionizing them. That releases additional electrons leading to an electron avalanche and the voltage decreases several times. The resistivity decrease is so great that the current increases with the decreasing voltage. The switching effect has a sequence. If the sample is again subjected to an increasing voltage, the current starts increasing from 100 V on. The initial course of the process is restored following a 24-h "rest" during which the ionized atoms restore their normal state. The cyclicity observed when alternating voltage application and "rest" reveals the lack of electrical breakdown at the switching. In the case of an electrical breakdown in the bulk of a material the processes are irreversible. The "switching effect" has been observed by other authors for aluminum powder and for vulcanizates thus filled.[45]

The volt–ampere characteristics of NBR-based composites filled with SiC, B$_4$C, and TiB$_2$ presented Figures 9.48–9.50, have a nonlinear character.

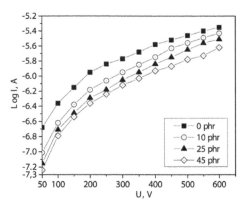

FIGURE 9.48   Volt–ampere characteristics of NBR-SiC composites.

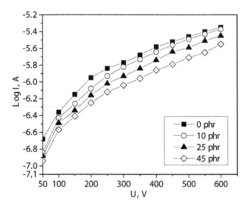

FIGURE 9.49   Volt–ampere characteristics of NBR-B$_4$C composites.

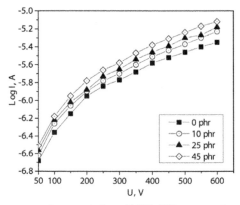

FIGURE 9.50   Volt–ampere characteristics of NBR-TiB$_2$ composites.

The arrangement of the curves representing the volt–ampere characteristics for NBR-based composites filled with SiC and B_4C is in a descending order under the curve for the unfilled NBR composite. The results are due to the volume resistivity of the fillers which is higher than that of the matrix. The volume resistivity of TiB_2 is higher than that of the matrix and of NBR-TiB_2 composites and their plot is the same as those for the carbides filled composites. In the case, the curves are arranged in an ascending order above the curve for unfilled NBR (Figure 9.50).

The change in the arrangement of the curves for volt–ampere characteristics of the vulcanizates regarding the unfilled sample is determined by the own resistivity of the fillers. It proves that the charge carriers transport proceeds owing to their particles. The transport generated by the charge carriers in the matrix is hindered by the high ohmic fillers and favored by the particles having a resistivity lower than that of the matrix.

The nonlinear course of the curves for the composites leads to the important conclusion that, the contacts between the filler particles are nonohmic, that is the contact between them is not a direct one. When the contact is direct, as in the case of "pure" fillers (Figure 9.47), the curves are straight lines. The results are related to the theory of particles distribution in the polymer matrix. Some researchers state that, the fillers build in the matrix their independent (own) network of contacting particles.[23,41,46,47] According to other opinions, the particles and macromolecules are localized in associates, forming domains with mosaic distribution in the matrix. The particles in the associates are insulates by polymeric layers; therefore, the conductivity is nonohmic, proving erroneous the statement of existing chains of particles contacting between each other.

Due to the high resistivity of NR-based composites, requiring a much higher voltage, their volt–ampere characteristics have not been investigated.

9.3.9 STRUCTURE AND PROPERTIES OF NBR AND NR-BASED COMPOSITES FILLED WITH SIC, B_4C, AND TIB_2

9.3.9.1 BOUND RUBBER

The amount of bound rubber is the structure element of the composites determining the adhesion, that is the elastomer–filler interaction. The amount of bound rubber in NR-based composites was determined using toluene as a solvent, and dichloroethane—for NBR-based composites. The amount of bound rubber was calculated with respect to the amount of elastomer in the sample. The results obtained are presented in Table 9.3:

TABLE 9.3   Bound rubber amount in NBR- and NR-based composites filled at 45 phr.

	Bound rubber amount (%)
NBR-SiC	5
NBR-B$_4$C	10
NBR-TiB$_2$	8
NR-SiC	15
NR-B$_4$C	18
NR-TiB$_2$	16

The mean square error of the experiment was ±0.6% for NBR-based composites and ±0.8% for NR-based composites.

The results obtained reveal the elastomer-filler adhesion interaction. The highest adhesion to both elastomers is exhibited by B$_4$C, then that by TiB$_2$ and SiC. The repeating gradation of bound rubber for both elastomers leads to the conclusion that the bound rubber amount is determined by the adhesion properties of the elastomers and the fillers specifics. The effect of those factors is seen when composites morphology is studied.

9.3.9.2   STRUCTURAL SPECIFICS OF COMPOSITES

The microscope images of the particles and agglomerates in SiC, B$_4$C, and TiB$_2$ are presented in Figures 9.51–9.53.

The filler SiC is less prone to agglomeration having in mind that its agglomerates are the smallest, while the distinct particles prevail in number. The latter are definitively in a small amount in B$_4$C and TiB$_2$ which are predominantly in a state of agglomeration with large enough agglomerates.

Obviously the cohesive forces in those two fillers are stronger than those in SiC. This morphologic characteristic of the fillers is due to their being dispersed in the dry state over the glass slide. That allows measuring the filler particles and their agglomerates. The size of distinct SiC particles is about 1 μm, and of its agglomerates—3–6 μm. The particles of B$_4$C are ~ 2 μm, and of its agglomerates 4–9 μm. The particles of TiB$_2$ are 2–4 μm in size, and its agglomerates—5–10 μm. The advantage of "dry dispersion" over the usage of emulsifiers is the fact that, one can evidence the natural agglomeration proneness of fillers particles. It is hard to calculate the percent of agglomerate formations as the process lacks regularity and reproduction. By rule, there are soft and solid agglomerates in all fillers.[48] The soft agglomerates are known to decompose during the compounding, while only a small part of the solid ones may decompose. That is why agglomeration proneness of fillers has been of particular interest.

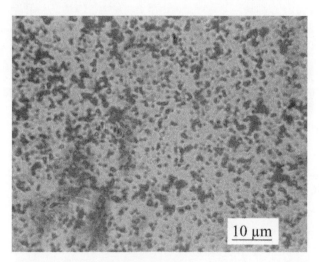

FIGURE 9.51   Particles and agglomerates in SiC.

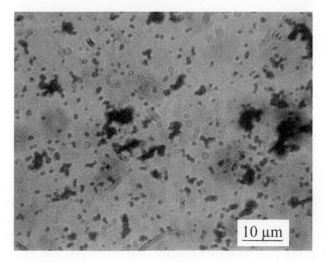

FIGURE 9.52   Particles and agglomerates in B_4C.

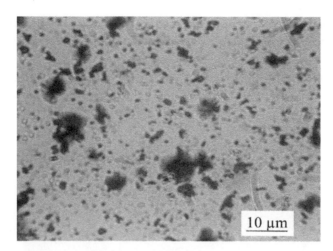

FIGURE 9.53   Particles and agglomerates in TiB$_2$.

The structure of unfilled NBR compound is presented in Figure 9.54, and that of a NR compound is in Figure 9.55.

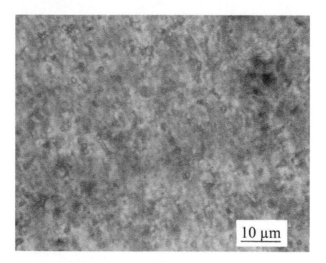

FIGURE 9.54   Unfilled NRB matrix.

The structure of nitrile butadiene rubber has a mosaic character—it consists of more compact and looser formations, which might be supposed to result from the presence of a macro- and microgel. The structure of NR also has a mosaic structure. The difference is in the greater amount of available gel formations.

The fillers used are of differently compatible with the nitrile butadiene rubber and NR matrixes. The findings of morphological studies show that B$_4$C has the best adhesion with both elastomers.

The structure of NBR-SiC composite is presented in Figure 9.56. It resembles the structure of the very filler—there is a great number of distinct particles and not quite compact associates of filler particles with NBR macromolecules.

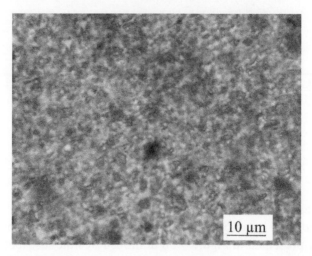

FIGURE 9.55   Unfilled NR matrix.

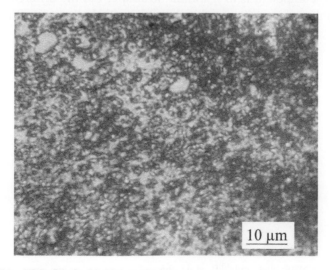

FIGURE 9.56   NBR filled with SiC at 45 phr.

The compatibility of SiC with NR is much better. As seen from Figure 9.57, there are well defined compact filler–matrix associates separated by domains of unfilled elastomer.

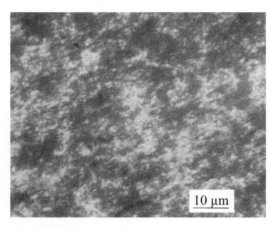

FIGURE 9.57   NR filled with SiC at 45 phr.

The adhesion of TiB_2 to the two elastomers is higher than that of SiC. That is manifested by associates of considerable size and compactness. The morphology presented in Figure 9.58 demonstrates the distribution of NBR and TiB_2 amongst rather large unfilled domains of the matrix. The structure of NR-TiB_2 compound is of a similar structure, but the difference is in the greater number of associates and in their higher polydispersity (Figure 9.59).

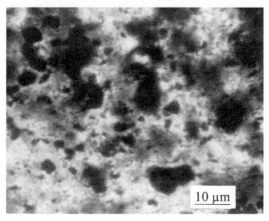

FIGURE 9.58   NBR filled with TiB_2 at 45 phr.

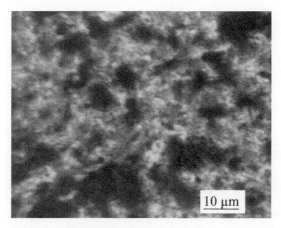

FIGURE 9.59   NR filled with TiB$_2$ at 45 phr.

The structure of B$_4$C differs from those described above by the higher content of associates formed in the matrix and by the much larger size of those associates. The structure of a NBR-B$_4$C composite is presented in Figure 9.60. The well-defined domains of unfilled matrix are typical of that structure. As seen in Figure 9.61, the associates between NR are B$_4$C are more numerous and one can speak of their forming a network in the elastomer matrix.

The described specifics of the morphology are in accordance with the calculated amounts of bound rubber (Table 9.3), which show that the more compact structures formed in NR-based vulcanizates are due to the better adhesion of NR and its higher amount of bound rubber.

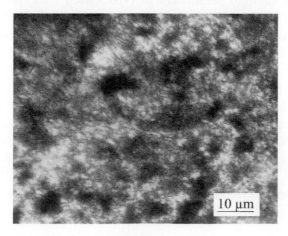

FIGURE 9.60   NBR filled with B$_4$C at 45 phr.

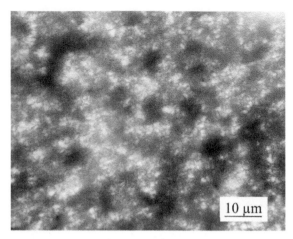

FIGURE 9.61   NR filled with B₄C at 45 phr.

9.3.9.3   TENSILE STRENGTH OF COMPOSITES

The reinforcement of noncrystallizing rubber, to which class NBR belongs, is related to the disentangling of macromolecules during the compounding process. As a result the cohesion forces increase considerably. Like SBR, NBR is very sensitive to reinforcement. Its properties improve even at reinforcement with less active fillers. The reinforcement of NBR with carbon black increases the tensile strength of its vulcanizates several times.[49]

The tensile strength parameters of NBR-based composites filled with TiB_2, B_4C, and SiC at different concentrations are presented in Figure 9.62.

As seen from the figure, the tensile strength (, MPa) of NBR-based composites increases with the increasing amount of the three fillers (TiB_2, B_4C, and SiC). The tensile strength values of NBR-SiC composites are the highest, while those for NBR-TiB_2 composites are the lowest. The highest tensile strength values of the former composites are due to SiC being the finest and succeeding best in disentangling NBR macromolecules. The adhesion of NBR macromolecules to SiC is weakest, having in mind the lowest amount of bound rubber, if compared with that in NBR-B_4C and NBR-TiB_2 composites (Table 9.3). As a result, the filler facilitates greater freedom of orientation and arrangement of the macromolecules at deformation.

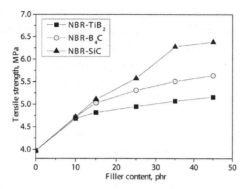

FIGURE 9.62   Tensile strength of NBR-based vulcanizates as a function of the concentration of SiC, B₄C, and TiB₂.

The results obtained are in agreement with the particles size of the fillers and their structure presented in Figures 9.51–9.53.

The reinforcement of crystallizing rubber, to which class belongs NR, as well is related to the disrupted orderness of macromolecules during compounding. As a result the cohesion forces weaken and the mechanical parameters of NR-based composites worsen.

The tensile properties of NR-based composites filled with TiB_2, B_4C, and SiC at different concentrations are presented in Figure 9.63.

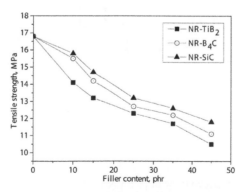

FIGURE 9.63   Tensile strength of NR-based vulcanizates as a function of the concentration of SiC, B₄C, and TiB₂.

As the figure show, the tensile strength of NR-based composites increases with the increasing amount of the three fillers (TiB_2, B_4C, and SiC). The tensile strength values of NR-SiC composites are the highest, then come those for NR-B$_4$C, NBR-TiB$_2$ composites have the lowest values. The order like in the case of unfilled NRB is related to fillers dispersion. The particles of SiC being the finest are the smallest hindrance to macromolecules orientation and worsen the least the tensile properties of NR. The particles of TiB$_2$ are larger (up–4 µm) what predetermines the orderness disruption of NR macromolecules. Hence, NR-TiB$_2$ composites have tensile parameters lower than those of the composites filled with B$_4$C, whose particles are about 2 µm. The tensile strength values for the composites based on the two types of rubber are determined at a relative error of ±2%.

9.3.10 THERMAL AGING OF COMPOSITES

9.3.10.1 AGING-RELATED CHANGES IN THE ELECTRIC PROPERTIES OF NBR AND NR-BASED COMPOSITES

The studied composites were subjected to 72 h aging at 100°C (NBR-based composites) and to 72 h at 70°C (NR-based composites) in order to evaluate the changes in their volume resistivity. The changes in the volume resistivity as a function of pressure for composites NBR-SiC, NBR-B$_4$C and NBR-TiB$_2$ after their aging are presented in Figures 9.64–9.66.

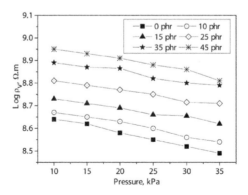

FIGURE 9.64 Pressure dependence of ρ$_v$ for NBR-SiC composites upon aging for 72 h at 100°C.

The comparison of Figures 9.6–9.8 with Figures 9.64–9.66, respectively, reveals ρ$_v$ values for the composites after aging to be higher than prior to their aging with the preserved tendency of their changes caused by filling and pressure. That is due to the degradation of NBR macromolecules what disturbs the normal amount and transport of charge carriers, and preserving the electronic character of conductivity.

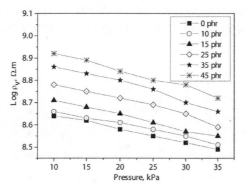

FIGURE 9.65   Pressure dependence of ρ_v for NBR-B$_4$C composites upon aging for 72 h at 100°C.

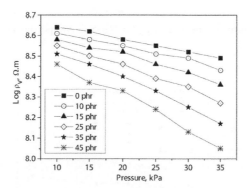

FIGURE 9.66   Pressure dependence of ρ_v for NBR-TiB$_2$ composites upon aging for 72 h at 100°C.

After aging for 72 h at 70°C, the volume resistivity for NR-based composites (Figures 9.67–9.69) is lower than that of the composites prior to their aging (Figures 9.9–9.11). The changes in ρ_v values as a function of pressure are the same as prior to composites aging.

The lower resistivity values of NR-based composites after their aging are explained by the low molecular weight fractions formed due to heating. They are an additional source of ions and electrons to which NR being poor in charge carriers is rather sensitive. Due to the increased amount of charge carriers, the resistivity decreases. The reason for the ascending course of the curves for composites filled with SiC and B$_4$C (Figures 9.67 and 9.68), as explained earlier, is due to the hindered electron transport. The descending course of the curve for NR- TiB$_2$ is because of the prevailing electronic conductivity in the presence of TiB$_2$, which is favored by increasing pressure (Figure 9.69). The same increase in conductivity upon aging has also been established by References[50,51] for composites with ethylene–propylene–diene polymer.

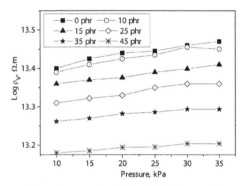

FIGURE 9.67   Pressure dependence of ρ_v for NR-SiC composites upon aging for 72 h at 70°C.

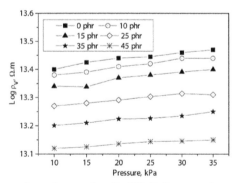

FIGURE 9.68   Pressure dependence of ρ_v for NR-B$_4$C composites upon aging for 72 h at 70°C.

The compression factor of resistivity for all composites upon aging changes negligibly and is in accordance with the changes in vulcanizates resistivity.

The aging coefficients (K_ρ, %), determined according to the volume resistivity for NBR- and NR-based composites, filled with TiB$_2$, B$_4$C, and

SiC at 45 phr and for the unfilled rubber matrixes are presented in Tables 9.4 and 9.5.

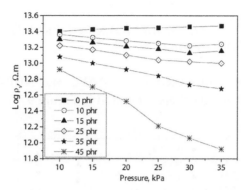

FIGURE 9.69   Pressure dependence of ρ_v for NR-TiB$_2$ composites upon aging for 72 h at 70°C.

TABLE 9.4   Aging Coefficient (K_ρ, %), Determined According to the Volume Resistivity for NBR-Based Composites Filled at 45 phr.

Aging coefficient (%)			
NBR	NBR-TiB$_2$	NBR-B$_4$C	NBR-SiC
140	157	151	140

TABLE 9.5   Aging Coefficient (K$_\rho$, %), Determined According to the Volume Resistivity for NR-Based Composites Filled at 45 phr.

Aging coefficient (%)			
NR	NR-TiB$_2$	NR-B$_4$C	NR-SiC
54	53	58	57

As seen from Tables 9.4 and 9.5, the change in ρ_v values for NBR-based composites upon aging for 72 h at 100°C is greater than that in ρ_v values for NR-based composites upon aging for 72 h at 70°C. In both cases ρ_v values of the composites reinforced with the different fillers are close to those of unfilled NBR and NR, what shows the changes in resistivity are on account mainly of the changes in ρ_v of rubber matrixes.

Having in mind the preserved linearity of the change in the volume resistivity as a function of pressure (up to 35 kPa) both for NBR- and NR-based composites prior to aging, as well as the changes of ρ_v values after that, one can draw the conclusion that the composites studied are still able to reflect relatively precisely the changes in resistivity. That is an advantage with regard to their application as pressure sensor.

9.3.10.2 CHANGES IN THE MECHANICAL PROPERTIES OF THE COMPOSITES AFTER AGEING

The tensile properties of NBR-SiC, NBR-B_4C and NBR-TiB_2 composites upon aging for 72 h at 100°C are presented in Figure 9.70.

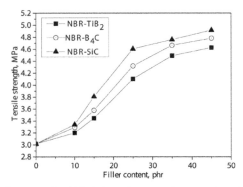

FIGURE 9.70   Tensile strength of NBR-based composites as a function of filler (SiC, B_4C, and TiB_2) concentration determined upon aging for 72 h at 100°C.

The changes in the tensile parameters of the composites with the increasing filler amount both before and after their aging have ascending curves (Figure 9.70). The difference is in the lower tensile strength value after aging. The tensile parameters of NBR-SiC composites upon aging remain the highest (with a small difference from those prior to aging), then come the values for NBR-B_4C and NBR-TiB_2 composites.

The aging coefficients (K_a, %), determined according to the tensile strength of the vulcanizates of unfilled NBR and those of NBR-composites filled at 45 phr with TiB_2, B_4C, and SiC are presented in Table 9.6.

TABLE 9.6   Aging Coefficient (K_a, %), Determined According to the Tensile Strength of NBR-Based Composites Filled at 45 phr.

Aging coefficient (%)			
NBR	NBR-TiB_2	NBR-B_4C	NBR-SiC
25	10	14	23

As seen from the table, the aging resistance of NRB-based composites increases. The lowest inhibition effect is produced by SiC, then by B_4C and TiB_2. The composites wherein the adhesion between the macromolecules and filler particles (B_4C and TiB_2) is higher the access of oxygen to the macromolecules is hindered. As a result the aging coefficient is lower. The tensile properties of NR-SiC, NR-B_4C and NR-TiB_2 subjected to aging for 72 h at 70°C are presented in Figure 9.71.

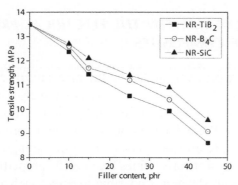

FIGURE 9.71 Tensile strength of NR-based composites as a function of filler (SiC, B_4C, and TiB_2) concentration determined upon aging for 72 h at 70°C.

The course of the curves for changes in the tensile properties as a function of filler concentration before and after aging is the same. The mechanical strength decreases with the increasing filler amount, but σ values are lower than prior to aging. That is explained by the appearance low molecular weight fragments of macromolecular chains and substances upon aging due to macromolecular degradation. Those have a plastification effect and worsen the tensile properties of NR-based composites.

The aging coefficients (K_σ, %), determined according to the tensile strength of unfilled NR-based and those of NR-composites filled at 45 phr with TiB_2, B_4C, and SiC are presented in Table 9.7:

TABLE 9.7 Aging Coefficient (K_σ, %), Determined According to the Tensile Strength of NR-Based Composites Filled at 45 phr.

Aging coefficient (%)			
NR	NR-TiB_2	NR-B_4C	NR-SiC
20	18	20	22

As seen from the Table, the fillers almost have no effect upon the aging of NR-based composites reinforced with TiB_2, B_4C, and SiC. Comparing the results for NBR- and NR-based composites, one sees that, the inhibition effect of the fillers for NBR-based composites is more pronounced than that for NR-based ones.

The mean square error in the aging coefficient determined according to the tensile strength for NBR- and NR-based composites was ±4%.

9.3.11 RELATION BETWEEN THE ELECTRICAL PROPERTIES AND MECHANICS OF COMPOSITES

As known, there are plenty of studies on the mechanical or electrical properties of elastomer composites that have given extremely thorough and useful results. But they all are strictly limited either on the mechanical or on the electrical properties. For unknown reasons, literature presenting an attempt to find a relationship (correlation) between tensile strength and conductivity is rare. Given that the structure defines the specific properties of the composites, it could be assumed that it is possible to detect such a dependency.

Our results allow the conclusion that the composites with better tensile characteristics have higher resistivity. It is more probable the mechanical and electrical properties of carbon black filled vulcanizates to be determined by the elastomer-carbon black structure and by the compatibility of those two phases. The effect of such a structure on the same properties (tensile strength, electrical properties), have been also observed for NBR- and NR-based composites comprising carbide and boride ceramics, regardless of the filler.

The resistivity values of NBR-based composites filled at 45 phr and their tensile strength also prove the relationship of those two parameters. The volume resistivity of NBR-SiC composites presented by its decimal logarithm is 8.57 Ω.m and the tensile strength—6.4 MPa. The values for vulcanizates filled with B_4C are 8.52 Ω.m and 5.6 MPa and for those with TiB_2—8.05 Ω.m and 5.2 MPa, respectively. In the case, the decreasing tensile strength values are in agreement with lower resistivity values. The order is the same for NR-based composites: for NR-SiC ρ_v = 13.55 Ω.m, σ = 11.8 MPa, for NR-B_4C 13.50 Ω.m and 11.1 MPa and for RN-TiB_2—13.25 Ω.m and 10.5 MPa.

The structure being an important factor determining the properties has also been proven by the morphological findings for NBR-based composites. The micrograph Figure 9.56 does not show any associates between the macromolecules and SiC. There are such associates formed between NBR and B_4C (Figure 9.60) but they are not compact enough, while Figure 9.58 presents well defined elastomer-filler associates and the structures of unfilled rubber.

The structural formations in NR-based composites are identical to those of NBR-based ones. In the former, however, the associates are larger, which is a result of the compatibility of the fillers with NR, which is greater than that with NBR. The elastomer–filler associates, increasing their size with regard to the initial one in the loading process, are the factor responsible for the change in composites conductivity. According to literature, that is due to the occurrence of a percolation threshold and ends with the asymptotic convergence of resistivity values to the abscissa where the increasing filler amount is plotted. The filler concentration at which the resistivity decreases sharply is its rate of reduction overlaps with the crucial indicator for elastomers reinforcement, known as the "optimum filling." That overlapping of critical points of the two parameters also determines the correlation between tensile strength and electrical properties of elastomer composites.

9.3.12 FUTURE INVESTIGATIONS: BUTYL RUBBER COMPOSITES COMPRISING SUPERCONDUCTING CERAMIC

The effect of Bi-containing superconducting ceramic on the volume resistivity of butyl rubber composites has also been investigated.[11] The microstructure of butyl rubber filled with a superconductor at different concentrations has been examined by scanning electron microscopy. The influence of the concentration of superconductor, time, applied static pressure and temperature on the volume resistivity has been investigated. The differences in volume resistivity—dependences on different filler amounts at varying temperature have also been measured. The effects observed are discussed viewing the ion and electronic mechanisms of electrical conductivity in rubber-based composites. It has been found that the volume resistivity of butyl rubber composites decreases with increasing ceramics concentration, applied pressure, and temperature. A change from nonlinear to linear behavior has been observed for volume resistivity with increasing superconductor concentration applied pressure; for volume resistivity-temperature dependences and for temperature hysteresis. Composites containing 30–35 phr of superconductor have good stability and reproducibility of volume resistivity and provide a wide range of relationships between the applied force and volume resistivity.

It has been also concluded[52] that a butyl rubber/boron carbide composite is a new prospective self-electrical heating composite with double (negative temperature coefficient of conductivity/positive temperature coefficient of conductivity) effect (i.e., V-shaped thermistor) and good for electromagnetic applications like enclosing computers, electronic devices and microwave shielding (they have been found to have high attenuation and reflection loss values). The investigations on composites filled with conductive ceramics are ongoing.

9.3.13 POSSIBLE ENGINEER APPLICATIONS OF THE COMPOSITES STUDIED

The results obtained from the studies on NBR- and NR-based composites, filled with carbide and boride ceramics (SiC, B_4C, and TiB_2), can be considered as characteristics defining the areas in those materials can find engineering applications. Regarding their sensitivity to external influences (mechanical, thermal, electrical, etc.) the composites belong to the group of semiconductors with all the ensuing positive effects. The composites studied offer an important convenience—they can be turned into electrical circuits and transform nonelectrical variables into electrical.

The nature of pressure and bending dependencies of volume resistivity for the vulcanizates studied allows their usage manufacturing of tensoresistors (tensotransducers) capable of responding to pressure, bending, and so forth, or pressure sensors. As such they exhibit good properties, namely sensitivity, resistance to a high degree of bending without disruption (a rubber tensotransducer could be deformed up to 100, 200, or 300%, while the classical one resists much smaller relative deformations of the order of 1×10^{-3}%), the possibility of making them in any size, ease of selection associated with resistance to environments of different aggressiveness, and so forth. A great sensitivity to minor impacts can be controlled by electrical devices, which are not a problem to create by modern technology. The elastomer-based pressure sensors are widely used in industry, especially when studying the phenomena of "rock pressure" on supporting structures in underground mines and civil engineering tunnels. Plenty of measuring detectors (sensors) should be installed both over the cross section of the gallery (tunnel) and along its length when conducting such experiments. Given the high price of the sensors used currently, this does not apply because of financial reasons. Since the manufacture of elastomer-based pressure sensors is related only to technological procedures typical of traditional rubber-industry, it can be expected that their price will be much lower than that of used till now. Those sensors have characteristics of long-lasting stability what allows their use in equipment with extended service life acting as long-term control sensors for preventing accidents. They can be used for reporting the stress state of dams, and so forth industrial and hazardous locations, incurring different loads. The above described results show that the investigated composites meet the requirements for this kind of materials, namely:

1. The relationship between volume resistivity and applied pressure should be linear over a wide range of variables.
2. The change in the resistivity with a change in pressure of the unit (e.g., 1 kPa) should be large enough to ensure measured value beyond the margin of errors of the adopted method for determining the resistivity.
3. The electrical parameters of the material should not change over time, that is the properties of the dielectric matrix, as well as the properties of the conducting phase should remain unchanged for a long time.

The change in the conductivity of the composites studied under the influence of changing temperature is a prerequisite for their application as thermistors with a negative temperature coefficient of resistivity. The usefulness of this feature is even more significant, taking into account the absence of a temperature hysteresis. Another advantage of the studied composites is their greater sensitivity (temperature coefficient of resistivity $\alpha = 0.0065°C^{-1}$), if compared with that of metal thermistors (for Ag $\alpha = 0.0036°C^{-1}$, for Al $\alpha = 0.0038°C^{-1}$). By this type of thermistors one is able to control and regulate not only the temperature changes, but the flow rate of fluids, rate of chemical reactions, and so forth. One of the most characteristic features of semiconductors is the strong temperature dependence of their resistivity. That is explained by the exponential increase in the concentration of charge carriers with increasing temperature. The semiconductors with high temperature sensitivity, that is, high negative temperature coefficient of resistivity are used for the production of thermistors. Termistors, including elastomer-based ones, may be used for measuring and regulating temperature, for making timers, for measuring the velocity of gases or liquids. On the basis of the altering intensity of cooling at various speeds of movement of the medium in which the thermistor is immersed, it can measure the liquid level, the power for thermal protection, and so forth.[53]

It has been found[39] that NR-based composites filled with SiC and B_4C ceramics may be used successfully as linear negative temperature coefficient thermistors and piezoresistive sensor materials.

The silicone rubber-based composites filled with SiC have been also found able to homogenize the electric field distribution of cable termination effectively. With the increase of nonlinear conductivity level, the electric field distribution of cable termination with nonlinear SiC/silicone rubber composite is more homogeneous. It has been proven by the electric field simulation that the more of nonlinear conductivity level increases, the more uniform the electric field distribution along the surface of silicone rubber composite insulators is.[54, 55]

A number of the applications of the composites based on silicone rubber and NR are related to the combination of properties "high thermoconductivity—excellent mechanical parameters."[56-58] The particular emphasis is on the improvement in tensile strength, elongation at break, hardness and density with adding the reinforcement phase (filling up to 10%). The same effect has been established with fluoroelastomer-based composites filled up to 20%.[59] The combination of properties "high thermal conductivity—excellent mechanical parameters" has also been established for NR-based composites filled with boride carbide at 6 phr.[60] It has been stated in Reference[61] that the thermoconductivity of NBR-based composites filled with boron carbide at 60 phr increases 20 times compared with that of the unfilled NBR composite. The effects of boron carbide content on bulk resistivity, voltage–current characteristic, thermal conductivity and thermal stability of boron carbide/butyl rubber (IIR) polymer composite have been investigated. The analysis results indicate that the bulk resistivity decreased greatly with increasing boron carbide content, and when boron carbide content reached to 60%, the bulk resistivity dropped to the minimum. Accordingly, electric heating behavior of the composite is strongly dependent on boron carbide content as well as applied voltage. The thermal stability test showed that, compared with pure IIR, the thermal stable time of composites was markedly extended, which indicated that the boron carbide can significantly improve the thermal stability of a boron carbide/IIR composite.

Various opportunities to use elastomer-based composites filled with boron carbide are related to the ability of this filler to absorb radioactive emissions effectively. In this connection it describes the use of elastomeric materials filled with boron carbide as shields for absorbing gamma radiation.[62] It is stated that an 18-mm-thick shield made of NR-based composite, filled with boron carbide at 20 phr can reduce the fast neutron beams about 50%. For slow neutrons, the reduction is about 46%. Another article[63] reported that, a 15-mm-thick ethylene-propylene diene rubber composite filled with boron carbide at 57 phr reduces sharply the initial direct slow neutron flux by about 85%. The possibilities to use elastomer-based composites as neutron shielding materials have been described in Reference.[64] The paper also presents results on the reinforcement with boron carbide leading to improvement in the mechanical properties of the composites: strength, moduli, wear resistance, tear resistance—what is of significant importance for the exploitation chrematistics of the protective shields.[69] The potential applications of elestomer composites comprising boron carbide in nuclear industry have been proven in Reference[65] as well.

The dependence of the volume resistivity on the frequency of the applied voltage opens an unusual opportunities to use NBR-SiC and NBR-B$_4$C composites—such as filters in high frequency circuits. It has been found that a decrease of the impedance in a certain frequency range unlocks current transport at those frequencies and makes conductive the element made of a polymer composite. For frequencies at which the resistivity is high the transport through the composite is hindered and no current circuit is observed at the given frequency. Constructively those filters are very simple compared with conventional high-frequency filters comprising resistors, capacitors and inductors.

The listed items are made with available compounds using a simplified rubber technology. They are easily replaced hence maintaining the equipment they are a part of is unproblematic.

9.4 CONCLUSIONS

1. The electrical properties of elastomer composites containing carbide and boride ceramics as functional fillers have been studied. It has been found that under the influence of fillers conductivity may transform from ionic into electronic.
2. The ion transport of charge carriers in composites with ionic conductivity of the matrix is retained when the volume resistivity of the fillers is not lower than 108 Ω.m. For matrixes with electronic conductivity the resistivity of the filler does not affect the type of conductivity of the composite. That indicates that the charge carrier transport depends on the electrical properties of both filler and matrix.
3. The electrical properties of the tested composites have been determined as a function of different factors: reinforcing, pressure, temperature, oxidation processes. It that has been found that by reinforcing, the volume resistivity of the tested composites increases or decreases smoothly, without a steep course of the dependences, as

observed in cases of a great difference in the resistivity of the matrix and the filler. Sudden changes in the volume resistivity have been registered with the application of high pressures.

4. Concerning the real existence of two structural phenomena of contacting and noncontacting filler particles in polymer composites: it has been proven that a direct contact between the particles does not take place. That has been demonstrated by the differences in resistivity of the very fillers and composites, by the type of changes in volt–ampere characteristics; by the changes in resistivity at high pressures and by optical microscopy findings.

5. The percolation effect in terms of contact between the particles of the fillers is not realized and probably results in the formation of associates between rubber macromolecules and filler particles that reach a certain size. Those, depending on their electrical salience and arrangement in the matrix define the conductive properties of composites.

6. The mechanism of charge carriers transport in the studied composites has been clarified by determining their characteristic parameters (activation energy, mobility, quantity and type of charge carriers). It has been found that, notwithstanding the Zone conductivity in the particles of the fillers, the transport of charge carriers in the polymer composites based on the restricted mobility (less than 1 cm^2/(V.s)), their small amount, the decrease of the measured resistivity with increasing frequency and lack of contact between particles, could most plausibly be explained by the hopping mechanism. According to it, hopping charge carriers in the localized state, do not depend on each other and are defined by the energy state and nature of the spatial location of the zones of localization (ions, impurities, substitutes in the molecular chains capable of exchanging electrons, etc.). It is preferable to determine the activation energy of the charge carriers by the Hopping transport equations.

7. The temperature hysteresis of the resistivity of the polymer conductive composites depends on the type of matrix and fillers. It decreases with the increasing conductivity of the composite components.

8. A relationship has been found between the optimum filling, strength and electrical properties of the composites, determined by the formed structure, where a higher electrical conductivity corresponds to reduced strength. Although such a link between those two parameters has not been discussed in the reviewed literature, it could establish express methods for determining the tensile characteristics.

9. The ranges wherein the tested composites can be used as high-frequency filters, sensors for temperature changes, registration of pressure and big bends, protective shields in nuclear industry, materials with high thermal conductivity, and so forth, have been defined experimentally.

10. The elastomer-based materials developed and investigated exhibit behavior of composites with typical semiconductor properties, which we consider as an important achievement of polymer materials science.

9.5 ACKNOWLEDGMENTS

The authors acknowledge the support provided by Dr Rashko Dimitrov and Dr Zornitsa Dimitrova in the experimental part and discussions on this work.

KEYWORDS

- **NBR and NR-based composites**
- **carbide ceramics**
- **electroconductive peculiarities**
- **compression factor of resistivity**
- **tensoeffects**
- **mobility and quantity of charge carriers**
- **volt–ampere characteristics**
- **structure and properties**
- **thermal aging**
- **possible applications**

REFERENCES

1. Peace, M. H. K.; Mitchell, G. R. Conductive Elastomeric Composites. *J. Phys.: Conf. Ser.* **2009,** *183,* 012011 Dielectrics 2009: Measurement Analysis and Applications, 40th Anniversary Meeting, IOP Publishing.

2. Ata, S.; Kobashi, K.; Yumura, M.; Hata, K. Mechanically Durable and Highly Conductive Elastomeric Composites from Long Single-Walled Carbon Nanotubes Mimicking the Chain Structure of Polymers. *Nano Lett.* **2012,** *12,* 2710–2716.

3. Thomas, S.; Stephen, R. Rubber Nanocomposites: Preparation, Properties and Applications, Wiley & Sons, 2010.

4. Park, M.; Park, J.; Jeong, U. Design of Conductive Composite Elastomers for Stretchable Electronics, Review. *Nano Today.* **2014,** *9,* 244–260.

5. Noh, J-S. Conductive Elastomers for Stretchable Electronics, Sensors and Energy Harvesters, Review. *Polymers.* **2016,** *8,* 123–141.

6. Zheng W.; Wong S. C.; Sue H. J. Transport Behavior of PMMA/expanded Graphite Nanocomposites. *Polymers.* **2002,** *43,* 6767–6773.

7. Chen, X. M.; Shen, J. W.; Huang; W. Y. Novel Electrically Conductive Polypropylene/Graphite Nanocomposites. *J. Mater. Sci. Lett.* **2002,** *21,* 213–214.

8. Das, T. K.; Prusty, S. Review on Conducting Polymers and Their Applications. *Polym.-Plast. Technol. Eng.* **2012,** *51,* 1487–1500.

9. Yi, X.; Shen, L.; Pan, Y. Thermal Volume Expansion in Polymeric PTC Composites: a Theoretical Approach. *Compos. Sci. Technol.* **2001,** *61,* 949–956.

10. Song, J.; Li, J.; Xu, J.; Zeng, H. Superstable Transparent Conductive Cu@Cu4Ni nanowire Elastomer Composites Against Oxidation, Bending, Stretching, and Twisting for Flexible and Stretchable Optoelectronics. *Nano. Lett.* **2014,** *14,* 6298–305.

11. Dishovsky, N.; El—Tantawy, F.; Dimitrov, R. Effect of Bi-containing Superconducting Ceramic on the Volume Resistivity of Butyl Rubber Composites. *Polym. Test.* **2004.** *23,* 69–75.

12. Xi, Y.; Ishikawa, H.; Bin, Y.; Matsuo, M. Positive Temperature Coefficient Effect of LMWPE-UHMWPE Blends Filled with Short Carbon Fibers. *Carbon.* **2004,** *42,* 1699–1706.

13. Obara, M.; Tajima, Y.; Suzuki. Y. Pressure-sensitive Electrically Conductive Composite Sheet. US Patent, 4,495,236, Jan 22,1985.

14. Kotani, T.; Arai, K.; Fukui, S.; Nagata, M. Pressure Sensitive Conductor and Method of Manufacturing the Same. US Patent, 4,292,261, Sept 29,1981.

15. Mashimo, S.; Nagayasu, S.; Yamaguchi, Y.; Noguchi, T.; Nakajima, M.; Kakiuchi, H.; Tanida, K. Pressure-responsive Variable Electrical Resistive Rubber Material. US Patent 4,765,930, Aug 23, 1988.

16. Kotani, T.; Nagato, M.; Arai, K. Pressure Sensitive Conductor. US Patent 4,302,361, Nov 24, 1981.

17. Sado, R.; Tahara, K. Pressure-sensitive Resistance Elements. US Patent, 4,145,317, Mar 20,1979.

18. Myers, T. E. Variable Resistance Material. US Patent 2,951,817, Sept 06,1960.

19. Neamen, D. *Semiconductor Physics and Devices*, 3rd ed.; McGraw-Hill: New York, 2003.

20. Yu, P. Y.; Cardona, M. Fundamentals of Semiconductors: Physics and Materials Properties, Springer-Verlag Berlin: Heidelberg, 2010.

21. Ferry, D. K. Semiconductors: Bonds and Bands, IOP Publishing: Bristol, UK 2013.

22. El-Tantawy, F. The Interrelation Among Network Structures, Molecular Transport of Solvent, and Creep Behaviors of TiB$_2$ Ceramic Containing Butyl Rubber Composites. *J. Appl. Polym. Sci.* **2005,** *98,* 2226–2235.

23. Roy, D.; Jana, P. B.; De, S. K.; Gupta, B. R.; Chaudhuri, S.; Pal, A. K. Studies on the Electrical Conductivity and Galvanomagnetic Characteristics of Short Carbon Fibre Filled Thermoplastic Elastomers. *J. Mater. Sci.* **1996**, *31*, 5313–5319.

24. Трухачев Б. С. Удалов Н. П., Полупроводниковые тензопреобразователи, М., Энергия, 1968.

25. Brown, R. *Physical Testing of Rubber*, 4th ed. Springer Science & Business Media, 2006.

26. Giancoli, D. Electric Currents and Resistance, in Philips, *J. Physics for Scientists and Engineers with Modern Physics*, 4th ed., Prentice Hall: Upper Saddle River, New Jersey, 2009.

27. Coropceanu, V.; Cornil, J.; da Silva Filho, D. A.; Olivier, Y.; Silbey, R.; Bredas, J. Charge Transport in Organic Semiconductors. *Chem. Rev.* **2007**, *107*, 926–952.

28. Lu, W.; Li, B.; Fu, M.; Huang, Z. Novel Thermistor of Bi-doped Ba(Sn,Sb)O3 with Linear Negative Temperature Coefficient. *Sens. Actuators, A.* **2000**, *80*, 38–41.

29. Blow C. M. Bound Rubber and Vulcanizate Properties. *Polymers.* **1974**, *15*, 814–815.

30. ISO 37:2002.

31. ISO 188:2002.

32. Pramanik P. K.; Khastgir D.; De S. K.; Saha T. N. Pressure-sensitive Electrically Conductive Nitrile Rubber Composites Filled with Particulate Carbon Black and Short Carbon Fibre. *J. Mater. Sci.* **1990**, *25*, 9, 3848–3853.

33. Blythe, T.; Bloor, D. *Electrical Properties of Polymers*, 2nd ed., Cambridge : Cambridge University Press, 2005, 186–216.

34. Aminabhavi, T. M.; Cassidy, P. E.; Thompson, C. M. Electrical Resistivity of Carbon-Black-Loaded Rubbers. *Rubber Chem. Technol.* **1990**, *63*, 451–471.

35. Todorova, Z.; El-Tantawy, F.; Dishovsky, N.; Dimitrov, R. Investigation of Conductivity Characteristics of Nitrile Butadiene Rubber Vulcanizates Filled with Semiconducting Carbide Ceramic. *J Appl. Polym. Sci.* **2007**, *103*, 2158–2165.

36. Todorova, Z.; El-Tantawy, F.; Dishovsky, N.; Dimitrov, R. Investigation of Electrical Properties of Elastomer Composites Filled with TiB_2. *J. Elastom. Plast.* **2007**,*39*, 69–80.

37. Todorova, Z.; El-Tantawy, F.; Dishovsky, N.; Dimitrov, R. Investigation of Some Electrical Properties of Natural Rubber Based Composites Containing as Fillers B_4C and SiC. *J. Chem. Technol. Metall.* **2006**, *41*, 285–290.

38. Todorova, Z.; El-Tantawy, F.; Dimitrov, R.; Dishovsky, N. *Electrical properties of NBR based composites containing B_4C and SiC, XVIIIth Congress of the Chemists and Technologists of Macedonia*, Society of Chemists and Technologists: Ohrid, 2004.

39. Todorova, Z.; Dishovsky, N.; Dimitrov, R.; El-Tantawy, F.; Aal, N. A. Natural Rubber Filled SiC and B_4C Ceramic Composites as a New NTC Thermistors and Piezoresistive Sensor Materials. *Polym. Compos.* **2008** *29*, 109–118.

40. Parris, D. R.; Burton, L. C.; Siswanto, M. G. Transient Electromechanical Behavior of Carbon-Black-Filled Rubber. *Rubber Chem. Technol.* **1987**, *60*, 705–715.

41. Dang, Z-M.; Shehzad, K.; Zha, J-W.; Mujahid, A.; Hussain, T.; Nie, J.; Shi, C-Y. Complementary Percolation Characteristics of Carbon Fillers Based Electrically Percolative Thermoplastic Elastomer Composites. *Compos. Sci. Technol.* **2011**, *72*, 28–35.

42. Bokobza, L. Multiwall Carbon Nanotube-Filled Natural Rubber: Electrical and Mechanical Properties. *eXPRESS Polym. Lett.* **2012**, *6*, 213–223.

43. Costa, P.; Silva, J.; Sencadas, V.; Simoes, R.; Viana, J. C.; Lanceros-Méndez, S. Mechanical, Electrical and Electro-Mechanical Properties of Thermoplastic Elastomer

Styrene–Butadiene–Styrene/Multiwall Carbon Nanotubes Composites. *J. Mater. Sci.* **2013**, *48*, 1172–1179.

44. Bhattacharya, P.; Fornari, R.; Kamimura H. ed. *Comprehensive Semiconductor Science and Technology, Six-Volume Set, Vol. 1 Physics and Fundamental Theory*, Elsevier, 2011.

45. Nassar, A.; Yehia, A. A.; El-Sabbagh, S. H. Evaluation of the Physico-Mechanical and Electrical Properties of Styrene-Butadiene Rubber/Aluminum Powder and Styrene-Butadiene Rubber/Cerium Sulfate Composites, *Polymers.* **2015**, *60*, 100–108.

46. Salaeh, S.; Boiteux, G.; Cassagnau, P.; Nakason, C. Conductive Elastomer Composites with Low Percolation Threshold Based on Carbon Black and Epoxidized Natural Rubber. *Polym. Compos*, Wiley Online Library. DOI: 10.1002 /pc.24136, (accessed July 12, 2016).

47. Grivei, E.; Probst, N. Electrical Conductivity and Carbon Network in Polymer Composites. *Kautsch. Gummi Kunstst.* **2003**, *56*, 460–464.

48. Schaefer, D. W.; Chen, C. Structure Optimization in Colloidal Reinforcing Fillers: Precipitated Silica. *Rubber Chem. Technol.* **2002**, *75*, 773–794.

49. Dick, J. S. *Rubber Technology, Compounding and Testing for Performance,* 2nd ed.; Carl Hanser Verlag GmbH & Co. KG: Munich, Germany, 2009.

50. Sau, K. P.; Chakai, T. K.; Khastgir, D. Conductive Rubber Composites From Different Blends of Ethylene–propylene–diene Rubber and Nitrile Rubber. *J. Mater. Sci.* **1997**, *32*, 5717–5724.

51. Mattson, B.; Stenberg, B., Electrical Conductivity of Thermo-Oxidatively-Degraded EPDM Rubber. *Rubber. Chem. Technol.* **1992**, *65*, 315–328.

52. El-Tantawy, F.; Dishovsky, N. Novel V-shaped Negative Temperature Coefficient of Conductivity Thermistors and Electromagnetic Interference Shielding Effectiveness From Butyl Rubber—Loaded Boron Carbide Ceramic Composites. *J. Appl. Polym. Sci.* **2004**, *91*, 2756–2770.

53. El-Tantawy F.; Bakry A.; El-Gohary A. R. Effect of Iron Oxide on The Vulcanization and Electrical Properties of Conductive Butyl Rubber Composites. *Polym. Int.* **2000**, *49*, 1670–1676.

54. Wang, F.; Zhang, P.; Gao, M.; Zhao, X. *Research on the Non-Linear Conductivity Characteristics of Nano-SiC Silicone Rubber Composites, 2013*; Annual Report Conference on Electrical Insulation and Dielectric Phenomena, 435–538, IEEE Xplore Digital Library, 2013.

55. Wang, F.; Peihong Zhang, P.; Gao, M. Improvement in the Electric Field Distribution of Silicone Rubber Composite Insulators by Non-Linear Fillers, Strategic Technology (IFOST) 2013 8th International Forum on. **2013**, *1*, 217–221.

56. He, Y.; Wu, X-S.; Chen, Z-C. Thermal Conductivity of Composite Silicone Rubber Filled with Graphite/Silicone Carbide. *Adv Mater Res.* **2011**, *221*, 382–388.

57. Kumfu, S.; Singjai, P.; Thamjaree, W.; Longkullabutra, H.; Nhuapeng, W. Mechanical Properties of Silicon Carbide Nanowires /Carbon Nanotubes/Rubber Composites. *NU. Int. Sci. J.* **2009**, *6*, 80–85.

58. Janyakunmongkol, K., Nhuapeng, W. Thamjaree, W. Effect of Added Silicon Carbide Nanowires and Carbon Nanotubes on Mechanical Properties of Natural Rubber Composites. *Jpn. J. Appl. Phys.* **2016**, *55*, 01AE21–01AE24.

59. Shuai, Z.; Liu, Z.; Wang, D.; Zhou, P.; Li, W.; Qiao, Y.; Liu, R.; Zhou, S. Preparation and Mechanical Properties of Micro- and Nano-Sized SiC/Fluoroelastomer Composites. *J. Wuhan Uni. Technol-Mater. Sci. Ed.* **2013**, *28*, 658–663.

60. Zakaria, M. Z.; Ahmad, S. H. Investigation on Thermal Conductivity and Mechanical Properties of Thermoplastic Natural Rubber Filled with Alumina and Boron Carbide Nanocomposites. *Energy. Environ. Eng.* **2013**, *J 2*, 11–14.
61. Meng, D.; Wang, N.; Li, G. Electric Heating Property From Butyl Rubber-Loaded Boron Carbide Composites. *J. Wuhan. Uni. Technol-Mater. Sci. Ed.* **2014**, *29*, 492–497.
62. Gwaily, S. E.; Hassan, H. H.; Badawy, M. M.; Madani, M. Study of Electrophysical Characteristics of Lead-Natural Rubber Composites as Radiation Shields. *Polym. Compos.* **2002**, *23*, 1068–1075.
63. Abdel-Aziz, M. M.; Gwaily, S. E.; Makarious, A. S.; El-Sayed Abdo, A. Ethylene-Propylene Diene Rubber/Low Density Polyethylene/Boron Carbide Composites as Neutron Shields. *Polym. Degrad. Stab.* **1995**, *50*, 235–240.
64. Salimi, M.; Amirabad, E. A.; Ghal-Eh, N.; Soltani, Z.; Etaati, G. Fabrication and Radio-characterization of Boron Carbide and Tungsten Incorporated, Rubber Shields. *Int. J. Innov. Appl. Studies.* **2013**, *4*, 437–440.
65. Subramanian, C.; Suri A. K.; Murthy, T.S.R.C. Development of Boron-based Materials for Nuclear Applications. *Barc. Newsletter.* **2010**, *313*, 14–21. http://www.barc.gov.in/publications/nl/2010/2010030403.pdf

CHAPTER 10

ELASTOMER-BASED COMPOSITE MATERIALS COMPRISING CALCINED KAOLIN

CONTENTS

ABSTRACT

The standard kaolin we use traditionally has been calcined at temperatures from 200 to 900°C. The fillers obtained have been characterized by X-ray, differential thermal analysis and infrared spectroscopy (IRS). Their properties have been compared with those of unmodified kaolin. In our opinion, the changes observed are due to the structural and chemical changes occurring in kaolin at its thermal modification. Electron-microscopic micrographs of the standard and kaolin calcined at different temperatures show that as a result of the thermal treatment the chemically-bound water in kaolin is released and the resulting metakaolin residue has some degree of disorderliness. The agglomerates are of undefined shapes and their size varies with increasing temperature.

Kaolin calcined at different temperatures has been used as a functional filler in elastomer composites based on styrene-butadiene rubber. The vulcanization characteristics, mechanical properties, resistance to heat aging, specific volume resistivity, and so forth, of the resulting vulcanizates have been determined. The analysis of the results shows the range of 400–600°C to be the most suitable for the calcination of the kaolin, viewing its activation as filler in elastomer composites. In this range the material undergoes dehydration and partial amorphizing, but its aluminosilicate architecture is retained. The surface of kaolin particles becomes coarser and therefore energetically more heterogeneous, which is a prerequisite for stronger "elastomer-filler" interactions.

Investigations have been performed on the properties of vulcanizates containing various amounts of calcined kaolin, introduced as a replacement for part of the carbon black in the composites for the production of structural elements of tires. It has been found that calcined kaolin can be used as a substitute of carbon black at 10–15 phr in composites intended for the production of tire carcasses and side walls.

10.1   INTRODUCTION

The term "high-performance fillers" has been introduced into the literature of recent years and annual international conferences have already been devoted to them.[1,2,3] According to the materials of those conferences, the annual market of fillers in Europe and the NAFTA (US, Canada, Mexico) amounts to 7.5 million tons worth about € 3.5 billion. Of these, the proportion of high-performance fillers amounts to about 1 million tons worth € 600,000,000.

Kaolin—an inorganic filler of natural origin is most used in rubber industry. About 54% of the consumption of this type of fillers account to it.[4] For the period of 1975–2003 its annual consumption only in rubber industry was in the range of 172–275 million tons.[5] The reasons are basically three: (i) low cost (price ratio of kaolin to that of carbon black N330 for the past 20 years ranges from 0.06 to 0.1); (ii) low energy consumption during production—14 MJ/kg (140 MJ/kg for the production of furnace carbon black) and (iii) relatively good mechanical parameters of the vulcanizates containing it. Therefore, in recent years, leading companies have been carrying out systematic research on the possibilities of using kaolin even in the production of tires[6]—in composites intended for the production of tire treads, carcasses, side walls, and so forth.

Kaolin is identified with the clay material kaolinite ($Al_2O_3.2SiO_2.2H_2O$), which is the constant component in its composition of Al_2O_3, SiO_2 and water. It belongs to the class of layered silicate minerals. The structure of kaolinite is a tetrahedral silica sheet alternating with an octahedral alumina sheet.[7] These sheets are arranged so that the tips of the silica tetrahedrons and the adjacent layers of the octahedral sheet form a common layer composed of $[(Si_2O_5)^{2-}]_\infty$. Schematically the hexagonal network of $(SiO_4)^{4+}$ is presented in Figure 10.1.

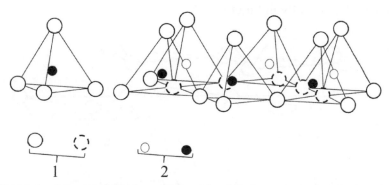

FIGURE 10.1 Scheme of $[SiO_4]$ tetrahedron (a) and the hexagonal network of tetrahedrons (b) 1-oxygen atoms, 2-silicon atoms.

The bases of the tetrahedrons lie on the same plane. The oxygen hexagonal network forms the first level of the layer. Silicon atoms are on the second level, located in the gap between the four oxygen atoms and also form a hexagonal network. The oxygen atoms located at the vertexes of the tetrahedrons princely above the silica atoms are at the third level.

The second structural unit consists of two layers of OH-groups, amongst which aluminum atoms in an octahedron coordination are located, so that they can be at the same distance from the six oxygen atoms or OH-groups which are at the vertexes of the octahedron. The octahedrons have common edges (Figure 10.2).

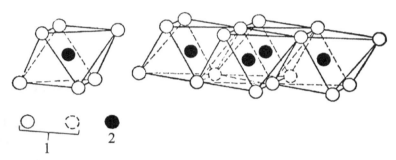

FIGURE 10.2   Scheme of an octahedron (a) and of an octahedron network (b) 1-OH-groups; 2-Al atoms.

The silicate packages of kaolin are two layer: one consisting of a tetrahedron and other with an octahedron network. Si^{4+} is in tetrahedrons while Al^{3+} is in the octahedron. Therefore, kaolin is dielectric. A kaolin layer is presented schematically in Figure 10.3.

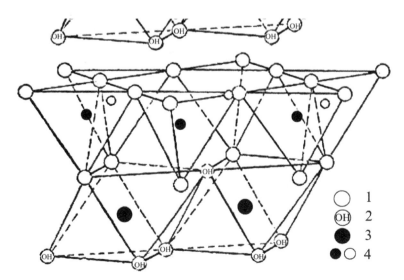

FIGURE 10.3   Scheme of the structure of a kaolin layer: 1-oxygen atoms; 2-hydroxyl groups; 3-aluminum atoms; 4-silicon atoms.

As the figure shows, the charges in the ideal kaolin lattice are balanced and it is electrically neutral. The strong bond between the layers of various surfaces of oxygen atoms and OH-groups occurs due to the hydrogen bond between the latter. That it presupposes that the bonds between the layers weaken upon OH-groups evolvement during dehydration. That is prerequisite for destruction of the crystal lattice and partial amorphization of the material.

The available literature data on thermal treatment effect (often named calcination) upon the reinforcing activity of kaolin and on the entire complex of the composites comprising it is neither complete nor systematic. Nedev et al.[8] have studied the effect that kaolin calcined at 850 and 1000°C has on the properties of SBR vulcanizates filled at 100 phr. However, they have not performed investigations in the 500–700°C range which is very important for kaolin. Having carried out experiments with kaolin calcined only at 850 and 1000°C, they draw the conclusion that with that product the properties of the vulcanizates are worse, if compared to those of vulcanizates comprising virgin kaolin.

Suito, Arakawa[9] suppose the reinforcing effect of kaolin which is due to the interaction of OH-groups located on the side faces of kaolin crystals with rubber macromolecules. If the thermal treatment is carried out up to 400C, it does not affect its reinforcing activity. According to the authors, heating at a temperature higher than 400°C leads to destruction of kaolin crystal structure and lessening of its reinforcing effect.

Calcination of kaolin changes its surface chemistry, optical and electrical properties.[10] It has also been noted that very rapid calcinations can lead to the formation of further voids. When heating above 550°C weight loss is also evident, that is the loss of water molecules from the lattice through dehydration and dehydroxylation. The scanning electron microscope (SEM) images of the studied samples reveal serious changes in the structure of the thermally-modified kaolin and a tendency of degradation of the substrate particles into much smaller ones as well as enhancing the "filler–filler" interaction what causes the formation of secondary structures by the particles (Figure 10.4).

Dannenberg in his profound review on mineral fillers compared some properties of standard kaolin and kaolin calcined at 550°C, produced by ECC International (Great Britain).[4] According to the author, substrate kaolin has a specific surface area of 20 m^2/g, upon calcination it is 30 m^2/g at an average particle size of about 1 μm. The oil absorption number increases twice, while its density decreases owing to the partial degradation of its structure and to the gaps formed therein. The investigations carried out on vulcanizates of SBR 1502 have shown a considerable shortening of optimal cure time T_{90} (11 min for the samples with virgin kaolin and 7.7 min for the samples with modified kaolin), an increase in the minimum torque in the vulcanization isotherm (0.57 N.m for virgin kaolin, 0.68 N.m for modified kaolin), and an increase in the values of the stress at 300% elongation for the samples with modified kaolin and a decrease of their tensile strength and elongation at break. The vulcanizates comprising calcined kaolin are more age resistant in the deformation state than those filled with virgin kaolin.

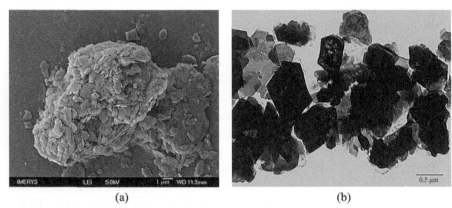

(a)   (b)

FIGURE 10.4 SEM images of (a) aggregated kaolin particles and (b) flash calcined kaolin particles.[10] (Reprinted from Mee, S. J.; Hart, J. R.; Singh, M.; Rowson, N. A.; Greenwood, R. W.; Allen, G. C.; Heard, P. J.;Skuse, D. R. The Use of Focused Ion Beams for the Characterisation of Industrial Mineral Microparticles. *Appl. Clay Sci.* **2008**, *39*, 72–77. © 2008 with permission from Elsevier.)

It has been shown[11] that, the addition of calcined kaolin instead of virgin kaolin (at 10 phr) to LDPE aimed for greenhouse folio, has a pronounced positive effect on resistance to air aging and to the thermal shielding (greenhouse) effect. The cost of heating the greenhouses drops about 25–35% and the crops from the grown cultures increase.

The calcined kaolin is whiter and more abrasive than the virgin kaolin and its surface chemistry, and physical properties completely changed. The largest utilization of calcined kaolin is in paints, rubber, and plastics.[12–16]

Using calcined kaolin, BASF has helped improve the electrical and physical performance—even at high stress with elevated moisture and at high temperature.[16]

Dannenberg[4] has turned the attention to the better results obtained when the already calcined kaolin is silanized with γ-mercaptopropyltrimethoxysilanes. Two commercial products of calcined silanized kaolin (Nicap 100 by Huber Inc and SCS by ECC International) that is virgin kaolin and carbon black N 762 (SRF) have been compared. The comparison has shown that the calcined silanized kaolin exceeds in its performance than the virgin kaolin as well as the only calcined one and in some cases these types of kaolin perform as the mentioned carbon black to a certain extent. Currently a number of companies offer calcined kaolin as well as silanized one, pointing out the significant opportunities these types of kaolin offer for improving the mechanical properties of the vulcanizates, their dielectric properties, thermal resistance, resistance to acids, bases, and so forth and chemical agents.[17]

It is also mentioned[18,19] that the companies producing tires conduct intensive research on the possibilities to use in their formulations other fillers than carbon black and silica—particularly kaolin (including calcined), zeolites, calcium carbonate, and so forth.

The aim of the study is to demonstrate the capabilities of thermal activation to modify the structure and chemical nature of kaolin surface, so that its reinforcing activity is improved.

The tasks related to the implementation of the above stated objective are:

- Optimizing the conditions for calcination of kaolin;
- Characterization of the resulting calcined kaolin;
- Comparison of the reinforcing effect of the filler before and after calcination, as well as the changes in the properties of the vulcanizates containing them; and
- Investigations and suggestion of opportunities for real-life applications of calcined kaolin.

10.2   EXPERIMENT

10.2.1   THERMAL TREATMENT OF KAOLIN

Standard milled kaolin produced by Kaolin Ltd., Bulgaria, was used as a substrate. The kaolin used had the following composition: SiO_2—54%; AL_2O_3—33.5%, Fe_2O_3—1%; TiO_2—0.3%; CaO—0.15%; MgO—0.25%; K_2O—1.1%; N_2O—0.2%, and so forth. The particle size determined by the Sedigraph 1500 particle size analyzer was: over 56 μm—0.02%; 10–56 μm—1.7%; 5—10 μm—17.6%; 1–2 μm—15.2%; 0.5–1 μm—17.2%; under 0.5 μm—44,9%. Regarding the characteristics of this type of kaolin (particle size, especially) it is close to the so called "hard kaolin"[17] It is more suitable than soft kaolin for the purpose of our investigations.

The calcination was conducted under laboratory conditions at 200, 300, 400, 500, 600, 700, 800, and 900°C in a muffle furnace at a constant temperature for 5 h. Temperature deviations varied within the range of ±10°C. Upon calcination the samples were stored in an exicator. For another group of experiments the calcination was run at 500°C and the heating lasted (retention time) for 1, 2, and 5 h.

The investigations in section 10.3.10 were performed on kaolin calcined under factory conditions of Kaolin Ltd., Bulgaria. The calcination was carried out in a rotating furnace in a non-stop regime which had the following parameters: length—60 m; diameter—3 m; slope—3.5°; rotation speed—0.64–1.23 min^{-1}. Calcination temperature was in the optimal temperature range established experimentally—500°C.

10.2.2 CHARACTERIZATION OF THE FILLER

The absorption of dibutylphthalate (DBP) was determined by the method described in BSS 15651-83. pH of the aqueous suspensions of the studied types of kaolin was determined on a laboratory digital pH-meter. The density was determined picnometrically by laboratory techniques. The X-ray phase analysis was carried out on a TUR-M62 apparatus with Cu-Kα radiation. The thermal gravimetric analysis was performed on an OD-102 *derivatograph* in the range 100–1000°C at a 5°/min heating rate. The infrared spectra (KBr pellets) were recorded on a SPECORD—71-IR in the 400–4000 cm^{-1} interval. The electron micrographs were taken on a TEM 400 Philips microscope. The samples were prepared for direct observation on a carbon support.

10.3 RESULTS AND DISCUSSION

10.3.1 EFFECT OF TEMPERATURE CALCINATION UPON KAOLIN MAIN CHARACTERISTICS

Table 10.1 summarizes the results from the determination of density, pH of the aqueous suspension and humidity of standard and thermally modified kaolin.

TABLE 10.1 Results from the Characterization of Standard and Thermally Modified Kaolin.

		Modification temperature (°C)							
	Non-modified	200	300	400	500	600	700	800	900
Density (g/cm³)	2.30	2.25	2.24	2.20	2.18	2.21	2.22	2.22	2.24
pH	7.20	7.46	8.04	8.01	7.96	7.26	7.09	6.54	6.62
Humidity (%)	1.16	0.60	0.54	0.41	0.33	0.28	0.27	0.24	0.22

The observed changes in the analyzed parameters can be rather commented more as trends than as absolute values, since in the majority of cases these are too close:

1. With increasing calcination temperature, the density decreases to about 500°C and in the range 600–900°C it remained almost constant and the values tended to increase slightly.
2. In the temperature range up to 300°C pH of the aqueous suspension increases initially, then there is a tendency toward a lasting and significant decrease in the 400–900°C range.
3. The humidity tends to last and there is significant decrease in the 200–900°C range.

The observed changes in the above mentioned characteristics of the fillers are apparently due to structural and chemical changes in kaolin occurring during its thermal modification. The dependence of DBP absorption on kaolin versus calcination temperature is shown in Figure 10.5.

The absorption of DBP on the modified kaolin is higher than that on the standard kaolin and increases at higher calcination temperature. This dependence is not valid only for kaolin calcined at 300 and 400°C, at which the absorption of DBP is almost identical.

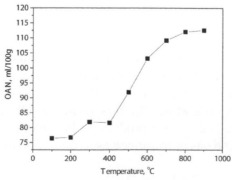

FIGURE 10.5 Absorption of dibutylphthalate on standard kaolin and kaolin calcined at various temperatures.

In the case of DBP absorption the calcination laves fully the rough surface of kaolin particles. Therefore, as the surface has more bumps, ridges and pores, it is energetically more heterogeneous and the absorption of DBP increases. When the filler particles merge, the free surface is reduced and the absorption of DBP is lower. It is the free surface of the filler that is open for interaction with rubber macromolecules. The latter being of larger size cannot penetrate into all the pores, dents and bumps on the surface of filler particles, where DBP molecules can reach, that is the free surface of the filler (which may be in contact with the rubber) does not coincide with the surface to which DBP adsorbs. Therefore, the increased absorption of dibenzofuran is not always a guarantee of improvement in the reinforcing activity and effect of a filler. A part of the derivatogram of standard kaolin is shown in Figure 10.6.

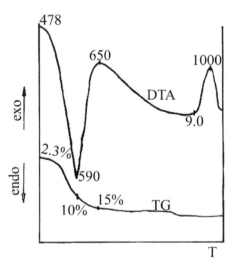

FIGURE 10.6   A part of the derivatogram of standard kaolin (DTA and TG are also presented).

Differential thermal analysis (DTA) curve presents an intensive endo-thermic reaction with a minimum of the differential curve at 590°C and an exothermic reaction with a maximum at 1000°C. The endothermic reaction starts at 478°C and ends at 650°C. It is related to OH evolving from the crystal lattice. Metakaolin is the product of the endothermic process in the 478–650°C range and has been formed as a result of kaolin dehydration:

$$Al_2O_3 \cdot 2SiO_2 \cdot 2H_2O = Al_2O_3 \cdot 2SiO_2 + H_2O$$

The exothermic effect starts at 900°C and corresponds to the degradation of metakaolin:

$$Al_2O_3 \cdot 2SiO_2 = \gamma - Al_2O_3 + 2SiO_2$$

TG curves plotted in Figure 10.6 show that in the beginning, dehydration mass losses are about 2.3% (that is humidity). At the temperature of maximum dehydration (590°C) mass losses are 10%, and at the completion of dehydration it is about 15%.

Figure 10.7 presents parts of derivatograms of kaolin calcined at various temperatures in the range typical of their dehydration.

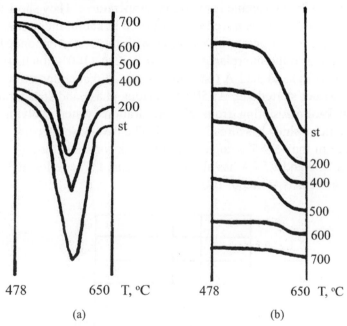

(a)   (b)

FIGURE 10.7 Parts of derivatograms (a—DTA curves; b—TG curves) in the temperature range typical of the dehydration of kaolin calcined at various temperatures ("st" stands for standard, standard kaolin).

As DTA curves show, with increase in the calcination temperature the endothermic effect is less pronounced, while the mass losses at the temperature of maximum dehydration decrease. The mass losses as detected by TG curves are the following:

- Standard kaolin—10.0%
- Kaolin calcined at 200°C—9.5%
- Kaolin calcined at 300°C—9.5%
- Kaolin calcined at 400°C—9.5%
- Kaolin calcined at 500°C—2.7%
- Kaolin calcined at 600°C—0.5%
- Kaolin calcined at 700°C—0.3%

According to literature data[7], dehydrated kaolin up to 600°C can slowly absorb (in the course of 70 h) water for filling the OH-groups in the lattice which is explained by a preserved structure fragment at this temperature. Figure 10.8 summarizes the results from the X-ray studies on standard kaolin and on kaolin calcined at various temperatures. They show that, up to 400°C no phase changes have been observed in kaolin inclusive. There are no changes in the structure of kaolinite with the best pronounced reflexes, corresponding to the interplanar space of 7.12 Å (100% intensity); 3.56 Å (100% intensity) and 3.32 Å (60% intensity). At that temperature the dehydration has not started yet. At 500°C serious phase and structural changes occur in kaolin resulting from its dehydration (water evolving from its crystal lattice) which has already started. The two most intensive reflexes of kaolin are in the 45–55% range. The reflexes corresponding to interplanar spaces of 2.33 and 2.55 Å has almost disappeared. The reflex at 3.32 Å is of increased intensity from 60–100%.

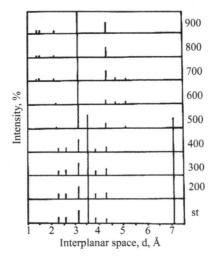

FIGURE 10.8 Results from X-ray diffraction studies of kaolin calcined at various temperatures in the 200–900°C range ("st" stands for the standard, virgin kaolin).

At 600°C the dehydration is at an advanced stage almost at its end— the two reflexes of 100% intensity typical of kaolinite have disappeared completely; the one at 3.32 Å is of 100% intensity and new reflexes appear (d= 4.50 Å, 5.03 Å); the structure corresponds to metakaolinite and there are some signs of structure amorphization. At 700°C the dehydration has already been completed; the structure is typical of metakaolinite and with increasing temperature the amorphization process continues.

The infrared spectra of the starting kaolin and of the samples calcined at different temperatures shows two areas in which changes occured in the process of calcination that is in 600–1500 cm⁻¹ range; wherein the vibrations of the aluminosilicate skeleton are in the area from 3000–4000 cm⁻¹. The symmetric valence vibrations of the hydroxyl group are[20] worth noting and is the absorption band at 1650 cm⁻¹ (deformation vibrations of the water molecules), which reveals whether the observed absorption bands are due to water molecules or hydroxyl groups. The results of infrared spectroscopy (IRS) confirm fully the results of X-ray and differential thermal analysis. By raising the temperature of calcination the absorption bands characteristic of the hydroxyl group change abruptly, their bandwidth is reduced (500–600°C) and disappear (700°C). The absorption band at 1650 cm⁻¹ is practically absent (in the spectra of some of the samples it is just hinted, which is probably due to secondary adsorption after the calcination of some quantity of moisture).

In the absorption band, characteristic of the aluminosilicate skeleton with increasing calcination temperature to 500°C there is a diffuse widening of the bands and reduction in their number. Closely located bands merge into one wide, so the two wide bands observed after 700°C cover totally an area of many narrow bands at 200–400°C. The fact that, with increasing the temperature of thermal processing of kaolin diffuses widening increases to the ongoing gradual amorphization of the material. The results presented (X-ray diffraction, DTA, IRS) allow the following conclusions about the changes occurring in thermally modified kaolin:

- The dehydration starts after 450°C, proceeds at a considerable rate in 500–600°C range and practically ends at 650°C;
- Its nature is demonstrated in evolving chemically bound water— hydroxyl groups of the crystalline lattice of the kaolinite and converting the latter into metakaolinite; and
- As a result of dehydration a metakaolinite residue is obtained with a certain degree of disorderliness (amorphization of the mineral), which determines its high chemical reactivity.

10.3.2 EFFECT OF CALCINATION TIME UPON KAOLIN MAIN CHARACTERISTICS

Figure 10.9 presents parts of derivatograms of standard kaolin and kaolin calcined 500°C at a different calcination time.

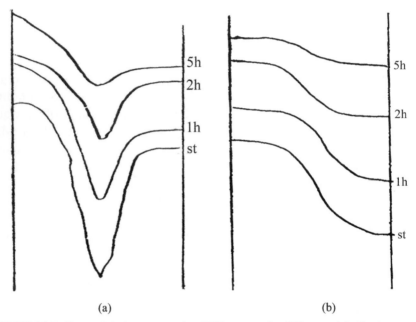

(a) (b)

FIGURE 10.9 Parts of derivatograms (a—DTA curves; b—TG curves) in the temperature range typical of the dehydration of kaolin calcinated at 500°C for different calcination time ("st" stands for standard, virgin kaolin).

As the DTA curves show, at longer calcinations time the endothermic effect is less pronounced, while the mass losses decreases at the temperature of dehydration at maximum rate. The mass losses calculated from the TG curves are the following:

- Standard kaolin—10%
- Kaolin calcined at 500°C for 1 h—7.2%
- Kaolin calcined at 500°C for 2 h—6.25%
- Kaolin calcined at 500°C for 5 h—2.75%

Figure 10.10 summaries the results from the X-ray studies on standard and kaolin calcined for a different time (1, 2 and 5 h) at 500°C. The results reveal serious phase and structural changes in kaolin because of the dehydration that has already started (evolving of water from its crystal lattice).

The changes mentioned are the greatest in the kaolin calcined for 5 h. The intensity of the two in which the most intensive reflexes of 61.5 and 53.3% correspond to interplanar distances of 7.12 and 3.56 Å. The reflexes corresponding to interplanar distance of 2.33 Å have disappeared while that of 3.32 Å has increased intensity (from 68.7 to 100%).

For the kaolin calcined for 2 h the reflexes corresponding to interplanar distances of 2.33 Å and 2.55 Å are preserved. The intensity of the reflex at 7.12 Å remains unchanged, while that at 3.56 Å decreases (from 100 to 73%).

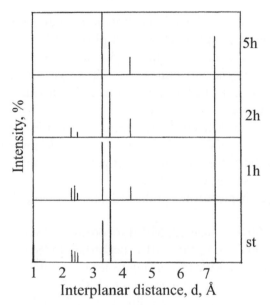

FIGURE 10.10   Results from X-ray diffraction studies of standard kaolin (st) and calcined at 500°C for 1, 2 and 5 h, respectively.

The X-ray diffraction result of standard kaolin is most similar to that of kaolin calcined for 1 h. The two most intensive reflexes characteristic of this type of kaolin almost do not change. The intensity of the reflex corresponding to an interplanar distance of 3.32 Å increases slightly (from 68.7 to 92.8%).

The changes observed are due to the different retention time which affects the dehydration in the 500–600°C range, hence the ratio of the phases of kaolinite and metakaolinite existing at a certain temperature are registered by the X-ray radiographs.

At longer time of calcination the reflexes characteristic of kaolinite are gradually reduced at the expense of increasing the reflexes typical of metakaolinite. Infrared spectra of the samples calcined for different time confirm those findings. The longer calcination time enables all processes to be realized at a higher degree. That applies both to changes in the alumino-silicate skeleton and to changes related to the hydroxyl group.

10.3.3 PROPERTIES OF SBR-BASED COMPOSITES COMPRISING CALCINED KAOLIN

Studying the influence of calcination temperature on the properties of vulca-nizates the following recipe was used: (in phr), SBR "Bulex 1500"—100; zinc oxide—5; stearic acid—2; filler—50; anti-aging agent (N-phenyl-N'-isopropyl-p-phenylene diamine)—1, 5; glycerol—3; accelerators (N-cyclohexyl-2-benzothiazolyl-1, 3-sulfenamid and zinc diethyldithiocar-bamate)—1; sulfur—2.

Kaolin calcined at different temperature (marked as CK-200; CK-300; CK-400; CK-500; CK-600; CK-700; CK-800; CK-900) was used as a filler. The number indicates the calcination temperature, standard virgin kaolin (marked SK) was used as a reference.

The following characteristics of the vulcanizates were determined:

- Stress at 100% elongation, MPa (according to ISO 37:1994)
- Stress at 300% elongation, MPa (according to ISO 37:1994)
- Tensile strength, MPa (according to ISO 37:1994)
- Elongation at break, % (according to ISO 37:1994)
- Residual elongation, % (according to ISO 2285:1988)
- Hardness shore A, (according to ISO 7619:2001)
- Thermal aging resistance with *forced air circulation at* 100°C (according to ISO 188:1998). The minus before the aging coefficient (%) shows the percentage of lowering the respective mechanical parameter.
- Specific surface area—laboratory method.[21]
- Equilibrium swelling degree[22]
- Microhardness—laboratory method[23,24]

The vulcanization characteristics of rubber compounds studied were calculated from the vulcanization isotherms. The minimum torque—M_L, which characterizes the viscosity of the uncured compound is lower than that of the compound containing standard kaolin. When mixed with modified kaolin M_L grew the least in CK-200. In compounds which contain as filler kaolin, calcined at temperatures from 300 to 600°C, M_L increases non-monotonously. This parameter reaches its maximum value with CK-500 and remains constant at CK-700, CK-800 and CK-900. The increase compared to the reference is 63.6%. Hence, by increasing the calcination temperature of kaolin (particularly at and above 500°C, when chemically bound water is evolved, parted by an increase in the number active adsorption centers on the surface of kaolinite particles) the interaction of *"elastomer–filler"* is enhanced. That finds a numerical representation in increasing the viscosity of the raw compound.

The maximum torque (M_H), which characterizes the hardness of vulcanizates is lower in the standard sample and their values are close to those of the samples containing CK-200, CK-300 and CK-400. By raising the calcination temperature of kaolin from 400°C its M_H increases and reaches maximum values at 700 and 900°C. That is due to an increase in the hardness of the filler and to its dehydration.

The difference between the maximum and minimum torque—ΔM is the measure of the curing network density in the vulcanizate. A 15–20% increase in ΔM with regard to the reference sample is observed for the composites filled with kaolin calcined at 500–900°C.

The optimal cure time for all rubber compounds containing fillers such as calcined kaolin is greater than that of the rubber compound filled with standard kaolin. The increase in the values of this parameter does not show a specific pattern depending on the calcination temperature. Probably kaolin calcined to 600°C adsorbs more active elements of the vulcanization acceleration group, if compared to standard kaolin. But running the kaolin calcination at temperature of 800–900°C which tends to lower due to the destruction of active adsorption centers as the final result of the amorphization and to the total destruction of the material.

The induction period of the vulcanization process is expressed by scorch time T_{s2}. It defines the thermoplasticity interval of a rubber compound. The compounds with calcined kaolin have an induction period longer (over 20%) than that of rubber compounds with standard kaolin. The maximum value of this parameter is reached with CK-800. Therefore, it is advantageous to use calcined kaolin as filler in compositions for thick-walled articles, as in the case the scorch time is prolonged (vulcanization isotherm has a broad plateau) compared to that of standard kaolin. Figure 10.11 shows the modification temperature dependence of Mooney viscosity ML (1 + 4) 100°C

of the rubber compounds containing standard kaolin and kaolin calcined at
different temperature.

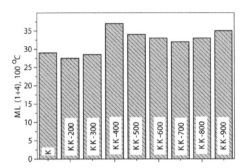

FIGURE 10.11   Mooney viscosity of rubber compounds containing standard kaolin (K) and
kaolin calcined at temperature in the 200–900°C range.

There are two groups of Mooney viscosity values, within these groups
the difference is negligible, but clearly noticeable. The first group includes
samples containing standard kaolin and those containing kaolin modified
at 200 and 300°C. The values are similar to those of standard kaolin. The
second group includes samples containing kaolin, calcined at temperature
in the 400–900°C range. Mooney viscosity of these samples is significantly higher than that of the samples from the first group. The highest
value is for the sample containing kaolin, calcined at 400°C, then in the
one in 500–700°C range. There is a tendency of a slight decrease and again
an increase for the kaolin, calcined in the 800–900°C range. In general,
however, there is a tendency of lower Mooney viscosity values after 400°C,
as compared with that of kaolin calcined at 400°C. The probable cause of
the effects observed is the enhanced "elastomer–filler" interaction, especially in the area of kaolin dehydration, wherein the active adsorption
centers are formed.

Figure 10.12 shows the reinforcing coefficient of standard kaolin and
kaolin calcined at 400, 500, 600 and 900°C kaolin.

As seen from Figure 10.12, the kaolin thermally modified in optimum
conditions (400–600°C) and used as a filler producing a reinforcing effect
better than that of virgin kaolin. That has been proven by mechanical properties (Table 10.2).

The calcinations temperature dependence of the amount of bound rubber
is presented in Figure 10.13.

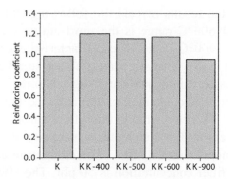

FIGURE 10.12 Reinforcing coefficient of standard kaolin (K) and kaolin calcined at various temperatures.

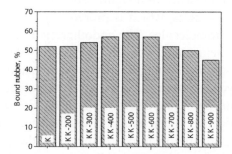

FIGURE 10.13 Calcinations temperature dependence of bound rubber amount.

If we accept bound rubber to be the criteria of the degree of "elastomer–rubber interaction",[17,25] then the most active kaolin is that calcined in the 400–600°C range. It has been confirmed by the investigations on the mechanical properties of the vulcanizates (Table 10.2) and by the obtained values of the reinforcing effect (Figure 10.13).[26,27]

TABLE 10.2 Results from Testing Vulcanizates Comprising Standard Kaolin and Kaolin Modified at Various Temperature.

	Calcination temperature, °C								
	K	**200**	**300**	**400**	**500**	**600**	**700**	**800**	**900**
M_{100} (MPa)	1.3	1.2	1.0	1.5	1.7	2.0	1.6	1.5	1.4
M_{300} (MPa)	2.6	2.5	2.3	3.2	3.4	4.7	3.2	–	–
σ (MPa)	4.8	4.0	4.6	5.0	5.7	5.2	4.3	4.0	3.8
ε_{rel} (%)	470	530	510	450	350	350	380	240	250
Shore A hardness	62	62	62	64	64	66	66	67	69

The analysis of the results reveals the following: the kaolin thermally modified in the 400–600°C range is the most active filler of all samples modified in the 200–900°C range. The characteristics of the vulcanizates containing them (M_{100}, M_{300}, tensile strength) are significantly better than those of vulcanizates containing standard virgin kaolin. The stress at 100% elongation of vulcanizates containing kaolin calcined under optimal conditions is improved by 30% compared to that of vulcanizates containing standard kaolin. The stress at 300% elongation increases also by 30% and the tensile strength by approximately 20%. The elongation at break decreases by about 25%. The filler concentration is 50 phr. The effects observed and described herein can be explained as follows:

It has been estimated[4,6] that the main reason for the enhancing effect of the standard kaolin is the presence of a significant number of hydroxyl groups on the surface of its particles, which is prerequisite for strong interactions with elastomeric macromolecules. The endothermic nature kaolin dehydration turning it into metakaolinite occurs in the 450–650°C range. That is connected with OH-groups leaving the crystal lattice and some amorphization. Removal of hydroxyl groups from the crystal lattice distorts its evenness and electroneutrality. Upon dehydration of kaolin its surface becomes rougher and energetically more heterogeneous. In accordance with the theoretical assumptions, adsorption forces increase in the presence of rough surfaces. The even surface of virgin kaolin is composed predominantly of smooth, low surface energy planes which facilitates relatively weak interactions with macromolecules (a decisive factor in the case is the presence of hydroxyl groups and their concentration). On the contrary, the rough surface of the calcined kaolin (that results in destruction of the smooth surface during dehydration) is characterized by a significant number of edges and other irregularities that are centers of high and uncompensated surface energy. That leads to stronger interactions with elastomeric macromolecules. Even simple physical models[28,29] confirm that the number of interactions with the surface greatly improves, when it is rough and energetically more heterogeneous. Four zones of adsorption energy over the filler surface have been defined[30] (Figure 10.14) and they are:

1st zone: zone of the planes with the lowest surface energy—16 kJ/mol;
2nd zone: zone of the amorphous material—20 kJ/mol;
3rd zone: zone of ridges and crumples over the crystal forms—25 kJ/mol;
4th zone: zone of cracks pores, and so forth (of highest energy)—30 kJ/mol.

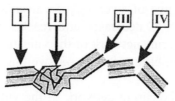

FIGURE 10.14 Adsorption energy zones over the surface of a filler.[30] (Reprinted with permission from Schröder, A., Klüppel, M., Schuster, R.H. . Characterisation of Surface Activity of Carbon Black and Its Relation to Polymer-Filler Interaction. Macromolecular Materials & Engineering 292, 885-916. © 2007 John Wiley and Sons.)

Obviously with calcination, especially in the temperature range of kaolin dehydration these areas reallocate, particularly the portions of the second and third zone increase at the expense of the decreasing portion of the first one. Accordingly, the fraction of high-energy sites decreases with the increasing primary particle size. The effect of increasing particle size is observed with raising calcination temperature. That is another reason for weakening the activity of reinforcing fillers at a calcination temperature over 650°C.

The tensile strength, elongation, resistance to aging and the amount of bound rubber are the highest at a calcination temperature of 500°C. In conclusion it can be said that, the most suitable temperature range for kaolin calcination, viewing its activation as a filler in elastomer composites is 400–600°C. That is the range of dehydration and partial amorphizing of the material, but with preserving the aluminosilicate skeleton. The surface of kaolin particles becomes coarser and therefore energetically more heterogeneous, which is a prerequisite for the strengthening of "elastomer–filler" interaction.

Figure 10.15 shows the equilibrium degree of swelling in an organic solvent of vulcanizates containing standard kaolin and kaolin calcined in the 200–900°C range.

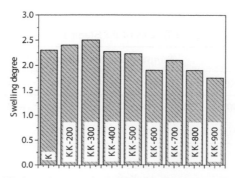

FIGURE 10.15 Equilibrium degree of swelling in an organic solvent of vulcanizates in toluene containing standard kaolin (K) and kaolin calcined at various temperature.

The equilibrium degree of swelling of vulcanizates containing kaolin calcined at 200 and 300°C decreases and reaches a minimum at 900°C. The parameters of the curing network of the vulcanizates studied have been determined by the equilibrium degree of swelling in an organic solvent (Table 10.3).

The parameters of the curing network of the vulcanizates studied have been determined by the equilibrium degree of swelling in toluene and are presented in Table 10.3.

TABLE 10.3 Parameters of the Curing Network of the Vulcanizates Determined by Equilibrium Swelling in Toluene.

	M_c	n_c
K	2100	95
CK-200	2200	91
CK-300	2400	83
CK-400	2100	95
CK-500	2150	93
CK-600	1900	105
CK-700	2000	100
CK-800	1600	125
CK-900	1000	200

The results in Table 10.3 correlate with the results presented so far. Table 10.4 shows the thermal aging coefficients regarding the tensile strength and the elongation at break of the vulcanizates comprising standard kaolin and kaolin calcined in the 200–900°C range.

TABLE 10.4 Thermal Aging Coefficients of the Studied Vulcanizates.

	Temperature of calcination (°C)								
	K	200	300	400	500	600	700	800	900
K_σ (%)	−29	−30	−32	−25	−2	−8	−15	−17	−20
K_ε (%)	−53	−52	−46	−10	−4	−10	−14	−18	−21

As the analysis of results obtained show, considerable improvement in the thermal aging resistance of the vulcanizates comprising kaolin calcined in the 500–600°C range. That is probably due to the better microstructure of the composite resulting from the additionally shattered calcined kaolin particles and to their higher activity. The better dispersion of the particles lowers the probability of local strains and stresses in the course of aging.

The described above effect of kaolin on the properties of SBR based vulcanizates has been fully confirmed by other authors.

In[31] kaolin and thermally-treated kaolin (especially m-kaolinite), which is the most active species of kaolin were added to SBR composites at different concentrations ranging between 10 and 50 phr to study their effect on SBR rheometric and its physical, mechanical, and thermal properties. The main aim of this study was to state whether the activity of calcined kaolin will allow its acting as a bifunctional additive (i.e. reinforcing filler and accelerator) or an adverse one (i.e. acting as an active radical breaking the polymer chains). The study revealed that SBR containing calcined kaolin yielded properties better than those of composites containing standard kaolin, indicating that calcined kaolin has acted as a bifunctional additive in SBR composites. Kaolin was thermally treated at temperatures ranging from 650 to 750°C, which is the temperature range after which kaolin can be activated to m-kaolinite species which is the active species of kaolin. Figures 10.16 and 10.17 shows the morphology of the different kaolin and thermally-treated kaolin using SEM and transmission electron microscope (TEM).

(a)   (b)

FIGURE 10.16   SEM micrographs of (a) kaolin and (b) calcined.[31]
(Reprinted with permission from Ahmed, N.M., and El-Sabbagh, S.H. The Influence of Kaolin and Calcined Kaolin on SBR Composite Properties. Polymer Composites, 35570–580. © 2014 John Wiley and Sons.

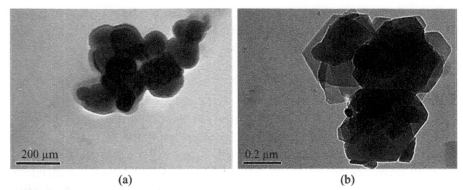

FIGURE 10.17   TEM micrographs of (a) kaolin and (b) calcined kaolin.[31] (Reprinted from Ahmed, N. M.; El-Sabbagh, S. H. The Influence of Kaolin and Calcined Kaolin on SBR Composite Properties. *Polym. Compos.* **2014**, *35*, 570–580. © 2014 with permission from John Wiley.)

As the micrographs show, kaolin particles have huge hexagonal platelet structures, while calcined kaolin structure appeared as smaller particles. These shapes were confirmed by TEM images presented in Figure 10.17, which showed that kaolin plates were homogenous hexagonal plates, and then random sizes appeared after the calcination process. In general, these plate structures of the clay help in slippage of rubber chains from the surface of clay particles leading to improved elasticity. This elasticity is attributed to the plasticizing and the conformational effects on the polymer at the clay–matrix interface, including that the clay particles are in the stacked condition and polymer chains are diffused inside kaolin.[32] The reinforcing index (RI) seen in Table 10.5 is an empirical parameter representing reinforcing effect and can be calculated from the mechanical properties according to the following equation:

$$RI = N / N_0 x \text{ (filler content \% / 100)} \qquad (10.1)$$

where N and N_0 are nominal values obtained by mechanical measurements of the sample in the presence and absence of filler, respectively. The obtained data was confirmed by some published results.[33,34] Table 10.5 indicates that, calcined kaolin was more effective than kaolin, because the value of RI for calcined kaolin is higher than RI for kaolin.

Authors concluded that rheological, mechanical, and thermal properties of SBR/calcined kaolin were better than those of SBR/kaolin. The swelling behavior for SBR containing calcined kaolin was better than that of SBR containing standard kaolin; that is due to the smaller particles of calcined kaolin which bar toluene penetration into the matrix. The morphological study showed that the smaller particles of calcined kaolin were distributed more homogeneously in SBR matrix than the huge ones of standard kaolin.

Publication[35] has shown that kaolin particles are dispersed uniformly in the calcined kaolin/SBR composite, and there is a stronger filler/matrix interfacial interaction, which has a significant effect on the mechanical properties of the composites. Stress–strain curves for SBR/kaolin and SBR/calcined kaolin composites at different filler loadings show that all curves are divided into three stages. The first stage concerns small strain (the stress increases slowly). The second stage is for medium strain (stress increases at a uniform rate). Finally, the third stage shows high strain (the stress increases drastically). Stress-strain properties were better for the composites with calcined kaolin. The stress at 100% elongation (M_{100}), tensile strength, elongation at break, Young's modulus, and hardness all exhibit a remarkable increase for the SBR/calcined kaolin composites. The tensile strength and M_{100} of SBR/calcined kaolin composites are greater than those of SBR/kaolin composites, which allows calcined kaolin particles to act as a reinforcing filler.

TABLE 10.5   The Value of Reinforcing Index (RI) for Kaolin and Calcined Kaolin.

Filler content (phr)	RI (%)
Kaolin	
0	–
10	20.55
20	9.49
30	6.37
40	4.44
50	3.61
Calcined kaolin	
10	25.35
20	14.11
30	11.48
40	9.66
50	8.41

The authors[36] studied the effect that various techniques for clay incorporation have on the mechanical properties of natural rubber (NR). The clay obtained from a local source within the south-eastern region of Nigeria was characterized using X-ray fluorescence and SEM. The calcined clay which contains more than 50% of silica with plate-like microstructure was used to produce NR/clay coagulum as a master batch for the master batch technique for reinforcing NR. The study showed that the NR compounds reinforced via the master batch technique had better tensile, hardness and abrasion properties. It was also evident that when using the master batch technique, the calcined clay filler dispersed better in the NR than when the conventional technique is implemented.

There are also data about the beneficial effect of calcined kaolin upon composites based on hydrogenated *acrylonitrile-butadiene rubber*[37], siloxane rubber[38], as well as low density polyethylene.[39]

10.3.4 POSSIBLE ENGINEER APPLICATION OF MODIFIED KAOLIN FOR PRODUCTION OF STRUCTURAL COMPONENTS (TREADS, CARCASSES, SIDEWALLS) FOR PNEUMATIC TIRES

The possibilities of using calcined kaolin for production of structural components for pneumatic tires (for agricultural, road constructor and building machinery) have been investigated. Kaolin calcined and milled under factory conditions was provided by the company Kaolin Ltd.—Bulgaria.

Kaolin industrially calcined under conditions close to the optimal ones substituted 5–15 phr carbon black in the real compositions. Vulcanizates comprising equivalent amounts of standard kaolin (K_{st}) and such comprising only carbon black (marked with "0") were used for comparison.

The ingredients in the recipes studied were the following:

- Compounds for carcasses (in phr): styrene butadiene rubber Bulex 1500 (produced in Bulgaria)—60; isoprene rubber SKI-3 (produced in Russia)—40; pyrolene—3; stearic acid—2; zinc oxide—5; anti-aging agent (N-isopropyl-N-phenyl-p-phenylenediamine)—1; 2,2,4-Trimethyl-1,2-Dihydroquinoline—1; carbon black N 550—50; processing oil—10; diphenylguanidine—0.2; *4-(2-benzothiazolylthio)-morpholine*—0.36; *N-cyclohexyl-2-benzothiazole sulfenamide*—1.1; sulfur—1.9.
- Compounds for protectors (in phr): Bulex 1500–75; butadiene rubber SKD (produced in Russia)—25; bitum—3; pyrolene—2; stearic acid—3; zinc oxide—3; anti-aging agent (N-isopropyl-N-phenyl-p-phenylenediamine)—1.2; 2,2,4-Trimethyl-1,2-Dihydroquinoline—1; carbon black N 330—65; processing oil—13; N-(Cyclohexylthio) phthalimide (Santogard PVI—0.3); diphenylguanidine—0.12; *N-cyclohexyl-2-benzothiazole sulfenamide*—0.95; sulfur—1.9.

- Compounds for side walls (in phr): isoprene rubber SKI-3 (produced in Russia)—50; butadiene rubber SKD (produced in Russia)—50; stearic acid—2; zinc oxide—5; anti-aging agent (N-isopropyl-N-phenyl-p-phenylenediamine)—1.5; carbon black N 550—50; processing oil—13; paraffin—0.5; *N-cyclohexyl-2-benzothiazole sulfenamide*—1.2; sulfur—1.2.

The vulcanization compositions designed for production of constructive elements for pneumatic tires, were subjected to further investigation to determine the following parameters:

- Fatigue resistance at multiple stretching, cycles till destruction according to Bulgarian State Standard BSS 14604:1978;
- Fatigue resistance at multiple and long lasting bending—laboratory method described[40,41];
- Heat build-up (at 17.5%, 20% and 25% deformation of the samples)—laboratory method described[40,41]; and
- Wear, mm³ (ISO 4649:1998) (performed only on the compound for treads).

10.3.5 EFFECT OF CALCINED KAOLIN UPON THE MECHANICAL CHARACTERISTICS

The introduction of calcined kaolin as a substituent of carbon black at an equimolar amount leads to the following changes in the mechanical parameters of the vulcanizates:

1. The stress at 100% elongation for the vulcanizates of the compound with calcined kaolin content of 5 phr intended for carcass production increased the most (by 42%). The values for the vulcanizates containing calcined kaolin at 10 and 15 phr slightly decrease, in comparison with those for the reference compound containing only carbon black. The lowest value is for the compound in the standard kaolin at 5 phr. The stress at 300% elongation of the vulcanizates with standard and calcined kaolin at 5 phr has increased compared to that of the control sample (over 13%). A decrease of the values for that parameter of the compound with calcined kaolin at a content of 10 and 15 phr has been observed. The most obvious improvement caused by introduction of calcined kaolin as a substitute for a part of the carbon

black is observed in the tensile strength, such as its magnification decreases with the increase in the amount of the calcined kaolin—at a kaolin content 5 phr it is 85%. The elongation at break also increases by 39% due to the introduced kaolin for vulcanizates with kaolin at 5 phr, by 44% for vulcanizates with kaolin at 10 phr and by 38% for the vulcanizates with kaolin at 15 phr. The effects are more pronounced for the vulcanizates containing calcined kaolin compared with those for the ones containing standard unmodified kaolin. Obviously, the calcined kaolin exhibits the strongest reinforcing effect on the compounds designed for carcasses regarding the modules at 100 and 300% elongation, the tensile strength and elongation at break, when it is at an amount of 5 phr (i.e. 5 phr of carbon black were substituted by 5 phr of calcined kaolin).

By increasing the amount of calcined kaolin introduced into the compound the values of the modules decrease. Given that they are a measure of the density of the curing network, it is clear that the impact of calcined kaolin on the vulcanization process is more favorable than that of carbon black. The probable reason for that is the greater adsorption activity of kaolinite particles which adsorb ingredients of the vulcanizing group. The strong increase in the residual elongation also evidences that the vulcanization degree of the compounds with calcined kaolin is lower than that of the compound filled only with carbon black.

Upon dehydration the surface of kaolin particles becomes rough (i.e. more energetically heterogeneous), therefore their adsorption activity increases. Owing thereto, the rubber macromolecules are attached more firmly to the surface of kaolin particles and at stress deformation they do not separate from the adsorption centers of the filler but break. That means the macromolecules gravity centers are moved, thus the stretched chains cannot be returned by the elastic forces to their previous places (they still fold, however, in new places), therefore the relative and residual elongation increase.

In the case of carbon black the situation is different. The macromolecules overflow the particles surface, what gives rise to continuous adsorption and desorption (in the course of deformation) and the macromolecules centers of gravity do not move, that is due to elastic forces the macromolecules shrink toward their centers of gravity. Therefore the residual elongation is low.

Of course, it is not clear how exactly the particles of kaolin interact with those of carbon black and how this interaction contributes to the reinforcing effect. It is thought, however,[42] that a compound with carbon black and kaolin has higher resistance to fatigue, lowered tear resistance (due to its anisotropic structure), and so forth. Particularly, kaolin combination with carbon black in SBR-based composites is good. Undoubtedly, there is a reinforcing effect on the tensile strength in the case of carbon black substitution with calcined kaolin of 5, 10, and 15 phr, as the optimum amount is 5 phr. An adverse effect in this case is the increased residual elongation due to the introduction of calcined kaolin into the compounds. That indicates reduced elasticity of the vulcanizates.

2. The module at 100% elongation for vulcanizates of composites intended for tire treads is increased by 22% (kaolin content at 5 phr), 16% (kaolin content at 10 phr) and 10% (kaolin content at 15 phr) compared to that of the control sample. Module at 300% elongation is slightly decreased, which is within the error range of the method (5 phr of kaolin—6%; 10 phr of kaolin—6%, and 15 phr of kaolin—10%). The same is observed for the tensile strength, the greatest decrease is for vulcanizates containing kaolin at 5 phr—15%, while the contents of 10 and 15 phr kaolin it is quite small—2%. The residual elongation for vulcanizates with 5 phr kaolin decreases slightly (7%) and slightly improves for vulcanizates with 10 and 15 phr kaolin—7 and 12%. In this case, the reinforcing effect of kaolin introduced as a substituent of carbon black on the physical and mechanical parameters of the calcined kaolin has been observed with introduction of 5 phr of calcined kaolin. Unlike in the case of vulcanizates for carcasses, here the residual elongation is low. It hardly changes (the differences are within the range of the method errors) compared to the residual elongation of the zero sample (i.e. the one containing only carbon black as filler).

3. The values for the stress at 100 and 300% elongation of vulcanizates of composites designed for side walls decrease with the increasing amount of calcined kaolin introduced as a substituent of carbon black. At a kaolin content of 5phr the decrease of the module at 100% elongation is 12% and of module at 300% elongation—9%. When kaolin content is of 10 phr, the decrease is 29% for the stress at 100% elongation and 26% for the stress at 300% elongation. When kaolin content is of 15 phr—it is 43% for the stress at 100% elongation and 29% for the stress at 300% elongation. The greatest decrease of the tensile strength is for vulcanizates containing kaolin at 10 phr—41%, the least is for those with kaolin of 5 phr—19% and

vulcanizates with kaolin of 15 phr—32%. The residual elongation is slightly changed. When kaolin content is of 5 phr, it decreases by 6%, and kaolin contents of 10 and 15phr the increase is negligible.

The substitution of a part of carbon black with calcined kaolin leads to deterioration of the modules and tensile strength of the vulcanizates of compounds for side walls. The relative and residual elongation hardly changes.

The changes in shore hardness for all vulcanizates (for carcasses, treads and side walls) are insignificant and are close to the precision of the method—about 1 relative unit.

10.3.5.1   EFFECT OF CALCINED KAOLIN UPON HEAT AGING RESISTANCE

In all tested cases of heat aging (for 24 and 72 h) the vulcanizates from zero compounds (without calcined kaolin) for carcasses, tread and sidewalls have positive values of the aging coefficient with respect to the stress at 100 and 300% elongation, that is, the modules increase. Such a trend is also exhibited by the vulcanizates of the compounds containing calcined kaolin except those for carcasses containing calcined kaolin at 5 phr. The increase in the modules make the vulcanizates tougher most likely due to the process of cross-linking taking place simultaneously with the processes of degradation during the thermal aging of the elastomeric materials. These processes depend on the chemical nature of the elastomer. Cross-linking predominates in the polymers, in which the carbon atom adjacent to a methylene group comprising at least one hydrogen atom, because cross-linking favors the double bonds.[43] For example, during the thermal aging of butadiene rubber-based vulcanizates the processes of structuring are dominant, due to the presence of 1,2—structure in rubber macromolecules.[43]

The introduction of calcined kaolin into the rubber compounds for carcasses, tread and sidewalls at all amounts mentioned improve the resistance to thermal aging, regarding both the tensile strength and elongation at break. The improvement is most pronounced at a calcined kaolin amount of 15 phr in the compounds for carcasses, 5 phr in the compounds for treads and 10 phr in the compounds for sidewalls (regarding both the tensile strength and elongation at break).

10.3.5.2 EFFECT OF CALCINED KAOLIN UPON DYNAMIC FATIGUE RESISTANCE

The effect that calcined kaolin has upon the resistance to dynamic fatigue at multiple stretching of vulcanizates of the compounds for treads, carcasses and sidewalls of pneumatic tires at different dynamic and static loads is presented in Figures 10.18–10.20.

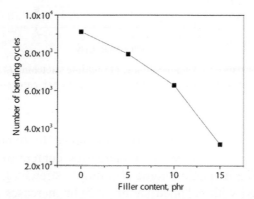

FIGURE 10.18 Resistance to dynamic fatigue at multiple stretching (150% dynamic load, 150% static load) of vulcanizates of compounds designed for treads which comprise calcined kaolin.

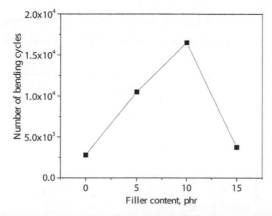

FIGURE 10.19 Resistance to dynamic fatigue at multiple stretching (150% dynamic load, 150% static load) of vulcanizates of compounds designed for sidewalls which comprise calcined kaolin.

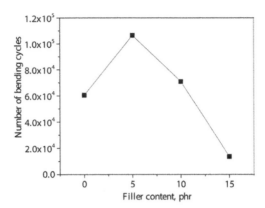

FIGURE 10.20   Resistance to dynamic fatigue at multiple stretching (150% dynamic load, 150% static load) of vulcanizates of compounds designed for carcasses which comprise calcined kaolin.

In the vulcanizates for treads the resistance to dynamic fatigue at multiple stretching decreases with the increasing content of calcined kaolin (Figure 10.18). The dynamic fatigue at multiple stretching in vulcanizates for carcasses filled with calcined kaolin at 5phr increases significantly, at 10 phr—increases less and decreases at 15 phr (Figure 10.20). In vulcanizates for sidewalls the resistance to dynamic fatigue at multiple stretching increases significantly at a calcined kaolin content of 5–10 phr, but at a content of 15 phr the increase is scarce.

The effect of calcined kaolin on the resistance to dynamic fatigue at multiple long-term bending is presented in Figures 10.21–10.23.

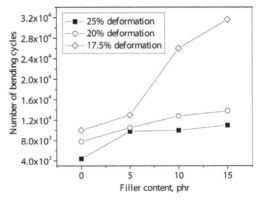

FIGURE 10.21   Resistance to dynamic fatigue at multiple long-term bending of vulcanizates for treads comprising calcined kaolin.

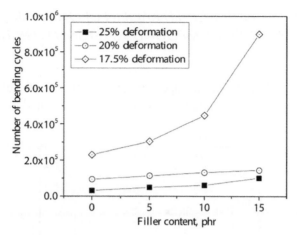

FIGURE 10.22 Resistance to dynamic fatigue at multiple long-term bending of vulcanizates for carcasses comprising calcined kaolin.

The resistance to dynamic fatigue at multiple long-term bending at all levels of deformation improves in all investigated cases. The improvement is most pronounced at calcined kaolin content of 10 and 15 phr and at 17.5% of deformation. For vulcanizates of the compounds for treads and carcasses the improvement is lesser at 20 and 25% of deformation.

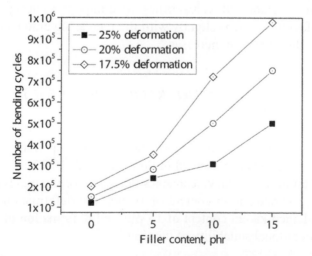

FIGURE 10.23 Resistance to dynamic fatigue at multiple long-term bending of vulcanizates for sidewalls comprising calcined kaolin.

10.3.5.3   EFFECT OF CALCINED KAOLIN UPON WEAR RESISTANCE OF VULCANIZATES OF COMPOUNDS DESIGNED FOR TIRE TREADS PRODUCTION

The effect of calcined kaolin upon wear resistance of vulcanizates of compounds designed for tire treads manufacture is presented in Figure 10.24.

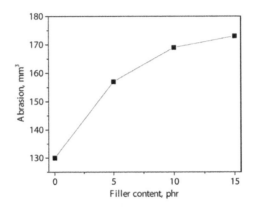

FIGURE 10.24 Wear resistance of vulcanizates of compounds with calcined kaolin designed for tire treads.

The wear resistance of vulcanizates of compounds designed for tire treads manufacture worsens slightly with the increasing amount of calcined kaolin that is their wearing increases.

10.3.5.4   EFFECT OF CALCINED KAOLIN UPON THE HEAT BUILD-UP OF VULCANIZATES

The effect that calcined kaolin has upon the heat build-up of the vulcanizates designed for tire treads is presented in Figures 10.25–10.27.

The heat build-up for the vulcanizates designed for tire treads production decreases at calcined kaolin content of 10 phr and deformation of 17% as well as at calcined kaolin content of 15 phr and deformation of 25%. That favors the operational suitability of the tires.

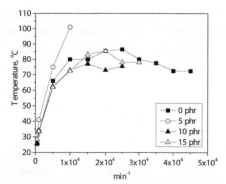

FIGURE 10.25   Heat build-up of the vulcanizates for manufacturing tire treads which comprise various amounts of calcined kaolin at deformation of 17.5%.

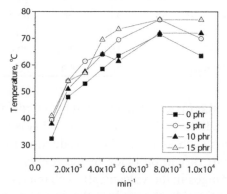

FIGURE 10.26   Heat build-up of the vulcanizates for manufacturing of tire treads which comprise various amounts of calcined kaolin at deformation of 20%.

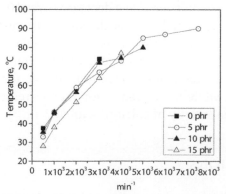

FIGURE 10.27   Heat build-up of the vulcanizates for manufacturing tire treads which comprise various amounts of calcined kaolin at deformation of 25%.

Effect of calcined kaolin upon the curing characteristics of compounds designed for the manufacture of construction elements for pneumatic tires.

The vulcanization characteristics estimated by the vulcanization isotherms recorded on a Monsanto rheometer are presented in Table 10.6.

TABLE 10.6 Vulcanization Characteristics of Rubber Compounds Designed for Production of Tire Carcasses, Treads, and Sidewalls (10 phr of carbon black are substituted with 10 phr of calcined kaolin).

	Compounds for carcasses		Compounds for treads		Compounds for sidewalls	
	0	10	0	10	0	10
M_L (dN.m)	10.5	9	11	10.5	7	8
M_H (dN.m)	48.5	42	42	44	33	32.5
$\Delta M = M_H - M_L$ (dN.m)	38	33	31	33.5	26	24.5
T_{s2} (min)	0.7	0.8	1	1.1	0.6	0.6
T_{90} (min)	13.6	16.6	23.6	29.4	8.1	8.3
Vulcanization rate V_c (%/min)	7.7	6.3	4.4	3.5	13.3	13

Vulcanization isotherms are characterized by a wide plateau, which does not change with the addition of calcined kaolin. That is beneficial for curing of thick walled items such as tires. The introduction of calcined kaolin as a substitute for a part of carbon black in the rubber compounds does not cause substantial changes in the vulcanization characteristics of the samples studied. The most noticeable change is in the optimum cure time T_{90}, which increases after the addition of calcined kaolin to the compounds intended for carcass and tread production. In the compounds for carcass production, calcined kaolin introduction leads to a decrease (over 10%) of the minimum and maximum torque (M_L and M_H); the density of the vulcanization network measured by ΔM decreases, respectively. The scorch time, albeit slightly increases under the influence of calcined kaolin for compounds intended for carcass and tread the production, which is favorable for the technical suitability of the latter, while the vulcanization rate decreases, probably due to increased adsorption activity of calcined kaolin.

10.4 CONCLUSIONS

The opportunities for thermal activation of kaolin, have been studied tracing the changes that occur in its structure and in its properties when heated in the 200–900°C temperature range. It has been found that kaolin modified at a temperature of 400–600°C produces the highest reinforcing effect.

The changes that occur in the mechanical, vulcanization, rheological, and so forth characteristics of the vulcanizates containing kaolin modified at different temperatures has been investigated. It has been found that the temperature of kaolin calcination has a significant influence on the entire range of properties of the vulcanizates containing modified fillers.

We believe that the main reason for the improvement in the reinforcing effect of thermally modified kaolin is the release of hydroxyl groups from the crystal lattice, resulting in a disturbed balance and electroneutrality, and its particle surface becoming rough and energetically more heterogeneous. That leads to an increase in the adsorption forces and to an enhanced interaction with the elastomer macromolecules.

The properties of vulcanizates containing varying amounts of calcined kaolin introduced as a replacement for a part of the carbon black in the recipes for production of structural elements (carcasses, treads and sidewalls) of tires for agricultural, road construction and building machinery, electro- and motorcars have been investigated. It has been found that, calcined kaolin can be used as a substitute of 10–15 phr carbon black in compounds intended for the production of carcasses and sidewalls.

10.5   ACKNOWLEDGMENT

The authors acknowledge the assistance of Dr. Petrunka Malinova in the experimental part and discussions of this work.

KEYWORDS

- calcined kaolin
- calcination conditions
- styrene–butadiene rubber
- reinforcing effect
- tires

REFERENCES

1. *High Performance Fillers. 2005*, Conference Proceedings—Compiled by Smithers Rapra: Cologne, Germany, March 8–9,2005; Smithers Rapra Publishing, 2005.
2. *High Performance Fillers. 2006*, Conference Proceedings—Compiled by Smithers Rapra: Cologne, Germany, March 21–22, 2006; Smithers Rapra Publishing, 2006.
3. *High Performance Fillers. 2007*; Conference Proceedings—Compiled by Smithers Rapra: Cologne, Germany, March 14–15, 2007.
4. Dannenberg, E. M. Filler Choices in the Rubber Industry. *Rubber Chem. Technol.* **1982,** *55*, 860–880.
5. U.S. Geological Survey.2014, Kaolin end-use statistics [through 2003; last modified September 15, 2005], in Kelly, T. D., and Matos, G. R., comps., Historical statistics for mineral and material commodities in the United States (2014 version): U.S. Geological Survey Data Series 140, http://minerals.usgs.gov/minerals/pubs/historical-statistics/. US Geological Survey, Data Series 140, (2005).
6. Waddell, W.; Evans, L. Use of Nonblack Fillers in Tire Compounds. *Rubber Chem. Technol.* **1996,** *69*, 377–423.
7. Герасимов, Е.; Атанасов, А. и др.; под ред. на С. Бъчваров, Технология на керамичните изделия и материали, София, ИК "Сарасвати", (2003).
8. Nedev, M.; Dazkova, T.; Kortschakova, M. Über die Thermische Behandlung von Kaolin und Die Thermoaktivierung Seiner Mischungen Mit Styrol-Butadienkautschuk (Thermal Treatment of Kaolin and Thermoactivation of its Components with Styrene-Butadiene Rubber). *Kaut Gummi Kunstst.* **1991,** *44*, 671–673.
9. Suito, E.; Arakawa, M. The Mechanism of Reinforcement with the Hard Clay. Infrared Studies of Rubber-Filler System. *Nippon Gomu Kyokaishi.* **1963,** *36*, 704–709.
10. Mee, S. J.; Hart, J. R.; Singh, M.; Rowson, N. A.; Greenwood, R. W.; Allen, G. C.; Heard, P. J.;Skuse, D. R. The Use of Focused Ion Beams for the Characterisation of Industrial Mineral Microparticles. *Appl. Clay Sci.* **2008,** *39*, 72–77.
11. Espí, E.; Salmerón, A.; Fontecha, A.; García-Alonso, Y.; Real, A. I. New Ultrathermic Films for Greenhouse Covers. *J. Plast. Film Sheeting* **2006,** *22*, 59–68.
12. Fanselow, J. R;Jacobs, D. A. Calcined Kaolin Clay Pigment. U.S. Patent 3,586,523 A, June 22, 1971.
13. Sare, E. J.; Adkins, T. L.; Raper, S. C.; Julie Figlar, J. High Whiteness Metakaolin and High Whiteness Fully Calcined Kaolin. Patent WO 2,005,019,349 A2, March 03, 2005.
14. Sullivan, R. E. Calcination Process for Kaolin-Containing Clay. U.S. Patent 3,476,511 A, Nov 04, 1969.
15. Grace, W. R.; Co.-Conn. Calcined Kaolin Product Stewardship Summary. https://grace.com/en-us/environment-health-and-safety/ProductStewardship/Documents/Calcined_Kaolin.pdf.
16. Kaolin Applications, BASF, http://www.kaolin.basf.com/applications.
17. Дишовски, Н.; Ценков, Г. Справочник по каучук, София, ЕС принт, (2006).
18. Scarlett, M. Reinforcement for Tires. *Tire Technol. Int.* **2001,** June, 23–25.
19. Wood, P. After Natural Rubber. *Tire Technol. Int.* **2004,** March, 42–45.
20. Болдырев, А.И. Инфракрасные спектров минералов, Москва, Недра, (1976).
21. Dishovsky, N.; El-Tantawy, F.; Dimitrov, R. Effect of Bi-Containing Superconducting Ceramic on the Volume Resistivity of Butyl Rubber Composites. *Polym. Test.* **2004,** *23*, 69–75.

22. Бергштейн, Л. Лабораторный практикум по технологии резин, Химия, Ленинград, (1978).
23. Zamfirova, G.; Dimitrova, A. Some Methodological Contributions to the Vickers Microhardness Technique. *Polym. Test.* **2000,** *19,* 533–542.
24. Perena, J. M.; Lorenzo, V.; Zamfirova, G. Microhardness of Polyethylene Surface Modified by Chlorosulphonic Acid. *Polym. Test.* **2000,** *19,* 231–236.
25. Mark, J. E.; Erman, B.; Eirich, F. Eds. *Science and Technology of Rubber,* 3rd ed.; Elsevier Academic Press: Amsterdam, 2005.
26. Perena, J. M.; Zamfirova, G.; Gaydarov, V.; Malinova, P.; Dishovsky, N. Rubber Filled with Thermally Activated Kaolin. World Polymer Congress MACRO 2004, *Proceedings of the 40th International Symposium on Macromolecules,* Paris, France 2004.
27. Zamfirova, G.; Gaydarov, V.; Malinova, P.; Dishovsky, N.; Perena, J. M. Mechanical Behavior of Rubber Filled with Thermally Modified Kaolin. *e-Polym.* **2004,** No. E_002.
28. Luchow, H.; Breier, E.; Gronski, W. Characterization of Polymer Adsorption on Disordered Filler Surfaces by Transversal 1H NMR Relaxation. *Rubber Chem. Technol.* **1997,** *70,* 747–801.
29. Evans, M. S. *Tyre Compounding for Improved Performance,* Vol. 12; Smithers Rapra Publishing: London, United Kingdom, 2002.
30. Schröder, A.; Klüppel, M.; Schuster, R. H. Characterisation of Surface Activity of Carbon Black and its Relation to Polymer-Filler Interaction. *Macromol. Mater. Eng.* **2007,** *292,* 885–916.
31. Ahmed, N. M.; El-Sabbagh, S. H. The Influence of Kaolin and Calcined Kaolin on SBR Composite Properties. *Polym. Compos.* **2014,** *35,* 570–580.
32. Ahmed, N. M.; El-Nashar, D. E.; Abd El-Messsieh, S. L. Utilization of New Micronized and Nano-CoO MgO/Kaolin Mixed Pigments in Improving the Properties of Styrene-Butadiene Rubber Composites. *Mater. Des.* **2011,** *32,* 170–182.
33. Haghighat, M.; Zadhoush, A.; Nouri Khorasani, S. Physicomechanical Properties of α-Cellulose–Filled Styrene–Butadiene Rubber Composites. *J. Appl. Polym. Sci.* **2005,** *96,* 2203–2211.
34. Kohjiya, S.; Ikeda, Y. Reinforcement of General-Purpose Grade Rubbers by Silica Generated In Situ. *Rubber Chem. Technol.* **2000,** *73,* 534–550.
35. Ahmed, N.; El-Sabbagh, S. H. Effect of Kaolin and Calcined Kaolin Loading on Styrenebutadiene Rubber Composites. Plastics research online. Society of Plastics Engineers (SPE), 2014. http://www.4spepro.org/pdf/005234/005234.pdf.
36. Ewulonu, C. M.; Obele, C. M.; Arukalam, I. O.; Odera, S. R. Effects of Local Clay Incorporation Technique on the Mechanical Properties of Natural Rubber Vulcanizates. *Int. J. Appl. Sci. Eng. Res.* **2015,** *4,* 307–318.
37. Anmin, H.; Xiaoping, W.; Demin, J.; Yanwei, L. Thermal Stability and Aging Characteristics of HNBR/Clay Nanocomposites in Air, Water and Oil at Elevated Temperature. *e-Polym.* **2007,** *7,* 588–598.
38. Pędzich, Z.; Bieliński, D. M.; Anyszka, R.; Zarzecka-Napierała, M. Influence of Boron Oxide on Ceramization of Silicone-Basing Composites.*Proceedings of the 12th Conference of the European Ceramic Society,* Stockholm, Sweden, 1–4, 2011.
39. Mallik, A.; Barik, A. K.; Pal, B. Comparative Studies on Physico-Mechanical Properties of Composite Materials of Low Density Polyethylene and Raw/Calcined Kaolin. *J Asian Ceram. Soc.* **2015,** *3,* 212–216.

40. Brown, R. P. (Ed.) *Handbook of Polymer Testing, Short term Mechanical Tests*; Rapra Technology, Smithers Rapra Press, Shawbury, Shrewsbury, Shropshire, SY4 4NR, United Kingdom, 2002.

41. Marsa, W. V.; Fatemi, A. A Literature Survey on Fatigue Analysis Approaches for Rubber. *Int. J. Fatigue* **2002,** *24*, 949–961.

42. Whelan, A.; Lee, K. S. Eds. *Developments in Rubber Technology 2—Synthetic Rubbers, Series: Polymer Science and Technology Series,* Vol. 37; Springer Science & Business Media: Springer Netherlands, 2013.

43. Дишовски, Н. Стареене и стабилизация на еластомерните материали, София, Информа, (1998).

CHAPTER 11

ELASTOMER-BASED COMPOSITES WITH REDUCED ZINC OXIDE LEVEL

CONTENTS

ABSTRACT

Recently, the industries related to rubber production and processing have been showing concern about environment protection and making efforts to reduce the amount of zinc oxide in rubber compounds. That has evoked our interest in the full or partial replacement of ZnO with other zinc salts that act as multifunctional ingredients. Zinc resinate is a salt that could be used in a natural rubber based compounds as a multifunctional additive of low zinc content with activating (in respect to the vulcanization process), plasticizing, dispersing, and antireversion effects.

The influence of zinc resinate on the main characteristics (vulcanization, mechanical, physicochemical, and dynamic) of compounds and vulcanizates based on unfilled and reinforced natural rubber in the presence of accelerators differing in their chemical nature (mercaptobenzothiazole, tetrametiltiuram disulfide, N-cyclohexyl-2-benzothiazolyl sulfenamide, and N-tert-butyl-2-benzothiazyl sulfenamide) has been investigated.

In vulcanizates based on unfilled rubber, substitution of zinc oxide by zinc resinate at equivalent amounts lowers the minimum and maximum torque, leading to an increase in the scorch time and the optimal curing time. When the vulcanizates of rubber compounds are filled with carbon black, the effects of lacking zinc oxide are less pronounced and the parameters are much closer to those of the control sample with zinc oxide. When the silica-filled compounds contain bifunctional organosilane—a combination of zinc oxide and zinc resinate—the antireversion effect is more pronounced. Vulcanizates containing zinc resinate or a combination of zinc oxide and zinc resinate have better dynamic properties than the respective control ones.

The results of our studies have demonstrated that when using zinc resinate as a polyfunctional additive, the content of zinc oxide in rubber compounds can be reduced, in some cases even eliminated, and certain properties of the vulcanizates improved.

11.1   INTRODUCTION

Rubber industry has always taken into account both latest achievements in the field of synthesis of polymers and ingredients and the ever-increasing demands for environment protection.

In this regard, over the last 15–20 years, the following trends have been observed:

11.1.1 REDUCING THE CONTENT OR COMPLETE REPLACEMENT OF ZINC OXIDE, DESCRIBED AS ECOTOXIC

The total consumption volume of zinc oxide in the European Union for 1995 is about 230,000 tonnes. The imported and exported volume of zinc oxide in the European Union is 32,000 tonnes and 16,600 tonnes, respectively. With these figures, the volume of available zinc oxide at the EU market could be calculated as approximately 250,000 tonnes per year (information from industry). The compound has a great range of applications, for instance, in rubber manufacture—tires and general rubber goods (36%), glass and ceramics (27%), ferrites, varistors and catalysts (12%), animal feed (9%), raw material for the production of zinc chemicals (4.5%), fuel and lubricant additives (4.5%), paints (4.5%), and cosmetics and pharmaceuticals (2%).[1] The concern about environment protection has led to the realization of the ECOZINC project funded by the European Commission, which aims at maximum reduction of zinc oxide in rubber compounds.[2] A number of restrictions have been imposed by the European Commission regarding the dose of ZnO in rubber compounds. The potential harmful environmental impact caused by the release of ecotoxic zinc species from rubbers lead to an increasing interest in potential substitutes.

11.1.2 WIDE APPLICATION OF SOME ZINC SALTS OF SATURATED AND UNSATURATED ORGANIC ACIDS (SO-CALLED ZINC SOAPS) AS MULTIFUNCTIONAL INGREDIENTS AND OFFERING THEM AS A MARKET PRODUCT

In fact, that is a different approach to reducing the content of zinc in the vulcanizates. Zinc soaps having a short aliphatic chain/C6–C8 exhibit an activating effect in the absence of zinc oxide. Replacement of zinc oxide in the rubber compound with zinc soaps leads to a lower amount of zinc ions, to better dispersion of accelerators and fillers in the rubber matrix, and to improved mechanical properties of the composites obtained.[3] Zinc salts of organic acids with a long aliphatic chain/C16–C18 produce a less pronounced activating effect, but their addition to rubber compounds comprising zinc oxide and thiazole accelerator improves compounds plasticity as well as the tensile strength, hardness, and elasticity of vulcanizates.[3]

Depending on the structure of the fatty acids, their zinc salts exhibit different efficiency in the rubber mixtures. Various soaps of organic acids have been studied in mixtures of natural rubber (NR) to show the dependence of their properties on their structure. The influence of the chain length, branching, and the content of the aryl group on the density of transverse links and resistance to reversion has been investigated.[3]

11.1.3 REPLACEMENT OF CARBON BLACK WITH SILICA, PARTICULARLY IN THE PRODUCTION OF TIRES, IMPOSED BY THE ENVIRONMENTAL CONCERNS AND REQUIREMENTS FOR PASSENGERS' SECURITY

When using silica in rubber compounds, the presence of bifunctional organosilanes is a must. The latter not only improve the dispersion of the filler during compounding, like zinc soaps, but also facilitate a stronger "polymer-filler" interaction, resulting in improved mechanical and dynamic properties of vulcanizates. The disadvantage of NR, however, is that vulcanization at temperatures above 150°C leads to undesirable reversion. That disadvantage could be overcome by the use of different multifunctional additives (MFAs), zinc salts of organic acids, inclusive. On the one hand, these additives may serve as an activator of the accelerated sulfur vulcanization by which to reduce the content of zinc oxide in the silica-filled compounds and on the other as softeners, dispersing agents, and substances which reduce reversion observed in the case of NR vulcanization at temperatures above 150°C.

Regarding the above said, the thesis of the study can be formulated as follows: According to theoretical prerequisites, zinc resinate, which is a zinc salt of monocarboxylic resin acids from the group of terpenoids with isocyclic structure, could be used as a MFA of low zinc content having activating, plasticizing, dispersing, and antireversion action in rubber compounds. Our interest is focused first of all on its use in NR, polyterpene being the most important for us. We have not found in the literature such an approach, namely, the search for a correspondence between the related terpenoid (zinc resinate) and polyterpenes (NR). The relationship and good technological compatibility should not be overlooked, as it is well known that due to the limited compatibility of certain ingredients in the elastomeric matrix, the so-called migration can be observed. That has motivated us to carry out further research in the particular area.

The aim is to investigate the effect of zinc resinate on the main characteristics (vulcanization, physical, mechanical, physicochemical, dynamic, etc.). Compounds and vulcanizates based on unfilled and filled NR in the presence of accelerators differing in their chemical nature, which are used most often in the current technology of rubber. The assessment of the possibilities for practical application of the results obtained and on the basis to propose routes to lowering the content of zinc oxide in the articles produced has also been a research goal.

Theoretical prerequisites for our attention to zinc resinate can be classified as follows:

1. Resinate is a zinc salt of the resin acids, which are monocarboxylic acids having isocyclic structure—mainly abietic, levopimaric, and dextro-pimaric, and can be represented by the following structure:

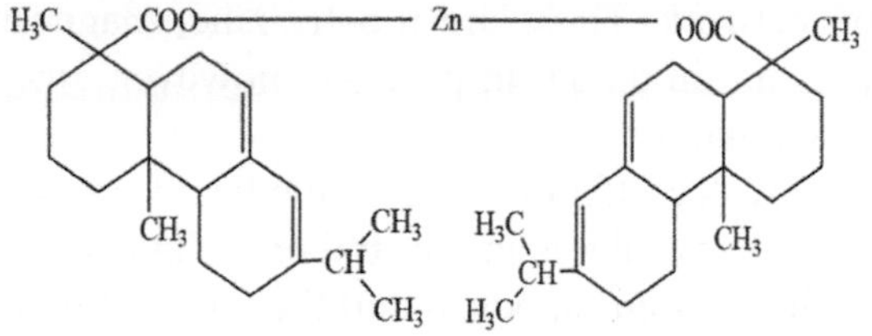

2. The compounds of isocyclic structure can be viewed as dimers (terpenes), trimmers, or polymers (polyterpenes) of isoprene and their oxygen derivatives (terpenoids). The polyterpene most important for us is NR.
3. Abietic acid and other resin acids can be regarded as isoprenoids, as seen from the following interpretation of its formula:[4]

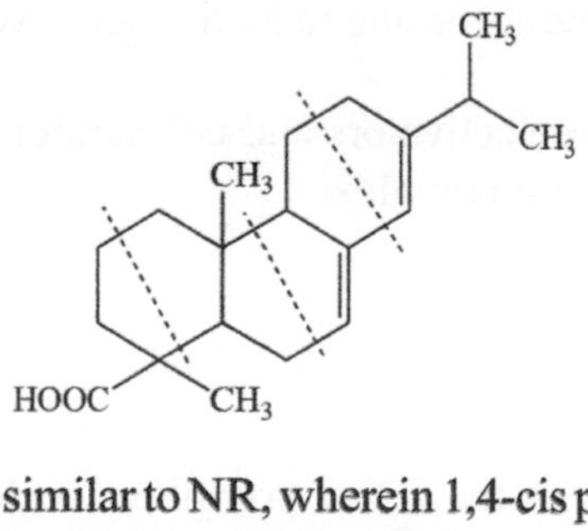

Thus, it is again similar to NR, wherein 1,4-cis polyisoprene dominates.
4. The specified chemical structure of zinc resinate hints at a considerable mobility of the zinc cation, which is of importance regarding its use as an activator of the curing process.
5. The percentage of zinc in ZnO is about 80%, whereas in the resinate, it is around 7.5%, that is, it is 10 times lower.[5]

As seen, the zinc content in zinc resinate is much lower than that in zinc oxide. On the other hand, a good technological compatibility with the NR based on their similar chemical nature as well as good dispersion in the elastomer matrix could be expected.

11.2   EFFECT OF ZINC OXIDE ON THE ACCELERATED SULFUR VULCANIZATION OF RUBBER

Activators of vulcanization are preferably metal oxides in the presence of accelerators, which exhibit maximum their activity during the vulcanization process. Their role is primarily associated with an increase in the density of the vulcanization network. These are mostly zinc, magnesium, cadmium, bismuth, mercury, and tin oxide. In practice, however, zinc oxide and less magnesium are used mainly.[6,7]

In an earlier study, the effect of zinc oxide has been associated with its ability to form complexes soluble in the rubber with the accelerators, stearic acid, and sulfur.[8,9] It has been supposed that these complexes dissociate with release of sulfur in an active form, capable of creating cross-links between the molecules of the rubber.

In other earlier studies, the activating effect of ZnO was associated with the formation of salt-like products from the reaction of zinc oxide and accelerators, appearing also as active accelerators such as carbamates.[10–16]

The theories of Kresja and Koening[17] and Nieuwenhuizen[18] can be mentioned as more recent. The main stages of the vulcanization process in the presence of zinc oxide according to Reference[19] would appear as follows:

- interaction between activators and accelerators with the formation of an active accelerator complex:

$$Acc. - \underset{\underset{Act.}{\uparrow}}{\overset{\overset{Act.}{\downarrow}}{Zn}} - Acc.$$

- interaction between the active complex accelerator and sulfur, with the result to form a sulfiding complex, which is the actual curing agent:

$$Acc - \underset{\underset{Act.}{\uparrow}}{\overset{\overset{Act..}{\downarrow}}{Zn}} - Acc. + S_8 \rightarrow Acc. - S_x - \underset{\underset{Act.}{\uparrow}}{\overset{\overset{Act.}{\downarrow}}{Zn}} - Acc.$$

- reaction of a sulfiding complex with the molecules of the elastomer by the formation of the so-called "cross-linking precursor"

- reactions for the formation of vulcanization network
- reactions of desulfurization of the cross-links

According to Coran,[20] the accelerator reacts with sulfur to give monomeric polysulfides of the type Acc-Sx-Acc in which Acc is an organic radical derived from the accelerator (e.g., benzothiazyl-). The monomeric polysulfides interact with rubber to form polymeric polysulfides (e.g., rubber-Sx -Acc). During this reaction, 2-mercaptobenzo-thiazole (MBT) is formed, if the accelerator is a benzothiazole derivative and if the elastomer is NR. When MBT itself is the accelerator in NR, it first disappears then transforms with the formation of rubber-Sx-Acc. Finally, the rubber polysulfides react, either directly or through an intermediate, to give cross-links, rubber-Sx-rubber.

There are different views on the role of zinc oxide at the different stages of the vulcanization process. Coran thinks the use of ZnO as an activator accelerates the early reactions of forming cross-linking precursors,[20] whereas in Reference,[21] there is a supposition of a greater activity of ZnO at the subsequent stages of the curing process—the opening of sulfur ring and conversion of sulfur into an active form.

Again, according to Coran,[20] the increase in the concentration of available Zn++ causes an increased overall rate in the early reactions (during the delay period), which leads to the formation of rubber-Sx-Acc. However, it gives rise to a decrease in the rate of cross-link formation, but also to an increase in the extent of cross-linking. The increase in the rates of the early reactions has been explained by the interaction as follows:

in which the chelated form of the accelerator is more reactive than the free accelerator during the early reactions:

Here, I-Sy is an ionized form of linear sulfur.

Zinc chelation changes the position of the S–S bond most likely to break. As a stronger bond must break, the rate is slower.[20] Though the rate of cross-linking is slower, the extent of cross-link formation is increased as less sulfur is used in each cross-link. That is, the cross-links are of lower sulfidic rank.

Despite the proposed mechanisms of activators action, there is no theory integrating the abundance of different concepts. Indisputable, however, remain some basic conclusions.

The action of an activator depends on the following:

- the type of elastomer
- the type of accelerator used in combination with the activator
- the nature of the filler
- the vulcanization conditions

The actuators affect predominantly the density of the curing network, not the rate of vulcanization.

11.3   APPROACHES TO DECREASING/REPLACING ZINC OXIDE AMOUNT IN RUBBER COMPOUNDS AND VULCANIZATES

According to several studies, the amount of zinc oxide can be lowered to 1–2 phr without significant performance deterioration of the produced rubber and rubber articles.[22,23]

Studies on the influence of calcium oxide, metal complexes of thiodiglycol and nanosized zinc oxide as activators of sulfur vulcanization of NR- and styrene–butadiene rubber (SBR)-based composites are reported in Reference.[24] According to the authors, the zinc complexes of thiodiglycol are acceptable substitutes of zinc oxide because they disperse better in the rubber matrix and tend to release higher zinc amounts. Calcium complexes of thiodiglycol prove inadequate, whereas nanosized zinc oxide improves the mechanical properties and resistance to abrasion of the vulcanizates due to its high-specific surface area. Nanosized zinc oxide is also recommended in Reference.[25] Some zinc chelate complexes assessed as suitable have been suggested as substitutes for zinc oxide.[26] The authors claim their usage allows to drop to 40% the amount of zinc ions if compared with the use of zinc oxide.

Investigations have been performed with MgO, CaO, and BaO in SBR-based composites and sulfenamide accelerator.[27] Reichle[28] has suggested an interesting approach to the synthesis of new activators which has obtained hydroxides of the type:

$M_a^{2+}M_b^{3+}(OH)_{2a+2b}(X^-)_{2b} \cdot xH_2O$, where
M^{2+} = Mg, Ni, Co, Zn, Cu, so forth.
M^{3+} = Al, Cr, Fe, Sc

$M^{2+}/M^{3+} \sim 1$–5;
$X^- = H_2O$ and a stable anion base
$X = 0$–6

The method included mixing an aqueous solution of M^{2+} and M^{3+} with an aqueous solution of carbonate hydroxide that yielded an amorphous gel. The latter was then subjected to crystallization in a wide temperature range of 60 to 325°C. The ratio of metal ions, pH of the solutions, and crystallization temperature were altered. The temperature and time were reported to affect the particles morphology and appearance of side phases.

B. Vega has recommended usage of mixed oxides instead of ZnO.[29]

The mixed oxides obtained had the general formula as follows:

$$[Zn_{1-x}Mg_xAl_{0,25}(OH)_2](CO_3)_{0,125} \cdot nH_2O$$

where: $x = 0$; 0,19; 0,38; 0,55 and 0,75

The ratio (Zn+Mg)/Al was not changed and was of constant value 3. The Zn^{2+}/Mg^{2+} ratio changed from 100–0% of Zn. The synthesis was performed using a coprecipitation method that consisted of the addition drop by drop of a solution containing salts of the cations (ZnCl, MgCl, and AlCl) in the desired proportions into a NaCO solution at room temperature. pH was maintained constant at a value of 10 with a pH-burete by adding, if necessary, solution of NaOH.[29]

The synthesis of defined Zn-containing organic compounds that can be potential donors of zinc ions to facilitate an active complex curing is an alternative to the methods described so far. The authors[30] reported studies on zinc-m-glycerol, zinc-2-ethylhexanoate as activators of the vulcanization process, indicating that glycerol is a better substitute of the reference zinc oxide. One of the most thorough studies on lowering the content of zinc oxide in sulfur vulcanization is that done by G. Heideman.[31] It has been found that, depending on the exact requirements for specific compounds, certain metal oxides are an alternative route to reducing zinc amount and therefore, to minimize the environmental impact. Although many alternative metal oxides and zinc complexes have been studied as activators for sulfur vulcanization, at present, no viable alternative has been found to eliminate ZnO completely from rubber compounds, without significant impediment to processing as well as to performance characteristics. Heideman reports on a fundamentally different approach: the application of MFAs. MFAs are amines complexed with fatty acids, developed to function both as activator and accelerator for sulfur vulcanization. Good physical properties can be obtained in s-SBR compounds using the MFA/s cure system. It is shown that amines play a crucial role in the vulcanization process, hence various amine complexes have been synthesized and investigated as zinc-free curatives in s-SBR compounds. It has been observed that the scorch time of the MFAs is related to the basicity of the amines used as components of the MFAs. The conclusion is that the chemistry involved in MFA systems is fundamentally different, indicated by the very fast decomposition of TBBS accelerator, fast formation as well as breakdown of the cross-link precursor and, above all, some major differences in the distribution of the cross-linked products. Overall, the results indicate that, dependent on the intended applications (e.g., tires or roofing foils), there exists a potential to significantly or even completely reduce ZnO via MFA-containing vulcanization systems. In Reference[31] the application of a new activator has also been discussed. A mineral clay, viz. montmorillonite, is used as a carrier material and loaded with Zn^{2+}-ions via an ion exchange process. The application in a wide range of natural and synthetic rubbers has been explored. It is demonstrated that this Zn clay can substitute conventional ZnO, grossly retaining the curing and physical properties of the rubber products, but reducing zinc concentration with a factor 10–20. It is concluded that systems with Zn^{2+}-ions on a support are a novel route to reducing zinc amount, and therefore to minimizing significantly its environmental impact. Details about the effectiveness of those activators have been summarized in.[32]

A detailed and comprehensive study on the effects of zinc stearate as an activator of the accelerated sulfur vulcanization has been made by Djagarova and Tipova.[33] The authors have investigated the effect of zinc stearate in a wide range of concentrations (0.5–10.0 phr) on the vulcanization characteristics and properties of rubber compounds and vulcanizates. Particularly successful and motivated in this study is the selection of accelerators: MBT, tetrametiltiuram disulfide—(TMTD) and N-cyclohexyl-2-benzothiazolyl sulfenamide—(CBS).

It has been found that zinc stearate can be an effective activator of sulfur vulcanization when the accelerator and an equivalent substitute for ZnO have been chosen effectively, while reducing the zinc content in the vulcanizates of 8–20 times.[33]

Other authors have developed new compositional activators for the vulcanization of rubber.[34] They called the resulting product "VULKATIV—FC" containing 30–35% of saturated fatty acids and 19% of zinc oxide, respectively.

Brazilian scientists reported even a new class of zinc complexes of coordination type with participation of sulfur atoms for activating and accelerating the vulcanization of NR as a result of systematic and thorough studies.[35,36]

Taiwanese researchers have prepared amorphous nanoparticles of Zn-complexes with sulfur by the hydrothermal method and studied their effect on vulcanization process.[37]

11.4 EXPERIMENT

11.4.1 RUBBER AND INGREDIENTS

11.4. 1.1   NATURAL RUBBER

The NR used were SMR10 and SVR10.

11.4.1.2   ZINC RESINATE

Zinc resinate $(Zn(R)_2)$ was purchased from Chemos GmbH, Germany.

The analysis we performed showed zinc content in the commercial product to be 6.68%, what is close to the theoretic one.

11.4.1.3   PLASTIKOL

Plastikol is a commercial product comprising zinc salts of fatty acids. It is produced by the Bulgarian company Industrial Chemistry, Ltd. Plastikol comprises mainly unsaturated fatty acids—about 50% of olein and 15% of linoleic acid. It also comprises saturated fatty acids—30% of stearic acid and 5% of palmitic acid. Zinc content is 8%, whereas the free zinc oxide does not exceed 1.5%.

All other ingredients used—MBT, TMTD and CBS, *N-tert*-butyl-2-benzothiazyl sulfenamide (TBBS), zinc oxide were produced by Lanxess.

11.4.1.4   VULCANIZATION AGENT

The used technical sulfur was of density 2.07 g/cm^3 and m.p., 112.8°C, could be assigned to the so-called "λ-form."

11.4.1.5   CARBON BLACK

The carbon black used as filler was N330 made in Russia. Its characteristics were: particles size 38–42 nm, specific surface area 75–82 m²/g; specific adsorption surface 75–85 m²/g, oil number 95–105 ml/100 g, density 1.86 g/cm³, aqueous suspension pH = 7–9.

11.4.1.6   SILICA

The used Ultrasil 7000 GR produced by Evonik Industries had the following characteristics: density 2.0 g/cm³; content of volatile compounds (moisture) 0.2%; aqueous dispersion pH = 6.5, specific surface area (S_{BET}) 175 m²/g and electroconductivity about 1,300 μS/cm.

11.4.2   RUBBER COMPOSITES AND VULCANIZATES

The rubber compounds were prepared on an open two-roll laboratory mill (L/D 320 × 160 and friction 1.27). The speed of the slow roll was 25 min⁻¹. The ready compounds in the form of thicker sheets stayed for 24 h prior to their analyses, tests and vulcanization. The vulcanization was carried out on a hydraulic press at electric heating of 160°C and 10.0 MPa.

11.4.3   CHARACTERIZATION OF THE RUBBER COMPOSITES AND VULCANIZATES BASED ON THEM

11.4.3.1   STANDARDIZED METHODS

Mooney viscosity of the rubber composites was determined according to: ISO 289-1:2002.

The vulcanization characteristics were determined by the isotherms recorded on a moving die rheometer MDR 2000, Alpha Technology at 160°C, according to ISO 3417:2002.

The following vulcanization characteristics were determined:

M_L—minimum torque moment, which is the measure of the effective viscosity of the rubber composites studied; M_H—maximum torque moment as a measure of vulcanizates hardness; the difference $\Delta M = M_H - M_L$—a parameter which is a measure for the curing network density; time to scorch—T_{s2} and optimal curing time—T_{90}.

The vulcanizates obtained were characterized by the following parameters:

1. Stresses at 100% and 300% elongation (M_{100} and M_{300})—ISO 37:2002
2. Tensile strength (σ)—ISO 37:2002
3. Elongation at break (ε_{rel})—ISO 37:2002
4. Residual elongation (ε_{res})—ISO 2285:1996
5. Shore A Hardness (Sh A)—ISO 7619:2001
6. Resistance to heat aging (72 h at 70°C)—ISO 188:2002
7. Wear resistance—ISO 4649:1998.

11.4.3.2 NONSTANDARDIZED ANALYTICAL METHODS

The molecular weight of the segment between two cross-links (M_c) was calculated by Eq. 11.1:

$$M_c = \frac{-\rho_r V_s \left(V_r\right)^{1/3}}{\ln\left(1 - V_r\right) + V_r + \chi V_r^2} \tag{11.1}$$

The average number of cross-linked units per molecule was determined by the formula:

$$n_c = \frac{M}{M_C}, \tag{11.2}$$

where M is molecular mass of the rubber; M_c is molecular weight of the segment between two cross-links.

11.4.3.3 DETERMINING THE DYNAMIC PROPERTIES OF THE VULCANIZATES STUDIED

The complex dynamic modulus, the equilibrium heat build-up and the residual deformation of the vulcanizates studied was determined on a Goodrich flexometer at a deformation rate of 850 min⁻¹. The temperature dependence of tan δ of the studied vulcanizates was determined by a Dynamic Mechanical Thermal Analyser Mk III system—Rheometric Scientific. The conditions of the test were: one point bending; deformation frequency 5 Hz; amount of deformation—64 μm.

11.5 RESULTS AND DISCUSSION

11.5.1 UNFILLED ELASTOMER COMPOSITES WITH MBT ACCELERATOR, COMPRISING ZINC RESINATE

Table 11.1 presents the formulations of the studied composites.

TABLE 11.1 Formulations of the Composites with Mercaptobenzothiazole (MBT) Accelerator, Comprising Zinc Resinate.

	M-0	M-1	M-2	M-3
NR(SMR10)	100.0	100.0	100.0	100.0
Zinc oxide	5.0	–	–	–
$Zn(R)_2$	–	5.0	10.0	15.0
Stearic acid	1.5	1.5	1.5	1.5
MBT	1.0	1.0	1.0	1.0
Sulfur	2.0	2.0	2.0	2.0

(Reprinted from Todorova, N., Radulov, I., and Dishovsky, N. Evaluating the effect of zinc resinate upon the properties of natural rubber vulcanizates. Journal of Chemical Technology and Metallurgy, 47, 505-512, 2012.)

11.5.1.1 VULCANIZATION CHARACTERISTICS OF THE STUDIED RUBBER COMPOSITES

The vulcanization isotherms of the studied elastomer composites are presented in Figure 11.1, and their vulcanization characteristics in Table 11.2.

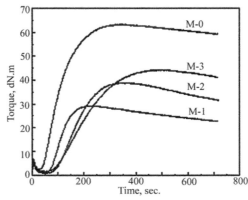

FIGURE 11.1 Vulcanization isotherms of rubber composites with Mercaptobenzothiazole accelerator. (Reprinted from Todorova, N., Radulov, I., and Dishovsky, N. Evaluating the effect of zinc resinate upon the properties of natural rubber vulcanizates. Journal of Chemical Technology and Metallurgy, 47, 505-512, 2012.)

The figure shows a considerable difference in the scorch time and cure time and vulcanization rate of the vulcanizates with and without zinc oxide when using an accelerator of a thiazole type.

TABLE 11.2 Vulcanization Characteristics of Composites with MBT Accelerator.

	M-0	M-1	M-2	M-3
M_L, dN.m	0.23	0.08	0.08	0.13
M_H, dN.m	6.33	2.92	3.90	4.43
$\Delta M = (M_H - M_L)$	7.49	3.0	3.82	4.3
T_{s2}, min.s	1:10	1:34	2:50	3:10
T_{90}, min.s	3:21	2:42	4:17	5:41

(Reprinted from Todorova, N., Radulov, I., and Dishovsky, N. Evaluating the effect of zinc resinate upon the properties of natural rubber vulcanizates. Journal of Chemical Technology and Metallurgy, 47, 505-512, 2012.)

The minimal torque M_L of the composites with MBT accelerator for the samples filled with Zn(R)$_2$ at 5.0 and 10.0 phr is much lower than the control ones with ZnO. That means their plasticity is higher.

It can be assumed that in both cases, Zn(R)$_2$ in larger quantities acts as a softener, that is, reduces the viscosity of the uncured rubber compound. That can be explained by the fact that, the organic portion of Zn(R)$_2$ molecule which is very bulky, sets apart the rubber macromolecules, and thus reduces the intermolecular forces between them. A similar effect is observed in the case of zinc stearate.[13] On the other hand, Zn salts of both aliphatic and cyclic organic acids exhibit properties of surfactants, what also influences the solubility of the formed sulfidizing complex between the accelerator, activator and sulfur in the elastomer matrix and the vulcanization mechanism, respectively.

The maximum torque M_H, which is a measure of the hardness of the vulcanizate is significantly smaller, thereby with increasing the amount of Zn(R)$_2$ in the rubber mixtures M_H increases to values close to the reference mixture of ZnO. The course of vulcanization isotherms (Figure 11.1) shows that reversal of vulcanization is slight and similar to that of the control sample with ZnO.

Time to scorch T_{s2}, is prolonged from 1:34–3:10 min.s with the increasing Zn(R)$_2$ amount, whereas for the control sample it is 1:10 min.s.

For the composites with MBT accelerator, with the increasing Zn(R)$_2$ amount to 5.0, 10.0, and 15.0 phr T_{90} is prolonged from 2:42–4:17 and 5:41 min.s., whereas for the control sample with zinc oxide it is 3:21 min.s.

11.5.1.2   PARAMETERS OF THE CROSS-LINK NETWORK

The obtained parameters—molecular weight of the segment between two cross-links (M_c) and the number of the cross-links (n) give additional information about the density of vulcanizates cure network (Table 11.3).

TABLE 11.3   Parameters of the Cure Network of Vulcanizates with MBT Comprising Zinc Resinate.

	M-0	M-1	M-2	M-3
M_c	6500	13,300	10,400	8500
n_c	60	30	40	45

As seen from Table 11.3, at the minimal $Zn(R)_2$ content in the rubber composite—5 phr—M_c has the highest value which decreases with the increasing $Zn(R)_2$ content, the number of cross-links increases, respectively, but in none of the cases the values of the control sample are reached.

If the data from the two tables are summarized, we find that the values of the molecular weight of the segment between two cross-links M_c for the composites with $Zn(R)_2$ activator in the best case are about 1.3 times higher than those of the control composite with ZnO. Meanwhile, the number of cross-links in the studied compounds—n_c accounts for about 75% of the number of cross-links in the composites with ZnO. The main conclusion is that the use of zinc resinate leads to the formation of a vulcanization system (what is confirmed by the measured M_H values) looser than that of the composite with zinc oxide, that is, the effectiveness of resinate as an activator of the vulcanization process is lower. Though, such efficiency does exist judging from our experiments with composites, containing no zinc oxide or zinc resinate whose $M_c = 27,000$ with MBT accelerator, whereas for the composites comprising zinc resinate and zinc oxide at 5 phr the corresponding M_c values are 13,300 and 6500, respectively. With the increasing amount of zinc resinate M_c values decrease significantly.

11.5.1.3   MECHANICAL PARAMETERS AND RESISTANCE TO HEAT AGING

The mechanical properties of the vulcanizates with MBT comprising zinc resinate are presented in Table 11.4.

TABLE 11.4 Mechanical Parameters of Vulcanizates with MBT Accelerator Comprising Zinc Resinate Before Heat Aging.

	M-0	M-1	M-2	M-3
M_{100} (MPa)	1.0	0.6	0.7	0.8
M_{300} (MPa)	2.3	1.2	1.6	1.8
σ (MPa)	14.7	9.7	11.0	14.1
ε_{rel} (%)	640	640	690	650
ε_{res} (%)	15	10	15	10
Shore A hardness	41	39	40	43

(Reprinted from Todorova, N., Radulov, I., and Dishovsky, N. Evaluating the effect of zinc resinate upon the properties of natural rubber vulcanizates. Journal of Chemical Technology and Metallurgy, 47, 505-512, 2012, Permission from University of Chemical Technology and Metallurgy.)

The mechanical properties of some of the vulcanizates with MBT accelerator comprising $Zn(R)_2$ are expectedly poorer than those of vulcanizates with ZnO. That is particularly valid for the stress at 100 and 300% elongation. The tensile strength of the control composite is 14.7 MPa, whereas of the composites with $Zn(R)_2$ at 5.0, 10.0, and 15.0 phr the values are 9.7, 11.0, and 14.1 MPa, respectively. Hence, the vulcanizates with $Zn(R)_2$ at 15 phr have almost the same tensile strength and elongation at break as the vulcanizates with ZnO.

The residual elongation and Shore A hardness do not change significantly.

The data of the accelerated heat aging of the vulcanizates for 72 h at 70°C are presented in Table 11.5.

TABLE 11.5 Aging Coefficients of the Vulcanizates with MBT Accelerator Comprising Zinc Resinate After the Heat Aging.

	M-0	M-1	M-2	M-3
K_σ (%)	−7	−9	−15	−17
K_ε (%)	−12	−17	−19	−17

In the results presented it is worth noting that at higher resinate amounts the vulcanizates age more what is probably related to the prevailing polysulfide bonds formed in its presence.

11.5.2 UNFILLED ELASTOMER COMPOSITES WITH TMTD ACCELERATOR COMPRISING ZINC RESINATE

The formulations of the elastomer composites are presented in Table 11.6.

TABLE 11.6 Formulations of Composites with Tetrametiltiuram Disulfide (TMTD) Accelerator Comprising Zinc Resinate.

	T-0	T-1	T-2	T-3
NR (SMR10)	100.0	100.0	100.0	100.0
Zinc oxide	5.0	–	–	–
Zn(R)$_2$	–	5.0	10.0	15.0
Stearic acid	1.5	1.5	1.5	1.5
TMTD	1.0	1.0	1.0	1.0
Sulfur	2.0	2.0	2.0	2.0

(Reprinted from Todorova, N., Radulov, I., and Dishovsky, N. Evaluating the effect of zinc resinate upon the properties of natural rubber vulcanizates. Journal of Chemical Technology and Metallurgy, 47, 505-512, 2012.)

Vulcanization characteristics of the studied rubber composites.

The vulcanization isotherms of the studied elastomer composites are presented in Figure 11.2, and their vulcanization characteristics in Table 11.7.

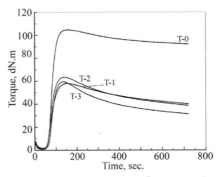

FIGURE 11.2 Vulcanization isotherms of rubber composites with tetrametiltiuram disulfide accelerator. (Reprinted from Todorova, N., Radulov, I., and Dishovsky, N. Evaluating the effect of zinc resinate upon the properties of natural rubber vulcanizates. Journal of Chemical Technology and Metallurgy, 47, 505-512, 2012.)

As seen from the figure, when using TMTD accelerators no significant differences in the scorch time and cure times of the control composite with ZnO and those with different amounts of Zn(R)$_2$. The same is valid for the optimal cure time (Table 11.7).

TABLE 11.7 Vulcanization Characteristics of Vulcanizates with TMTD Accelerator.

	T-0	T-1	T-2	T-3
M_L (dN.m)	0.17	0.15	0.12	0.07
M_H (dN.m)	10.48	5.95	6.34	5.81
$\Delta M = (M_H - M_L)$	10.31	5.8	6.22	5.74
T_{S2} (min.s)	1.11	1.18	1.12	1.13
T_{90} (min.s)	1.45	1.44	1.43	1.49

The increasing $Zn(R)_2$ amount in the rubber composites with TMTD leads to lower M_L what is related to the improved thermoplastic properties. Both M_H and ΔM of the rubber compounds comprising zinc resinate are much lower what prompts that the density of cross-link network is also lower.

11.5.2.1 CURING NETWORK PARAMETERS

The obtained parameters—molecular weight of the segment between two cross-links of the macromolecule—M_c and the number of intermolecular cross-links—n_c reveal additionally the density of curing network of the vulcanizates. Table 11.8 presents the values of those parameters of the type of vulcanizates with TMTD.

TABLE 11.8 Parameters of the Curing Network of Vulcanizates with TMTD Accelerator, Comprising Zinc Resinate.

	T-0	T-1	T-2	T-3
M_c	4,500	10,400	5,700	5,600
n_c	90	40	70	72

The analysis of the data from Table 11.8 for composites with TMTD accelerator reveals that the composites with $Zn(R)_2$ 10.0 phr and 15.0 phr have M_c =5700 and MC = 5600, respectively. Those values are very close to those of the control sample with zinc oxide. The number of cross-links in the aforesaid composites is 70 and 72, respectively, whereas in the composite with ZnO—90.

11.5.2.2 MECHANICAL PARAMETERS AND HEAT AGING RESISTANCE

The parameters of vulcanizates with zinc resinate (Table 11.9) are inferior to those of the control composite with ZnO at 5 phr, but increasing the amount of zinc resinate to 15 phr this difference decreases markedly.

The elongation at break of the vulcanizates with $Zn(R)_2$ is much higher than that of the control composite. The hardness is about 7–10 units lower, respectively.

TABLE 11.9 Physical and Mechanical Performance Before Aging of Vulcanizates with Accelerator, Containing Zinc Resinate.

	T-0	T-1	T-2	T-3
M_{100} (MPa)	1.5	0.8	1.0	1.1
M_{300} (MPa)	6.1	1.6	2.5	3.0
σ (MPa)	27.6	20.4	23.4	25.0
ε_{rel} (%)	580	700	660	630
ε_{res} (%)	15	15	15	15
Shore A hardness	51	41	43	44

The data of accelerated heat aging test (72 h at 70°C) of the vulcanizates with TMTD comprising zinc resinate, presented in Table 11.10 also deserve interest.

TABLE 11.10 Aging Coefficients of Vulcanizates with TMTD Accelerator.

	T-0	T-1	T-2	T-3
K_{σ}	−15	−12	−10	−15
K_{ε}	−25	−30	−32	−35

(Reprinted from Todorova, N., Radulov, I., and Dishovsky, N. Evaluating the effect of zinc resinate upon the properties of natural rubber vulcanizates. Journal of Chemical Technology and Metallurgy, 47, 505-512, 2012.)

The values of the coefficients of aging are negative for vulcanizates with thiuram, an indication of the reduced monitored parameters after aging. As seen, the vulcanizates with zinc resinate age somewhat more in terms of elongation, but less in terms of tensile strength.

The results of the study of the activating effect of zinc resinate in the presence of accelerators from the group of thiazoles and thiurams are described in detail in Reference.[38]

In general, it can be said that, with TMTD accelerator in the presence of 15 phr zinc resinate in the rubber composites instead of 5 phr zinc oxide yields vulcanizates with a very similar mechanical performance.

11.5.3 UNFILLED COMPOSITES WITH CBS ACCELERATOR, COMPRISING ZINC RESINATE

Table 11.11 presents the formulations of the elastomer compounds.

TABLE 11.11 Formulations of the Elastomer Compounds with *N*-cyclohexyl-2-benzothiazolyl Sulfonamide (CBS) Accelerator.

	C-0	C-1	C-2	C-3
NR (SMR10)	100.0	100.0	100.0	100.0
Zinc oxide	5.0	–	–	–
Zn(R)$_2$	–	5.0	10.0	15.0
Stearic acid	1.5	1.5	1.5	1.5
CBS	1.0	1.0	1.0	1.0
Sulfur	2.0	2.0	2.0	2.0

11.5.3.1 VULCANIZATION PROPERTIES OF THE RUBBER COMPOUNDS STUDIED

The vulcanization characteristics of the studied compounds are summarized in Table 11.12. Noteworthy is the fact that, the compound with 5.0 phr zinc resinate has T_{90} 3:33 min.s whereas the reference compound with of ZnO has T_{90} 3:48 min.s. The difference in the scorch time T_{s2} of the two compounds is minimal.

TABLE 11.12 Vulcanization Characteristics of Vulcanizates with CBS Accelerator Comprising a Different Content of Zinc Resinate.

	C-0	C-1	C-2	C-3
M_L, dN.m	0.11	0.21	0.12	0.11
M_H, dN.m	9.0	4.76	4.85	4.70
$\Delta M = (M_H - M_L)$	8.89	4.55	4.73	4.59
T_{s2} (min.s)	2.07	2.44	3.07	3.26
T_{90} (min.s)	3.48	3.33	4.40	5.32

The nature of sulfenamide accelerator CBS undoubtedly affects the vulcanization process, which is illustrated by the vulcanization isotherms in Figure 11.3—a delay in the curing process and a gradual prolongation of the optimal cure time T90 with increasing content of Zn(R)$_2$ in the elastomeric compounds—has been expected. After optimum cure, however, there is a strong reversion, which is probably related to the nature of the polysulfide nature of the formed vulcanization network.

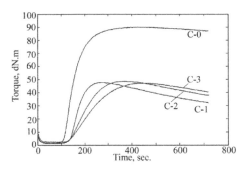

FIGURE 11.3 Vulcanization isotherms of rubber compounds with N-cyclohexyl-2-benzothiazolyl sulfonamide accelerator.

The difference $\Delta M = M_H - M_L$, which is a measure of the hardness of the vulcanizates, density of the vulcanization network, respectively (Figure 11.4), is a very important parameter for our investigations.

The dependence of ΔM on the amount of $Zn(R)_2$ for the rubber compounds when using various accelerators MBT, TMTD, and CBS is plotted in Figure 11.4. As seen, the curves representing the dependence above have a different course and a different location in the coordinate system. That allows the assumption that, under the same other conditions the parameters of the vulcanization network when using the three types of accelerators differ, that is, they have a different impact on the effectiveness of $Zn(R)_2$. As seen, the accelerator from the group of thiazoles (MBT) has the strongest influence on the amount of zinc resinate, whereas the accelerator from the thiuram type (TMTD)—the least. Thiuram has the highest ΔM values, whereas those for thiazoles are the lowest. The latter fact is in full agreement with our concept for the selection of accelerators of different chemical nature, chosen to clarify the presumed of activating mechanism of zinc resinate. Probably various coordination complexes with a sulfided Zn ion are formed, which have different solubilizing action in the elastomeric matrix respectively.

It is also evident that the structure of zinc resinate determines some of its specific characteristics, due to the structure of abietic acid. A number of zinc salts of carboxylic acids may have a surfactant action. They may form layers between the elastomeric chains,[3] thereby lowering the viscosity of the compounds and improve their flowability. In the cases studied that is revealed by the reduction of the minimum torque M_L, which is related to the effective viscosity, and to a decrease of Mooney viscosity which has also been observed, that is, to determining the softener and/or plasticizing activity of the product.

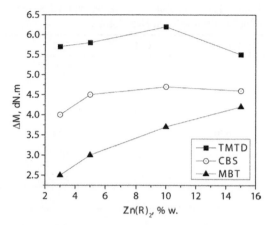

FIGURE 11.4   Dependence of ΔM on the amount of $Zn(R)_2$.

11.5.3.2   DETERMINATION OF THE MECHANICAL PARAMETERS AND HEAT AGING RESISTANCE

The mechanical parameters of the vulcanizates correspond to the data about their vulcanization characteristics, namely that a less dense vulcanization network is formed. The tensile strength of the vulcanizates comprising different amounts of zinc resinate is lower than that of the reference sample with ZnO, whereas the elongation at break is higher. Shore A hardness is reduced by about 10 units (Table 11.13). The results obtained so far evidence that direct replacement of equivalent amounts of zinc oxide with zinc resinate worsens the mechanical performance due to the less dense vulcanization network. It is obvious that the increasing the amount of $Zn(R)_2$ till 10–15 phr leads to almost equal mechanical characteristics with the control sample. According to the current understanding of the mechanism of accelerated sulfur vulcanization,[3] the role of zinc in the process as a transition metal is reduced to form a relatively strong coordination bonds and as a result, stable complexes (typical 4, 5, or 6 coordination complexes are formed due to the electronic structure of zinc). During vulcanization in the presence of thiazolidinedione and sulfenamides such complexes are even isolated.[3] It is assumed that zinc ions cause activation of these complexes by expansion and contraction of their electron shells, and thus they become the driving force of the vulcanization process as the activity of the complexes controls the formation of sulfur bonds during vulcanization.

TABLE 11.13   Mechanical Parameters of Vulcanizates with CBS Accelerator, Comprising Different Amounts of $Zn(R)_2$, Prior to Their Heat Aging.

	C-0	C-1	C-2	C-3
M_{100} (MPa)	1.2	0.8	0.9	1.0
M_{300} (MPa)	2.8	1.5	1.7	1.8
σ (MPa)	20.3	17.8	18.6	19.3
ε_{rel} (%)	630	710	690	670
ε_{res} (%)	15	10	10	10
Shore A hardness	49	38	39	37

As the said above shows, at a drastic decrease of zinc amount in the two activators (zinc resinate—7.5%, and zinc oxide—80%) a deficit of zinc ions occurs, of coordination complexes, respectively and smaller number of sulfur bonds as a final result. That has been observed in all cases studied. The higher resinate amount leads to an increase in zinc ions concentration, in the coordination complexes, respectively. That leads to a denser vulcanization network, higher vulcanization degree and better mechanical properties. Obviously, there should be another approach to substitution of zinc oxide. Namely, the substitution of zinc oxide should be partial and particularly by combinations of zinc oxide with zinc resinate or with other zinc salts of different structure.

The resistance to heat aging of the studied vulcanizates is presented in Table 11.14.

TABLE 11.14   The Aging Coefficient of Vulcanizates with CBS Sccelerator.

	C-0	C-1	C-2	C-3
K_σ (%)	−75	−5	−4	−21
K_ε (%)	−54	−15	−11	−23
K_{sh} (%)	+2	+7	+5	+5

A noteworthy fact from the table is that the resistance to heat aging of the vulcanizates comprising sulfenamide accelerator is quite higher than that of the control sample comprising zinc oxide, which is a very important result.

11.5.4 CARBON BLACK FILLED ELASTOMER COMPOSITES WITH TBBS ACCELERATOR COMPRISING ZINC RESINATE

To give a more applied orientation of our research, we compared the effect of zinc oxide, zinc stearate, and zinc resinate on the behavior of the composites and of the vulcanizates based on them when carbon black was used as filler.

The formulations of the studied composites are presented in Table 11.15.

TABLE 11.15 Formulations of the Carbon Black Filled Composites with *N-tert*-butyl-2-benzothiazyl Sulfenamide (TBBS) Accelerator Comprising Zinc Resinate.

	ZnO	Zn(St)$_2$	Zn(R)$_2$
NR (SMR10)	100.0	100.0	100.0
Zinc oxide	5.0	–	–
Zn(St)$_2$	–	5.0	–
Zn(R)$_2$	–	–	5.0
Stearic acid	2.0	2.0	2.0
TBBS	3.0	3.0	3.0
Sulfur	1.0	1.0	1.0
N330	50.0	50.0	50.0

11.5.4.1 VULCANIZATION PROPERTIES

The vulcanization characteristics of NR-based vulcanizates with sulfenamide accelerator TBBS and three different activators (ZnO-compound was denoted by O, Zn—stearate, the compound was denoted by S, Zn-resinate compound was denoted by R) in amounts of 5 phr are presented in Table 11.16. Data show that M_L increases gradually in the direction from ZnO to zinc resinate and M_H decreases in the same direction.

TABLE 11.16 Vulcanization Characteristics of Carbon Filled Vulcanizates with TBBS Accelerator.

	O	S	R
M_L (dN.m)	1.58	1.62	1.75
M_H, (dN.m)	29.40	19.97	18.65
$\Delta M = (M_H - M_L)$	27.82	18.45	16.90
T_{S2} (min.s)	2.29	2.31	2.31
T_{90} (min.s)	6.37	3.45	3.22

The ratio $\Delta M = (M_H - M_L)$, which is a measure for vulcanizates hardness, the density of the vulcanization network, respectively, also decreases in the case of stearate and resinate, having the highest values in the case of reference sample with ZnO—27.82 dN.m. Hence, zinc resinate as an activator produces the less dense vulcanization network. The closeness of the values in the cases of zinc stearate and zinc resinate is obvious.

The optimal cure time is shorter, whereas that for zinc stearate and resinate is very close. That means, at equal amounts of zinc oxide, zinc stearate and zinc resinate—5 phr and the same accelerator TBBS, the latter two composite are cured in a shorter time, which is likely due to the formation of a less dense vulcanization structure and less possibilities for the formation of new bonds after a specified period of time. Another possible explanation is that, according to the theory of Coran,[20] a chelate complex is formed between zinc oxide and thiazole ring of the accelerator, and the sulfur molecule attached to it. That retards the vulcanization rate but a vulcanization network is formed with less polysulfidity of the bonds and more thermostable vulcanization network, respectively. In the absence of zinc oxide stearate and resinate such a complex is not formed and vulcanization takes place quickly. The scorch time T_{s2} is almost the same.

11.5.4.2   MECHANICAL PARAMETERS AND HEAT AGING RESISTANCE

The mechanical parameters of the vulcanizates studied are presented in Table 11.17.

TABLE 11.17   Mechanical Parameters of the Vulcanizates Comprising Different Activators with TBBS Prior to Heat Aging.

	O	S	R
M_{100} (MPa)	7.8	4.6	4.1
M_{300} (MPa)	28.0	19.8	18.8
σ (MPa)	28.0	27.1	26.7
ε_{rel} (%)	300	410	450
ε_{res} (%)	10	15	15
Shore A hardness	64	60	57

The mechanical parameters of the vulcanizates show that at a same content of the different activators the modules are visibly different. The highest values are in the case of zinc oxide. Concerning the tensile strength the difference between the vulcanizates is minimal (only 1.3 MPa) in favor of those with zinc oxide. Vulcanizates with zinc resinate have the highest elongation values. The differences in the performance of zinc stearate and zinc resinate are minimal.

The data from the accelerated heat aging of the vulcanizates for 72 h at 70°C are presented in Table 11.18.

TABLE 11.18   Aging Coefficients of Vulcanizates with TBBS Accelerator.

	O	S	R
K_a (%)	−25	−33	−29
K_ε (%)	−13	−12	−16
K_{sh} (%)	3	5	0

As seen, the differences in the aging coefficients values of the different of vulcanizates are minimal. In general, in the case of carbon black filled vulcanizates with TBBS accelerator, the differences in the properties of the vulcanizates containing zinc oxide and zinc resinate are not so great regarding the unfilled ones. The values for zinc stearate and zinc resinate are too close. Given, however, that the zinc content in the zinc resinate is 10 times lower than in the zinc oxide, it can be considered that in many cases, the former may be used as a substitute of the latter.

11.5.5   SILICA-FILLED ELASTOMER COMPOSITES WITH TBBS ACCELERATOR COMPRISING ZINC RESINATE

The formulations of the composites are presented in Table 11.19.

Two series of four compounds were prepared. The first four blends (Ctrl, St, Rs, and Pl) do not contain zinc oxide and stearic acid. Control compound (Ctrl) contains an activator of sulfur vulcanization and the other three contain zinc stearate (St), zinc resinate (Rs) or Plastikol (Pl) at 5 phr. In the second series, the control compound (Ctrl + ZnO) contains zinc oxide at 4 phr and stearic acid at 1 phr, the other three compounds (St + ZnO, Rs + ZnO, Pl + ZnO) contain the same amounts of zinc salts of saturated fatty acids, as well as zinc oxide at 4 phr and stearic acid at 1 phr.

Viewing the possible practical applicability of these combinations of rubber filler and accelerator in the production of some tire treads special attention has been paid to them and much serious research has been carried out.

TABLE 11.19 Formulation of Silica-Filled Elastomer Composites with TBBS Accelerator Comprising Zinc Resinate.

	Ctrl	St	Rs	Pl	Ctrl+ ZnO	St+ ZnO	Rs+ ZnO	Pl+ ZnO
NR-SVR 10	100.0	100.0	100.0	100.0	100.0	100.0	100.0	100.0
SiO_2, Ultrasil 7000 GR	60.0	60.0	60.0	60.0	60.0	60.0	60.0	60.0
Zinc oxide	–	–	–	–	4.0	4.0	4.0	4.0
Stearic acid	–	–	–	–	1.0	1.0	1.0	1.0
Zinc stearate	–	5.0	–	–	–	5.0	–	–
Zinc risinate	–	–	5.0	–	–	–	5.0	–
Plastikol	–	–	–	5.0	–	–	–	5.0
TESPT	5.0	5.0	5.0	5.0	5.0	5.0	5.0	5.0
TBBS	1.5	1.5	1.5	1.5	1.5	1.5	1.5	1.5
Sulfur	2.0	2.0	2.0	2.0	2.0	2.0	2.0	2.0

11.5.5.1 VULCANIZATION CHARACTERISTICS AND MOONEY VISCOSITY OF THE STUDIED RUBBER COMPOSITES

Figure 11.5 presents the Mooney viscosity of the rubber composites studied. As seen from the figure, the composites containing zinc salts have a Mooney viscosity of about 50 MU, whereas the presence of zinc oxide and stearic acid has no effect on this parameter. In the absence of zinc oxide, zinc salts tend to reduce the viscosity to about 15 MU and somewhat better plasticizing effect of zinc resinate in comparison with zinc stearate and plastikol could be observed. The latter fact confirms the prerequisites motivated earlier for the role of zinc resinate structure, in particular its relation to the elastomeric matrix, when considering its functions in the rubber composite. In the presence of zinc oxide the plasticizing effect of zinc salts is slightly more pronounced (Mooney viscosity decreases by about 20 units). In the case, there is a significant influence of the type of zinc salt on the viscosity of the tested rubber composites.

The vulcanization curves of the studied compositions without zinc oxide and stearic acid are presented in Figure 11.6. The curves were taken on a Monsanto rheometer at 160°C.

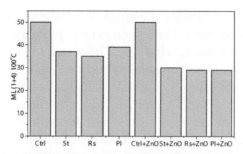

FIGURE 11.5 Mooney viscosities of the studied rubber composites dependent on the presence of different zinc salts.

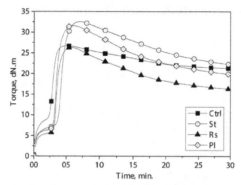

FIGURE 11.6 Vulcanization characteristics of rubber composites comprising different zinc salts and no zinc oxide.

The figure shows that for all composites investigated curing starts very quickly, so there was no thermoplasticity. The scorch time T_{s2} is of the order of 30 s, which is very short for compositions accelerated by a sulfenamide accelerator. A strong reversal in the course of the curves after optimum cure time has been observed. The most pronounced reversion is for the control composite free of zinc salt (Ctlr). Figure 11.7 shows the vulcanization curves for composites containing various zinc salts of organic acids in the presence of zinc oxide.

As Figure 11.7 shows, the control composite (Ctrl + ZnO), wherein only zinc oxide and stearic acid are present, the vulcanization also starts very quickly, that is, without thermoplasticity. The maximum torque of this mixture is the highest (35 dN.m) and a reversion of the curve after reaching the maximum torque is observed. In this case, the reversion is significantly less pronounced than that of the composite containing zinc oxide and stearic acid or containing only zinc salts (Figure 11.6). A radical change in the course of the curves is observed when a combination of zinc oxide and zinc salts is present in the compound. The scorch time is prolonged to 6 min., which is normal for sulfenamide accelerators. The maximum torque is lower than that of the control composite containing no zinc salts (Ctrl + ZnO), but there is no reversion in the course of the curve, that is, we have a broad plateau of vulcanization. The reason is probably in the formation of a coordination complex between zinc oxide, zinc salts, and the sulfur accelerator, resulting in prolonged scorch time and to the formation of a more stable vulcanization structure. Perhaps the complex (ZnO + zinc salt) has a thermostabilizing effect preventing the destruction occurring along the main chain of the elastomer at the vulcanization temperature. As seen, the compounds with St and Rs have almost the same effect.

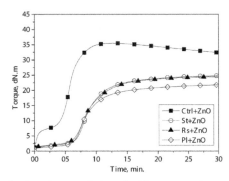

FIGURE 11.7   Vulcanization characteristics of rubber composites comprising different zinc salts in the presence of zinc oxide.

The vulcanization characteristics of rubber composites as determined by the above curves of vulcanization kinetics are presented in Table 11.20.

TABLE 11.20   Vulcanization Characteristics.

	Ctrl	St	Rs	Pl	Ctrl+ ZnO	St+ ZnO	Rs+ ZnO	Pl+ ZnO
M_{min} (dN.m)	3.7	3.3	2.2	3.0	3.1	1.3	1.4	1.3
M_{max} (dN.m)	26.7	32.6	26.3	31.6	38.1	24.7	24.4	23.7
ΔM (dN.m)	23.0	29.3	24.1	28.6	35.0	23.4	23.0	22.4
T_{s2} (min)	0:24	0:42	0:38	0:30	0:23	6:25	6:01	6:27
T_{90} (min)	3:27	5:15	4:16	4:22	8:36	14:28	14:25	15:16
V, (%/min)	33	22	29	29	13	12	13	10
Tan δ at 160°C	0.137	0.078	0.109	0.087	0.086	0.048	0.082	0.067

The table shows that the compositions containing zinc oxide have a higher vulcanization rate even for a very short scorch time (T_{s2}), but generally no significant differences are observed. In the presence of zinc oxide, these differences are greater, especially with regard to the scorch and optimal cure time.

11.5.5.2 VULCANIZATION NETWORK PARAMETERS

Figure 11.8 presents the molecular weight values of the segments between two cross-links (M_c), as determined by the equilibrium degree of swelling at a different cure time. As seen, the vulcanizates formed in the presence of zinc oxide or of a combination of zinc oxide and zinc salts form slightly denser vulcanization network in comparison with that of vulcanizates containing zinc oxide at the optimum cure time of the specimens (T_{90}). At a prolonged vulcanization $(4 \times T_{90})$ the density of the vulcanization network of those vulcanizates increases slightly, whereas for vulcanizates in which zinc oxide is absent a longer curing period loosens the vulcanization network. The results obtained are in accordance with the above vulcanization curves, with a strong reversal in the curves course for composites with no zinc oxide.

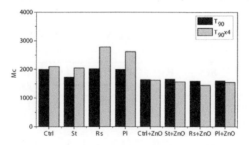

FIGURE 11.8 Molecular weight values of the segments between two cross-links of the studied composites at a different cure time T_{90} and $4 \times T_{90.}$

11.5.5.3 MECHANICAL PROPERTIES OF THE VULCANIZATES AND HEAT AGING RESISTANCE

The mechanical properties of the rubber composites subjected to vulcanization for the optimum cure time (T_{90}) are presented in Table 11.21. As seen from the table, the presence of zinc oxide leads to a slight tension increase of the stress at 300% elongation and to a decrease of the residual elongation ε_{res}. It should be noted that the specimens with zinc resinate, either on its own or in a combination with ZnO, gave tensile stress results better than the control one.

TABLE 11.21 Mechanical Properties of the Rubber Composites Subjected to Vulcanization for the Optimum Cure Time (T_{90}).

	Ctrl	St	Rs	Pl	Ctrl+ ZnO	St+ ZnO	Rs+ ZnO	Pl+ ZnO
M_{100} (MPa)	1.9	1.9	1.7	2.0	2.2	2.2	1.9	2.3
M_{300} (MPa)	7.6	7.2	6.5	8.9	9.0	9.1	8.0	9.1
σ (MPa)	19.1	19.2	20.1	22.8	20.8	24.3	22.8	22.6
ε_{rel} (%)	530	570	610	610	560	670	660	580
ε_{res} (%)	25	50	35	45	30	20	25	20
Shore A hardness	63	67	61	65	68	67	65	65

Table 11.22 shows the mechanical parameters of the vulcanizates obtained during a $4 \times T_{90}$ vulcanization. The study was done viewing the usage of these compositions in the production of massive rubber products. Again, we should mention the results for Rs + ZnO sample in terms of tensile elongation which are better than those for Ctrl samples of both variants.

TABLE 11.22 Mechanical Parameters of the Vulcanizates Obtained for a Four Times Longer Cure Time ($T_{90} \times 4$).

	Ctrl	St	Rs	Pl	Ctrl + ZnO	St + ZnO	Rs + ZnO	Pl + ZnO
M_{100} (MPa)	1.6	1.6	1.0	1.2	2.4	2.6	2.4	2.6
M_{300} (MPa)	5.4	5.5	3.1	4.6	9.0	12.1	10.8	10.8
σ (MPa)	14.5	14.7	12.6	14.3	15.5	22.7	23.1	20.0
ε_{rel} (%)	560	560	670	630	480	510	615	520
ε_{res} (%)	30	40	35	40	25	25	30	30
Shore A hardness	63	65	56	60	70	70	67	66

As seen, the presence of a combination of zinc oxide and a zinc salt results into an increase in the tensile strength, which is pronounced best for the combination of zinc oxide with zinc resinate. That, in our opinion, is another important and very interesting fact in favor of the use of $Zn(R)_2$, alone or in combination with classic ZnO. Its multifunctional effect upon elastomer composites and vulcanizates based on them is probably also owing to its unique relation to the elastomeric matrix. The results fully correlate with the observed reversion of vulcanization curves. The absence of reversion of the curves of composites containing a combination of zinc oxide and a zinc salt does not lead to a change in the physical and mechanical performance of these composites, regardless of the duration of vulcanization. Vulcanizates of the compositions, whose curves exhibit reversion during vulcanization (Ctrl, St, Rs, Pl, Ctrl + ZnO), significantly impair their mechanical properties at a longer time of vulcanization in comparison with the ones vulcanized for the optimum cure time.

The wear resistance values of the composites studied are presented in Figure 11.9. It is worth noting that the vulcanizates comprising a combination of zinc oxide and a zinc salt possess much higher wear resistance than those with no zinc oxide or only with zinc oxide. That is in agreement with the other mechanical properties of the vulcanizates.

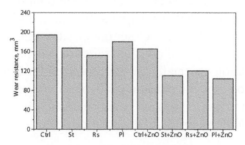

FIGURE 11.9   Wear resistance of the vulcanizates studied.

The heat aging coefficients of the samples vulcanized for the optimum cure time after aging for 72 h at 70°C are presented in Table 11.23.

TABLE 11.23   Thermal Aging Coefficients of the Composites Vulcanized for the Optimum Cure Time (T_{90}).

	Ctrl	St	Rs	Pl	Ctrl + ZnO	St + ZnO	Rs + ZnO	Pl + ZnO
K_{sh} (%)	+1	+3	−2	−2	−1	0	+2	+1
K_{σ} (%)	−39.3	−30.7	−34.3	−36.8	−15.4	−12.7	−2.2	−11.0
K_{ε} (%)	−15.1	−24.6	−16.4	−19.7	−16.1	−20.9	−17.2	−10.3

The mechanical characteristics of the composites vulcanized for four times the optimum cure time after aging for 72 h at 70°C are presented in Table 11.24.

TABLE 11.24 Mechanical Characteristics of the Composites Vulcanized for Four Times the Optimum Cure Time ($4 \times T_{90}$).

	Ctrl	St	Rs	Pl	Ctrl + ZnO	St + ZnO	Rs + ZnO	Pl + ZnO
K_{sh} (%)	−1	+3	+1	+2	−	−	+2	+1
K_a (%)	−80.0	−39.5	−31.7	−45.4	−12.2	−12.3	−9.9	−6.0
K_e, %)	43.8	−26.8	−20.9	−20.6	−14.6	−13.7	−15.0	−9.6

As seen from the table, the samples with zinc resinate both on its own and in a combination with zinc oxide exhibit high heat aging resistance. If the results of Tables 11.23 and 11.24 are compared, one sees that samples cured for four times T_{90}, are more prone to aging. The only exceptions are the composites containing $Zn(R)_2$, whose coefficient of aging is nearly the same regardless of the cure time. This fact once again confirms our thesis about the unique opportunities that the isoprenoid structure gives with regard to tailoring the properties of the elastomer composites and vulcanizates. Noteworthy, the vulcanizates containing zinc salts are generally more resistant to aging than the control, especially those vulcanized for four times T_{90}. Most likely, the formation of a vulcanization network of lower polysulfidity in the presence of the combination of zinc oxide and zinc salt is the reason for the more significant thermostability of those vulcanizates.

11.5.5.4 DYNAMIC PROPERTIES OF THE VULCANIZATES

The obtained good results from the study on vulcanizates containing zinc resinate or another zinc salt encouraged us to conduct some additional investigations related to their actual application in the production of tire treads.

Figure 11.10 presents the dependence of the equilibrium heat build-up on the dynamic deformation of vulcanizates with no zinc oxide, determined on a Goodrich flexometer. The figure shows that at 4–8% of deformation the tested vulcanizates have close values. At higher dynamic deformations the vulcanizates containing no zinc salt (Ctrl) have an equilibrium heat build-up significantly higher than the vulcanizates containing zinc salts. The vulcanizate containing zinc stearate has the lowest heat build-up. The vulcanizates containing Plastikol have a value close to that of the vulcanizates with zinc stearate. Vulcanizates containing zinc resinate have an equilibrium heat build-up higher than the latter two types, which is probably due to the bulky nature of its molecular structure.

Figure 11.11 presents the dependence of the equilibrium heat build-up on the dynamic deformation of vulcanizates containing zinc oxide or a combination of zinc oxide and zinc slats, as determined on a Goodrich flexometer. The course of the curves is similar to the one of the curves for vulcanizates with no zinc oxide. In this case the vulcanizate containing only zinc salt (Ctrl) has also the highest equilibrium heat build-up. The vulcanizates containing a combination of zinc oxide and zinc stearate have lower equilibrium heat build-up. It is worth noting that the vulcanizates containing zinc oxide or a combination of zinc oxide and zinc salts have equilibrium heat build-up much lower than that of the vulcanizates with no zinc oxide (Figure 11.10).

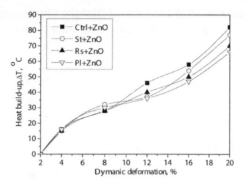

FIGURE 11.10   Dependence of the equilibrium heat build-up on the dynamic deformation for vulcanizates containing no zinc oxide.

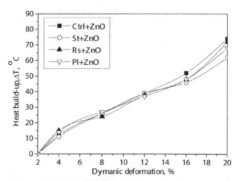

FIGURE 11.11   Dependence of the equilibrium heat build-up on the dynamic deformation for vulcanizates containing zinc oxide.

Figures 11.12 and 11.13 present the temperature dependences of the dynamic modulus for vulcanizates with no zinc oxide and for vulcanizates containing zinc oxide.

The vulcanizates containing zinc stearate have a higher dynamic storage modulus E', whereas E' for those containing zinc resinate is lower than the one of the control vulcanizates, at temperature higher than the glass transition temperature T_g (Figure 11.12). As Figure 11.13 shows, the vulcanizates with zinc oxide in the entire temperature range investigated have a higher dynamic storage modulus, if compared with that of vulcanizates containing zinc oxide. There is also a change in the dynamic modulus depending on the presence of zinc salts. The control sample containing only zinc oxide has the lowest dynamic modulus values, whereas the vulcanizates comprising a combination of zinc oxide and zinc resinate have the highest ones. In the case of composites with no zinc oxide, the vulcanizate containing zinc resinate has the lowest value, that is, it can be said that the combination of zinc oxide/ zinc resinate has a synergistic effect regarding the dynamic storage modulus.

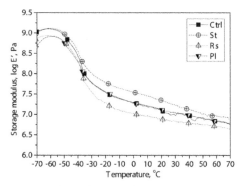

FIGURE 11.12 Temperature dependences of the storage modulus E' for vulcanizates without oxide vulcanized for the optimum cure time.

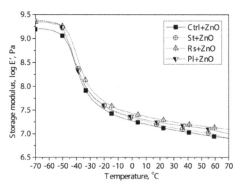

FIGURE 11.13 Temperature dependences of the storage modulus E' for vulcanizates containing zinc oxide or a combination of zinc oxide and zinc salts vulcanized for the optimum cure time.

The temperature dependences of mechanical loss angle tangent for the compositions with no zinc oxide vulcanized for the optimum cure time T_{90} is presented in Figure 11.14. As seen from the figure, T_g for the vulcanizates is about − 40°C. The lowest tan δ = 0.15 is at about 0°C. At a higher temperature the value of tan δ = 0.20 at 60°C revealing formation of a poor curing network in those vulcanizates.

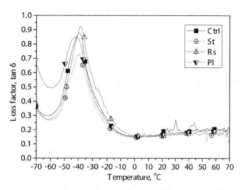

FIGURE 11.14   Temperature dependences of mechanical loss angle tangent (tan δ) for the compositions with no zinc oxide vulcanized for the optimum cure time.

As seen from Figure 11.15, the mechanical loss angle tangent of the vulcanizates containing zinc oxide or a combination of zinc oxide and zinc salts at 0°C is about 0.13 (lower than that of the compositions with no zinc oxide) and decreases with the increasing temperature. Exceptions are the vulcanizates containing only zinc oxide, whose tangent remains constant:

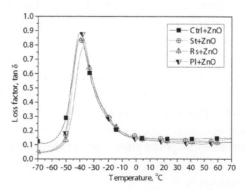

FIGURE 11.15   Temperature dependences of mechanical loss angle tangent (tan δ) for the compositions containing zinc oxide or a combination of zinc oxide and zinc salts vulcanized for the optimum cure time.

tan δ is about 0.10 to 0.11 at 60°C, what is about twice as low as the value of vulcanizates containing zinc oxide (Figure 11.14). The result would be that vulcanizates containing a combination of zinc oxide and zinc salts have formed rolling resistance two times lower compared with that of vulcanizates with no zinc oxide. There is no noticeable difference in tan δ values at 60°C depending on the type of zinc salt used.

The above studies have been described in details in Reference.[39]

11.6   CONCLUSIONS

1. The influence that zinc resinate has on vulcanization, physicochemical, mechanical and dynamic properties of unfilled composites and vulcanizates based on NR containing MBT, TMTD and CBS as accelerators as well as on composites filled with carbon black and silica comprising *N-tert*-butyl-2-benzotiazolil sulfenamide as an accelerator have been investigated. It has been found that:

2. The substitution of zinc oxide with zinc resinate in equivalent amounts in unfilled vulcanizates lowers the minimum and maximum torque, prolongs the scorch time and the optimal cure time. The vulcanization network is less dense, reaching a maximum of 70–80% as compared with that when using zinc oxide, which is the reason for the lower mechanical parameters. The increase in the amount of resinate improves the parameters studied and there is a smaller difference between them and the parameters when zinc oxide is used.

    • The chemical nature of the accelerator affects the effectiveness of zinc resinate as a vulcanization agent. The observed effects on vulcanizates containing accelerator such as MBT are much more pronounced, compared with those upon vulcanizates containing TMTD and CBS.
    • The formation of a less dense vulcanization network is probably due to the zinc content in zinc resinate which is considerably lower compared with that in zinc oxide (7.5 vs. 80%).
    • Vulcanization with zinc resinate in the role of an activator could be run by an appropriate choice of accelerator and quantity of zinc resinate (about 10 phr), so to achieve acceptable performance of the vulcanizates.

3. In the case of rubber composites filled with carbon black the conclusions about unfilled vulcanizates have been confirmed, but the effects

of zinc oxide absence are less pronounced, that is, the parameters are much closer to those of the control sample with zinc oxide.

4. The comparative study on the effect of zinc stearate, zinc resinate and the commercial product Plastikol (a combination of unsaturated and saturated fatty acids) by itself or in a combination with zinc oxide in silica filled composites, containing bifunctional organosilane, has shown:

- The presence of zinc resinate at 5 phr leads to a significantly lower Mooney viscosity of the rubber composites; in the presence of zinc oxide the plasticizing effect of zinc salts and in particular of zinc resinate is more pronounced than when using zinc oxide by itself.

- The combination of zinc oxide and zinc resinate has a strong anti-reversion effect. Probably the reason is the formation of a coordination complex between zinc oxide, zinc resinate, the accelerator and sulfur, which results in a prolong scorch time and in the formation of a more stable vulcanization structure.

- The absence of reversion results into the preservation of the mechanical parameters of the vulcanizates, irrespective of the vulcanization duration.

- Vulcanizates, containing zinc oxide are significantly less resistant to heat aging than vulcanizates containing zinc salts; most resistant to aging are vulcanizates containing zinc oxide and zinc salts, zinc resinate, inclusive. Probably a complex having a thermostabilizing effect is formed between them.

- The vulcanizates containing the combination of zinc oxide and a zinc salt are significantly more wear resistant than vulcanizates containing zinc oxide, or those containing only zinc oxide and no zinc salts.

- Vulcanizates containing zinc resinate, or a combination of zinc oxide and zinc resinate have better dynamic properties than the control ones.

- No significant differences have been observed in the dynamic modulus E' and the mechanical loss angle tangent of the vulcanizates containing zinc resinate and of the composites with other zinc salts.

The combinations "terpenoid-polyterpenes," that is, "NR-resinate" and "zinc oxide–zinc resinate" used in NR-based composites filled with silica open opportunities to overcome a number of problems related to the processing and use of this elastomer. We believe that with the use of zinc resinate acting as a MFA, the content of zinc oxide in rubber compounds may be reduced, in some cases even eliminated, and some of the properties of the vulcanizates—improved.

11.7 ACKNOWLEDGMENTS

The authors acknowledge the assistance of Dr Nina Todorova and Dr Ilia Radulov in the experimental part and discussions of this work.

KEYWORDS

- **vulcanization**
- **natural rubber**
- **zinc oxide**
- **properties**

REFERENCES

1. RISK ASSESSMENT ZINC OXIDE CAS-No.: 1314-13-2 EINECS-No.: 215-222-5 Final report, May 2008. https://echa.europa.eu/documents/10162.
2. Chapman, A. V. Safe Rubber Chemicals: Reduction of Zinc Levels in Rubber Compounds, TARRC/MRPRA, 1997; Reducing zinc in rubber compounds. In Proc. Int. Rubber Conference IRC '05, Maastricht, The Netherlands, June 6–9, 2005.
3. Struktol Company of America. Struktol Rubber Handbook. 2004. www.struktol.com.
4. Roberts, J. D.; Caserio, M. C. *Basic Principles of Organic Chemistry*, 2nd ed.; W. A. Benjamin, Inc.: Menlo Park, CA, 1977.
5. De, S. K.; White, J. R. *Rubber Technologist's Handbook*; Smithers Rapra Publishing: London, UK, 2001; Vol 1.
6. Wypych, G. *Handbook of Fillers*, 4th ed.; Chem Tec Publishing: New York, 2016.
7. Blackley, D. C. *Synthetic Rubbers: Their Chemistry and Technology*; Springer Science & Business Media, Dec 6 2012.
8. Barton, B. C.; Hart, E. J. Variables Controlling the Cross-Linking Reactions in Rubber. *Ind. Eng. Chem.* **1952**, *44*, 2444–2448.
9. Adams, H. E.; Johnson, B. L. Cross Linking in Natural Rubber Vulcanizates. *Ind. Eng. Chem.* **1953**, *45*, 1539–1546.

10. Блох, Г. А. Органические ускорители вулканизации каучуков. Под ред. П. И. Захарченко, М.—Л., Химия, (1964).(In Russian)

11. Coran A. Y., Vulcanization. Part V. The Formation of Crosslinks in the System: Natural Rubber-Sulfur-MBT-Zinc Ion; Part VI. A Model and Treatment for Scorch Delay Kinetics. *Rubber Chem. Technol.* **1964,** *37,* 679–688; 689–697.

12. Auerbach, I., Mercaptobenzothiazole Vulcanization Using Sulfur-35. *Ind. Eng. Chem.* **1953,** *45,* 1526–1538.

13. Tsurugi, J.; Fukuda H., The Chemistry of Vulcanization. VI. Action of the Zinc Salt of 2-Mercaptobenzothiazole on the Reaction of Diphenylmethane with Sulfur. *Rubber Chem. Technol.* **1960,** *33,* 211–216.

14. Tsurugi, J.; Fukuda H., The Chemistry of Vulcanization. VII. Role of Zinc Butyrate in the Reaction of Diphenylmethane, Sulfur and 2-Mercaptobenzothiazole. *Rubber Chem. Technol.* **1960,** *33,* 217–228.

15. Blokh, G.A. The Use of Radioactive Sulfur in Vulcanization. *Rubber Chem Technol.* **1958,** *31,* 1035–1054.

16. Донцов, А.А., Никаноренкова, А.В., Догадкин, Б.А., Исследование реакции полиэтилена с серой в присутствии меркаптобензотиазола в окси цинка. *Высокомол соед* 6, 2023–2029 (1964).(In Russian).

17. Kresja, M. R.; Koenig, J. L. A Review of Sulfur Crosslinking Fundamentals for Accelerated and Unaccelerated Vulcanization. *Rubber Chem. Technol.* **1993,** *66,* 376–410.

18. Nieuwenhuizen, P. J.; Ehlers, A. W.; Haasnoot, J. G.; Janse, S. R.; Reedijk, J.; Baerends, E. J. The Mechanism of Zinc(II)-Dithiocarbamate-Accelerated Vulcanization Uncovered; Theoretical and Experimental Evidence. *J. Am. Chem. Soc.* **1999,** *121,* 163–168.

19. Morrison, N. J.; Porter, M. *Crosslinking of Rubbers, in the Synthesis, Characterization, Reactions and Applications of Polymers*; Allen, G. Ed., Pergamon press: Oxford, UK, 1984.

20. Coran, A. Y. Vulcanization. In *Science and Technology of Rubber*, 4th ed.; Mark, J. E., Erman, B., Roland, M., Eds.; Elsevier: Oxford, 2013; pp 337–381.

21. Morton, M. (Ed.) *Rubber Technology*, 3rd ed.; Springer: Netherlands, 1999.

22. Heideman, G., Datta, R. N., Noordermeer, J. W. M.; van Baarle, B. Activators in Accelerated Sulfur Vulcanization. *Rubber Chem. Technol.* **2004,** *77,* 512–541.

23. Heideman, G., Datta, R. N., Noordermeer, J. W. M.; van Baarle, B. Influence of ZnO in Different Stages of Sulphur Vulcanization. Elucidated by Model Compound Studies. *J. Appl. Polym. Sci.* **2005,** *95,* 1388–1404.

24. Taghvaei-Ganjali, S., Malekzadeh, M., Abbasian, A.; Khosravi, M. Effects of Different Activator Systems on Cure Characteristics and Physicomechanical Properties of a NR/SBR Blend. *Iran. Polym. J.* **2009,** *18,*415–425.

25. Roy, K.; Alam, M. N.; Mandal, S. K.; Debnath, S. C. Sol–gel Derived Nano Zinc Oxide for the Reduction of Zinc Oxide Level in Natural Rubber Compounds. *J. Sol-Gel Sci. Technol.* **2014,** *70,* 378–384.

26. Przybyszewska, M.; Zaborski, M.; Jakubowski, B.; Zawadiak, J. Zinc Chelates as New Activators for Sulphur Vulcanization of Acrylonitrile-Butadiene Elastomer, *eXPRESS Polym. Lett.* **2009,** *3,* 256–266.

27. Heideman, G.; Noordermeer, J. W. M.; Datta, R. N.; van Baarle, B. Effect of Metal Oxides as Activators for Sulphur Vulcanization in Various Rubber. *Kaut Gummi Kunstst.* **2004,** *57,* 30–42.

28. Reichle, W. T. Synthesis of Anionic Clay Minerals (Mixed Metal Hydroxides, Hydrotalcite), *Solid State Ionics.* **1986,** *22,* 135–141.

29. Vega, B. New Insights in Vulcanization Chemistry Using Microwaves as Heating Source. PhD Thesis, Institute Quimic de Saria, Barcelona, Spain, 2010.

30. Heideman, G.; Noordermeer, J. W. M.; Datta, R. N.; van Baarle, B. Various Routes for Reduction of ZnO Levels in Rubber Compounds. *Tire Technol. Inter.* (Annual Review) Nov. **2004**, 22–27.

31. Heidemann, G. Reduced Zinc Oxide Levels in Sulphur Vulcanization of Rubber Compounds. PhD Thesis, University of Twente, The Netherlands, 2004.

32. Heideman, G.; Noordermeer, J. W. M.; Datta, R. N.; van Baarle, B. Zinc Loaded Clay as Activator in Sulfur Vulcanization: A New Route for Zinc Oxide Reduction in Rubber Compounds, *Rubber Chem. Technol.* **2004**, *77*, 336–345.

33. Tipova, N. Investigation of Zinc Stearate as Activator of Accelerated Sulphur Vulcanization from the View Point of Decreasing of Zinc Content in the Rubber Vulcanizates, PhD Thesis, University of Chemical Technology and Metallurgy, Sofia, Bulgaria 2007.

34. Karmanova, O. V.; Tikhomirov, S. G.; Ososhnik, I. A., *Creating a New Compositional Activators for Rubber*, 3rd Scientific and Practical Conference with International Participation, November 2–4, 2012, Central Hotel Forum, Sofia, Bulgaria, 2012.

35. Mariano, R. M.; Oliveira, M. R. L.; Rubinger, M. M. M.; Visconte, L. L. Y. Synthesis, Spectroscopic Characterization and Vulcanization Activity of a New Compound Containing the Anion bis(4-methylphenylsulfonyldithiocarbimato)zincate(II). *Eur. Polym.* **2007**, *J 43*, 4706–4711.

36. Cunha, L. M. G.; Rubinger, M. M. M.; Oliveira, M. R. L.; Tavares, E. C.; Sabino, J. R.; Pacheco, E. B. A. V.; Visconte, L. L. Y. Syntheses, Crystal Structure and Spectroscopic Characterization of bis(dithiocarbimato) zinc(II) Complexes: A new class of vulcanization Accelerators. *Inorg. Chem. Acta.* **2012**, *383*, 194–198.

37. Wu, S. H.; Shih, C. F.; Pan, H. C.; Wang, Y. Y.; Chen, H. M.; Wu, C. S. Investigation of vulcanization of Non-crystalline Cu_2ZnSnS_4 Nano-particles. *Thin. Solid Films.* **2013**, *544*, 19–23.

38. Todorova, N., Radulov, I.; Dishovsky, N. Evaluating the Effect of Zinc Resinate Upon the Properties of Natural Rubber Vulcanizates. *J. Chem. Technol. Metall.* **2012**, *47*, 505–512.

39. Todorova, N., Mihaylov, M., Damyankin, I.; Dishovsky, N. Zinc Resinate Influence on the Properties of Silica Filled Composites Based on Natural Rubber. *J. Chem. Technol. Metall.* **2014**, *49*, 213–219.

INDEX

Printed and bound by CPI Group (UK) Ltd, Croydon, CR0 4YY

23/10/2024

01777704-0014